Semiconductor-Laser Fundamentals

Springer
*Berlin*
*Heidelberg*
*New York*
*Barcelona*
*Hong Kong*
*London*
*Milan*
*Paris*
*Singapore*
*Tokyo*

Weng W. Chow    Stephan W. Koch

# Semiconductor-Laser Fundamentals

## Physics of the Gain Materials

With 132 Figures and 3 Tables

Springer

Dr. Weng W. Chow
Sandia National Laboratories,
Albuquerque, NM 87185
USA

Professor Dr. Stephan W. Koch
Philipps-Universität Marburg,
Fachbereich Physik und
Wissenschaftliches Zentrum
für Materialwissenschaften
Mainzergasse 33
D-35032 Marburg, Germany

ISBN 978-3-642-08386-0

Springer-Verlag Berlin Heidelberg New York

Library of Congress Cataloging-in-Publication Data. Chow, W.W. (Weng W.), 1948 –
Semiconductor-laser fundamentals: physics of the gain materials / W.W. Chow, S.W. Koch. p. cm.
Includes bibliographical references and index.
1. Semiconductor lasers. I. Koch, S. W. (Stephan W.)
QC689.55.S45C45 1999 621.36'6-dc21 99-11674



Cover design: *design & production* GmbH, Heidelberg
Computer to film: Saladruck, Berlin

To our parents,
Ho Yin Hong, Hildegard and Friedrich Koch
and to
Ruth and Rita

# Preface

Since Fall of 1993, when we completed the manuscript of our book "Semiconductor-Laser Physics" [W.W. Chow, S.W. Koch, and M. Sargent III (Springer, Berlin, Heidelberg, 1994)] many new and exciting developments have taken place in the world of semiconductor lasers. Novel laser and amplifier structures were developed, and others, for example, the VCSEL (vertical cavity surface emitting laser) and monolithic MOPA (master oscillator power amplifier), made the transition from research and development to production. When investigating some of these systems, we discovered instances when device performance, and thus design depend critically on details of the gain medium properties, e.g., spectral shape and carrier density dependence of the gain and refractive index.

New material systems were also introduced, with optical emission wavelengths spanning from the mid-infrared to the ultraviolet. Particularly noteworthy are laser and light-emitting diodes based on the wide-bandgap group-III nitride and II–VI compounds. These devices emit in the visible to ultra-violet wavelength range, which is important for the wide variety of optoelectronic applications. While these novel semiconductor-laser materials show many similarities with the more conventional near-infrared systems, they also possess rather different material parameter combinations. These differences appear as band structure modifications and as increased importance of Coulomb effects, such that, e.g., excitonic signatures resulting from the attractive electron-hole interaction are generally significantly more prominent in the wide bandgap systems.

On the theoretical side, important progress was made concerning the long-standing problem of the semiconductor laser lineshape. The solution of this problem may be obtained on the basis of a systematic analysis of carrier damping and dephasing processes. This improved level of theoretical analysis led to quantitative agreement between experimentally measured and theoretically predicted gain/absorption and refractive index spectra for a wide variety of semiconductor-laser materials. Since it can be used directly in the engineering of laser and amplifier structures, the improved gain medium theory is of more than academic interest.

The success and usefulness of the new gain medium theory in explaining experiments and designing devices, combined with the complexity in implementing the calculations provided motivation for the present book,

"Semiconductor-Laser Fundamentals: Physics of the Gain Materials". To provide the background of introducing the new developments, we integrated into this book the material related part of our original "Semiconductor-Laser Physics" book. Besides some of the basic chapters, that have been updated and reorganized, we extensively cover band structure engineering aspects and the microscopic theory of the semiconductor gain materials in order to adequately account for the recent progress in materials and theoretical understanding. We included a wealth of examples, involving many different material combinations that are used in quantum-well laser systems. All of these results are obtained consistently at the level of the full microscopic many-body theory, and we expect a good degree of quantitative and predictive value from these numerical examples. As a guide for people interested in reproducing our numerical results, we included a variety of technical details involved with the coding of the set of many-body equations.

As always, this book could not have been written without the interaction and collaboration with numerous colleagues, including (in alphabetical order) K. Choquette, M. Crawford, A. Girndt, F. Jahnke, E. Jones, A. Knorr, J. Moloney and A. Wright. It is our pleasure to thank Murray Sargent III for his collaboration on the first book, and we are very sorry that his new job does not allow him the time to continue working in this area. Special thanks are due to Renate Schmid for her expert technical help in preparing the manuscript and handling the extensive e-mail exchanges between Marburg and Albuquerque. SWK thanks Sandia National Laboratories for the hospitality during the three months when this book was finished.

The research has been funded by the Deutsche Forschungsgemeinschaft, partly through the Leibniz prize, and by the U. S. Department of Energy under Contract DE-AC04-94AL85000.

Albuquerque, NM W.W. Chow
Marburg S.W. Koch
October 1998

# Contents

# 1. Basic Concepts

This chapter reviews some background material on semiconductor lasers and lays the theoretical foundation for the development of a theory for the gain media in these lasers. We begin with a brief summary of the historical background of semiconductor laser development in Sect. 1.1. Section 1.2 describes the basic laser structure and discusses how an inversion is created. Section 1.3 introduces the concept of heterostructures. Some basic aspects of the semiconductor band structure are presented in Sect. 1.4. A more detailed band-structure analysis including the modifications caused by quantum confinement and strain effects is presented in Chaps. 5, 6. Section 1.5 briefly discusses cgs and MKS units, both of which are used extensively in the semiconductor laser literature. The problem is that MKS has been used traditionally for lasers, while cgs is often used in semiconductor theory. Hence, the marriage of the fields requires a familiarity with both systems of units. Section 1.6 discusses the Fermi-Dirac distributions of the carrier probabilities. Section 1.7 introduces the concept of quantum confinement and Sect. 1.8 makes contact with the laser electric field by outlining a derivation of the slowly varying electromagnetic-field equations. This shows how the field amplitude and phase are influenced by an induced polarization of the medium. Section 1.9 begins our discussion of this induced polarization using a quantum mechanical description of the semiconductor medium. This lays the foundations for later chapters, which derive the polarization of the semiconductor medium with increasing levels of accuracy and complexity.

## 1.1 Historical Background

The concept of a semiconductor laser was introduced by *Basov* et al. (1961) who suggested that stimulated emission of radiation could occur in semiconductors by the recombination of carriers injected across a *p-n* junction. The first semiconductor lasers appeared in 1962, when three laboratories independently achieved lasing. After that, progress was slow for several reasons. One reason was the need to develop a new semiconductor technology. Semiconductor lasers could not be made from silicon where a mature fabrication technology existed. Rather, they require direct bandgap materials which were found in compound semiconductors. At the time compound semiconductors were less well understood. There were also problems involving high threshold currents for lasing, which limited laser operation to short pulses at cryogenic temperatures, and low efficiency, which led to a high heat dissipation. A big stride toward solving the above problems was made in 1969, with the introduction of heterostructures. In a heterostructure laser, one replaces the simple *p-n* junction with multiple semiconductor layers of different compositions. Constant wave (cw) operation at room temperature became possible because of better carrier and optical confinement. Laser performance continued to improve as more advanced heterostructures, such as quantum wells and strained quantum wells were used as gain media.

Two factors are largely responsible for the transformation of semiconductor lasers from laboratory devices operating only at cryogenic temperatures into practical opto-electronic components capable of running continuously at room temperature. One is the exceptional and fortuitous close lattice match between AlAs and GaAs, which allows heterostructures consisting of layers of different compositions of $Al_xGa_{1-x}As$ to be grown. The second is the presence of several important opto-electronic applications where semiconductor lasers are uniquely well suited because they have the smallest size (several cubic millimeters), highest efficiency (often as much as 50 % or even more) and the longest lifetime of all existing lasers. This enables the field of semiconductor lasers to draw the attention and resources that are necessary for its development.

One such application is optical-fiber communication, where device design is simplified by the fact that the laser output can be modulated simply by modulating the injection current. Gigahertz information transmission rates are now possible. Optical fiber communication also motivated the development of semiconductor lasers at 1.3 μm, where optical fiber loss is minimum, and at 1.5 μm, where dispersion in the fiber is minimum. The need for repeaters led to the development of laser amplifiers, the introduction of underwater optical communication lines necessitate improvements in device reliability, and frequency multiplexing of transmissions led to distributive feedback (DFB) and Bragg reflector (BR) lasers for frequency stability.

There are other applications of semiconductor lasers as well. The optical memory (audio and video discs) industry has generated a large enough demand of semiconductor lasers to help in reducing laser cost. High power semiconductor lasers are being used in printers and copiers. When even higher power is reliably available, the list of applications will expand to include, e.g., line-of-sight communications, laser radar and fuzing. Schemes for increasing laser power are plentiful. They involve widening the active region (broad-area lasers), phase locking many narrow active regions (arrays and external cavity lasers) and optically pumping a solid state laser with a stack of semiconductor lasers. In addition to high output power, external cavity lasers can also be designed to have very narrow linewidths. While a semiconductor laser by itself does not make a good high energy pulse laser because its gain medium performs badly in terms of energy storage due to a short carrier recombination time, it can be used to optically pump a Q-switched solid state laser. Diode pumped YAG lasers have produced high energy (over 1 joule) short (tens of nanoseconds) pulses. They also have the advantages over conventional flashlamp pumped ones by being more efficient and compact.

Advances are continually being made. For example, vertical-cavity surface-emitting lasers (VCSELs) now rival conventional edge-emitting laser diodes in efficiency, and surpass them by a wide margin in threshold current. Lasing has been demonstrated in II–VI and group-III nitride compounds. In addition to the many applications for these visible and potentially ultraviolet light sources, there are interesting material sciences and basic physics issues as well. So, we see that semiconductor lasers have many applications. Important applications drive the need for better and newer lasers, which generate more applications, which in turn support the development of even better and more versatile lasers.

## 1.2 Laser Device

The semiconductor laser looks different to different people. An electrical engineer may think of it as a forward biased *p-n* junction, while a crystal grower may see mainly the heterostructure. The laser fabricator's view probably involves too much engineering detail for the solid-state physicist. Since this book is primarily concerned with the theory of the semiconductor laser gain medium, we structure this and the following sections to provide the background for constructing useful theoretical models.

The basic features of the experimental device are shown in Fig. 1.1, which is a vastly simplified diagram of a semiconductor laser. A semiconductor laser is usually fabricated by growing a *p*-doped layer on top of an *n*-doped semiconductor substrate. The *n*- and *p*-doping are results of donor and acceptor impurities, respectively, in the semiconductor medium. In an *n*-doped

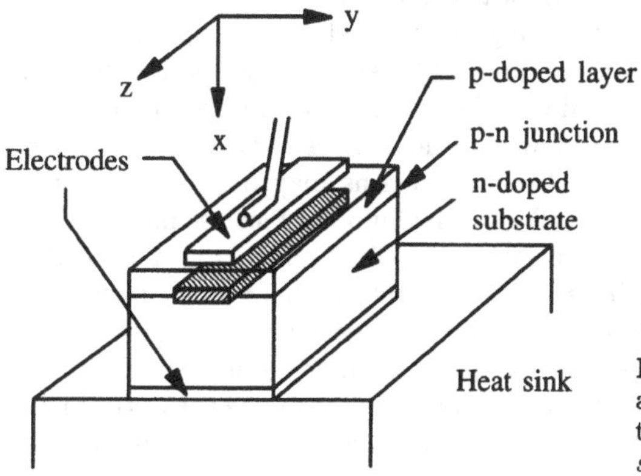


**Fig. 1.1.** Schematic diagram of a semiconductor laser. The active region is indicated by the *shaded area*

medium, the donor levels lie well within a thermal energy $k_BT$ of the conduction band, so that they are effectively ionized, yielding conduction electrons, i.e., negatively charged current carriers. (Here $k_B$ is Boltzmann's constant and $T$ is the absolute temperature.) $n$-doped media can conduct well, compared to intrinsic semiconductors, which do not have impurity levels and derive conductivity only by carriers that manage to bridge the whole bandgap, an energy large compared to the thermal energy. In the GaAs semiconductor, a popular donor impurity is Se from column VI in the periodic table, which has one more valence electron (6) than As, which is in column V. Similarly, in a $p$-doped medium, the acceptor impurity level lies well within a thermal energy of the valence band, allowing an electron from the valence band to fall into the level, thereby creating a hole in the valence band. This creates positively charged current carriers, which gives the name $p$-doped. In the GaAs semiconductor, a popular acceptor impurity is Zn from column II, which has one valence electron less (2) than Ga from column III. Note that the doped media by themselves are electrically neutral, that is, they are not charged positively or negatively, even though they have current carriers.

Current is injected via two electrodes, one of which is electrically connected to a heat sink. Lasing occurs in the active (gain) region between the electrodes as indicated by the shaded area in Fig. 1.1. This shaded area represents the depletion region in a simple $p$-$n$ junction or the specially fabricated intrinsic layer in a heterostructure laser (Sect. 1.3). The optical resonator is formed by two parallel facets that are made by cleaving the substrate along crystal planes. Owing to the high gain in the active region, resonator facets are often left uncoated, which gives a Fresnel reflectivity of about 30 %, since the semiconductor index of refraction is about 3.5 compared to unity outside. The laser in Fig. 1.1 is mounted p-side up. For more efficient heat removal, semiconductor lasers are sometimes mounted p-side down, so that the active region is closer to the heat sink. Typical sizes of the active region are 1000 Å thick by 10 μm wide by 250 μm long. Much smaller lasers can be fabricated easily with interesting special properties as we discuss in later chapters.

Several refinements may be incorporated into the laser of Fig. 1.1 to attain certain desirable characteristics: e.g., low threshold, cw operation, operation at high temperature, narrow linewidth (even single mode) spectra, or high output power. All semiconductor lasers are now fabricated with heterostructures in order to have the low thresholds necessary for cw or room temperature operation. The heterostructure widths may be chosen to produce either a bulk or a quantum-well gain medium. In most semiconductor lasers on the market, the lateral variations (y direction in Fig. 1.1) of the light are *gain guided*, that is limited in y extent to the region having appreciable current flow. Some lasers are fabricated more intricately to achieve index guiding, which allows them to operate in a single mode. Combinations of high and low reflection coatings are often used to optimize the optical resonator quality factor, Q. Nonabsorbing facet technology, which increases the facet damage threshold, is instrumental in the development of high-power

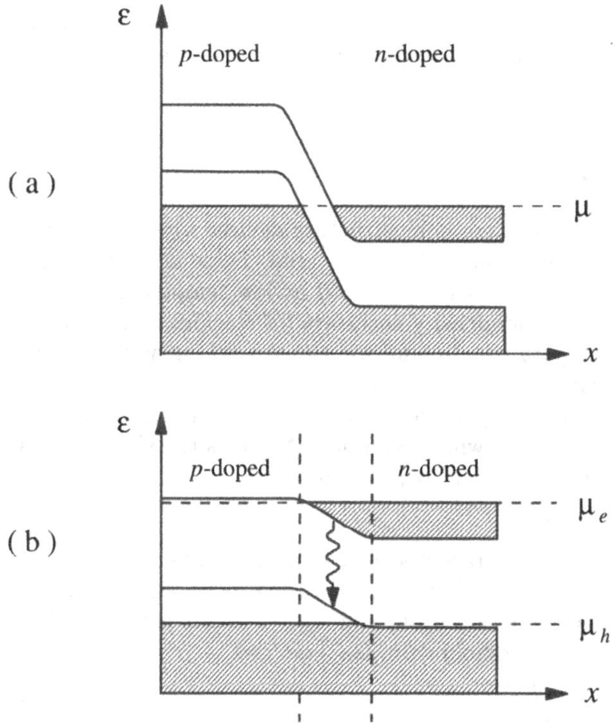


**Fig. 1.2a, b.** Electron energy and occupation perpendicular to the $p$-$n$ junction (**a**) without an applied voltage and (**b**) with a forward biased applied voltage

single-mode lasers. Some semiconductor lasers are fabricated with narrow stripes for single transverse-mode operation, while others are fabricated with broad active areas for high output power.

To see how an inversion is created at a $p$-$n$ junction, we plot in Fig. 1.2 the energy bands and electron occupation as functions of position in the transverse x direction, i.e., perpendicular to the junction plane. This figure shows that in the absence of an applied voltage across the electrodes, the chemical potential (Fermi energy) is constant throughout the entire structure, as shown in Fig. 1.2a, resulting in no net flow of carriers. More importantly, there is no region containing both electrons in the conduction band and holes in the valence band, which is necessary to obtain an inverted population. When a voltage is applied so that the $p$-doped region is positive relative to the $n$-doped region, the electron energies are altered as shown in Fig. 1.2b. The voltage drop across the junction is reduced (forward biased). When the forward bias approximately equals the bandgap potential, an inverted or active region is created within the junction. Inside this region, stimulated emission occurs due to electron-hole recombination. At steady state, the inversion is maintained by the injection of carriers, via the electrodes, by an external power supply. Lasing occurs when the rate of stimulated emission due to electron-hole recombination approximately equals the total rate of optical losses.

## 1.3 Heterostructures

The first semiconductor lasers were homostructure devices where each laser was fabricated with only one semiconductor material. These lasers had high threshold current densities, even when operated at low temperature where the gain is higher and the carrier density necessary for reaching transparency is lower than at room temperature. In addition to not working at room temperature, these lasers also could not operate cw.

The methods which drastically improve semiconductor laser performance may be understood with the following discussion. A laser's threshold gain is often estimated from the unity round-trip condition

$$R_1 R_2 \, e^{2(\Gamma G_{th} - \alpha_{abs})L} = 1 \ , \tag{1.1}$$

where $R_1$ and $R_2$ are the facet reflectivities, $\Gamma$ is the confinement factor, which is a measure of the overlap between the lasing mode and the active region cross section, $L$ is the laser length, and $\alpha_{abs}$ accounts for the optical losses. Solving (1.1) for the threshold gain, $G_{th}$, we have

$$G_{th} = \frac{1}{\Gamma}\left(\alpha_{abs} - \frac{\ln(R_1 R_2)}{2L}\right) \ . \tag{1.2}$$

For a bulk gain medium, and under very restrictive conditions, as will become clear later in this book, one can sometimes approximate the gain as

$$G = A_g(N - N_g) \ , \tag{1.3}$$

where $A_g$ is the gain coefficient and $N_g$ is the carrier density needed to reach transparency in the gain medium. $A_g$ and $N_g$ depend on both the gain material (via relaxation rates, band structure, etc.) and the laser configuration (via the lasing frequency, temperature, etc). In the more phenomenological approaches, one takes their values from experiment. In later chapters we derive more precise gain formulas, which when reduced to the form of (1.3) give the functional form of $A_g$ and $N_g$. Note that although (1.3) displays no tuning dependence, it clearly shows that with too few carriers, the medium absorbs radiation instead of amplifying it.

Another approximation, which can be used under some conditions, is to relate the injected carrier density to the injection current density by

$$N = \frac{J\eta}{e\gamma_{eff} d} \ , \tag{1.4}$$

where $e$ is the electron charge, $\gamma_{eff}$ is an effective carrier recombination rate, $d$ is the active region thickness (or depth), and $\eta$ is the quantum efficiency with which the injected carriers arrive in the active region and contribute to the inversion. Since $N$ is inversely proportional to $d$, the gain (1.3) increases as $d$ decreases for a given $J$. More specifically, solving (1.3) for $N$ using the threshold gain of (1.2), we find the threshold injection current density

$$J_{\mathrm{th}} = \frac{e\gamma_{\mathrm{eff}} d}{\eta} \left\{ N_{\mathrm{g}} + \frac{1}{A_{\mathrm{g}} \Gamma} \left[ \alpha_{\mathrm{abs}} - \frac{1}{2L} \ln(R_1 R_2) \right] \right\} \ . \quad (1.5)$$

We note that the threshold current density is a strong function of the active region thickness. In a homostructure laser, $d$ is the distance traveled by a conduction electron going from the $n$-doped region to the $p$-doped region before it recombines with a hole. In a homostructure GaAs laser, $d \approx 1$ μm. Reduction of this thickness reduces the threshold current density proportionately, unless $\Gamma$ is changed.

The method that is now generally adopted for decreasing the active layer thickness involves blocking the carrier flow with a layer of material that has a higher bandgap energy than the active region. The resulting structure is called single heterostructure if only one blocking layer is used, and double heterostructure if a blocking layer is used on either side of the active region (Fig. 1.3). With a heterostructure laser, the thickness of the active region is determined during growth, and active region thicknesses of 0.1 μm or less can readily be achieved.

Heterostructures may only be grown epitaxially with crystals that have sufficiently similar lattices. For example, one may use GaAs and AlAs because both are face-centered cubic crystals, with almost equal lattice constants of 5.652 and 5.662 Å at room temperature, respectively. One can then grow layers of $(\mathrm{GaAs})_{1-x}(\mathrm{AlAs})_x$ which is also written as $\mathrm{Al}_x\mathrm{Ga}_{1-x}\mathrm{As}$. For $x > .45$, this compound has an indirect bandgap, which is not useful as a gain medium. For smaller values of $x$, the bandgap energy of such a layer is given empirically by

$$\varepsilon_{\mathrm{g}0} = 1.424 + 1.266x + 0.266x^2 \ . \quad (1.6)$$

For two materials that can form a stable heterostructure, the larger bandgap material usually has a lower refractive index. According to experimental data, the refractive index of $\mathrm{Al}_x\mathrm{Ga}_{1-x}\mathrm{As}$ may be approximated by

$$n = 3.590 - 0.710x + 0.091x^2 \ . \quad (1.7)$$

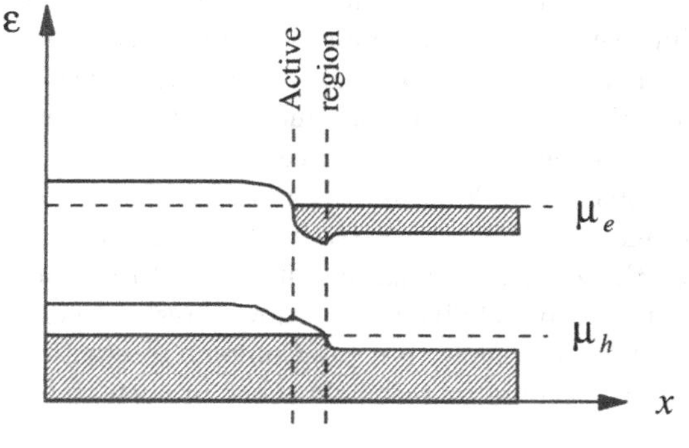


**Fig. 1.3.** Electron energy and occupation for a double heterostructure laser

Therefore, the double heterostructure also provides an optical waveguide for the laser field, resulting in a higher confinement factor. Equally important, because of their wider bandgap, the blocking layers are transparent to the laser field, thus reducing optical losses. The improvement in laser performance due to the introduction of heterostructures is largely responsible for making semiconductor lasers into practical devices.

Once lasing is achieved, the next goals are usually to increase output power and lower laser threshold current. For the double heterostructure laser just described, scaling to higher power is a problem. The reason is that both the carriers and the laser mode are confined to within the same thin region. While we would like the carriers to be in a thin layer to maximize the density, we would also like the radiation field to be in a thick layer to ensure that its intensity is below the material damage threshold. It turns out that we can have both with more complicated heterostructure configurations. Examples are the large optical cavity (LOC) structure shown in Fig. 1.4a, or the separate confinement heterostructure (SCH) shown in Fig. 1.4b. These heterostructures involve either one or two barrier layers for carrier confinement, and two cladding layers for optical confinement. The flexibility of the LOC and SCH designs makes them widely used in semiconductor lasers.

Present state of the art fabrication techniques allow one to reduce the active layer thickness even to the dimension of the order of or less than an electron de Broglie wavelength, which is about 120 Å in GaAs. We then have a quantum-well laser where the carriers are confined to a square well in the transverse dimension and move freely in the other two dimensions. As later chapters show, the change from a three-dimensional to a two-dimensional free-particle density of states causes a quantum-well gain medium to behave differently from a bulk gain medium. A useful property of a quantum-well layer is that it is thin enough to form stable heterostructures with semiconductors of noticeably different lattice constants. The necessary deformation (strain) in the quantum-well lattice structure produces stress in the neighborhood of the interface which significantly alters the band structure. The change in band structure can lead to a reduction of laser threshold current density. This and other features of strained layer quantum-well gain media are discussed in Chaps. 6, 7.

Additional improvements of the semiconductor laser performance appear possible if one reduces the dimensionality of the gain medium even further than in the quasi-two-dimensional quantum wells. Instead of having carrier confinement only in one space dimension, one may produce structures where the quantum confinement occurs in two or even all three space dimensions. These quasi-one-dimensional or quasi-zero-dimensional nanostructures are referred to as quantum wires or quantum dots, respectively. Simple density-of-state arguments indicate that the reduced dimensionality leads to a more efficient inversion and, hence, to the possibility of ultra low threshold laser operation. However, more recent studies show that the Coulomb interaction effects among the charge carriers become increasingly more important for

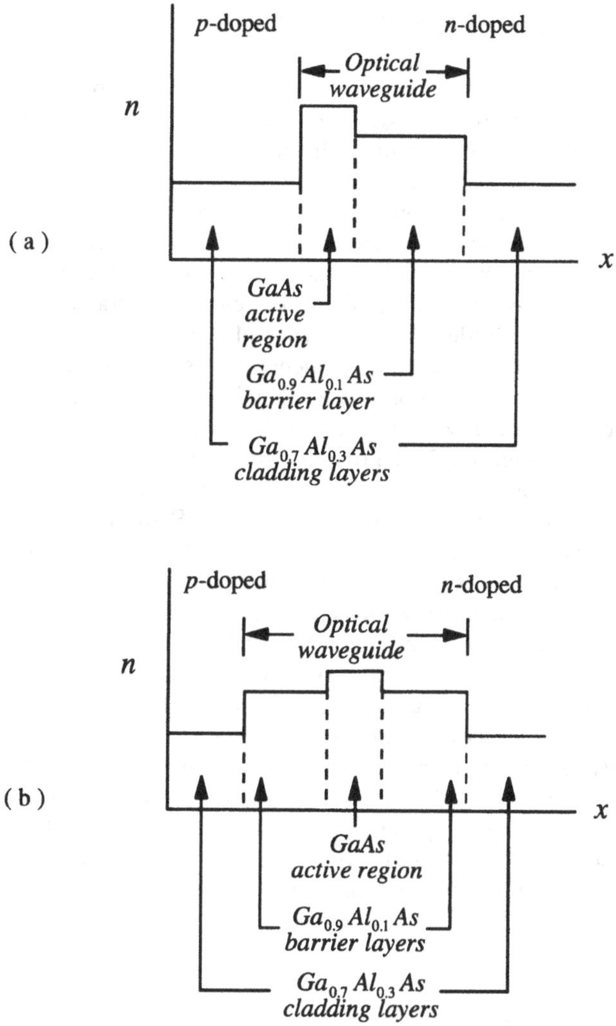


**Fig. 1.4a, b.** Schematic diagram of (**a**) the large optical cavity (LOC) structure, (**b**) the separate confinement heterostructure (SCH)

a decreased dimensionality of the semiconductor structure. These Coulomb effects seem to, at least partially, remove the advantages gained by the modified density of states. Furthermore, the manufacturing of quantum-wire or quantum-dot laser structures is still in its infancy. Therefore, we do not discuss the potentially very interesting quantum-wire or quantum-dot laser devices in this book. For more information we refer the interested reader to the literature at the end of this chapter.

## 1.4 Elementary Aspects of Band Structures

In a simple picture of a semiconductor, an electronic state is identified by its momentum, $\boldsymbol{k}$, and $z$-component of spin, $s_z$. The allowed electronic energies are a result of the interaction of the electron with the regular lattice of ions. The resulting band structure describes how the energy of an electron in the ionic lattice is related to the carrier momentum in the absence of other mobile carriers. Figure 1.5 shows, as an example, the electronic band structure for GaAs. We see that it is quite complicated. There are regions with continuous distributions (or *bands*) of energies and regions where electronic states are forbidden (*bandgaps*). The figure identifies two types of bands: conduction bands, which consist of unoccupied states; and valence bands, which consist of occupied states. GaAs is an example of a direct bandgap semiconductor, for which the conduction-band energy minimum and the valence-band energy maximum have the same momentum. If the band extrema occur at different momentum values, the semiconductor has an indirect bandgap. Most III–V and II–VI compounds (the numerals refer to columns in the Periodic Table) are direct bandgap materials. Examples of indirect bandgap materials are Si and Ge (both column IV), and AlAs (III–V). Direct bandgap materials tend to have high radiative transition rates, whereas indirect bandgap materials do not. Since the abscissa in Fig. 1.5 covers the full Brillouin zone, it gives the complete description of the electronic band structure. In general, a semiconductor electronic band structure has numerous bands with asymmetric shapes and sometimes several energy maxima and minima.

Fortunately, for optical transitions with frequencies in the visible or near infrared, it is often sufficient to consider only a small portion of the band structure shown in Fig. 1.5. This simplification is due to two factors. One is that optical transitions are direct transitions, i.e., the momenta of the initial and final electronic states are essentially equal. This is because from

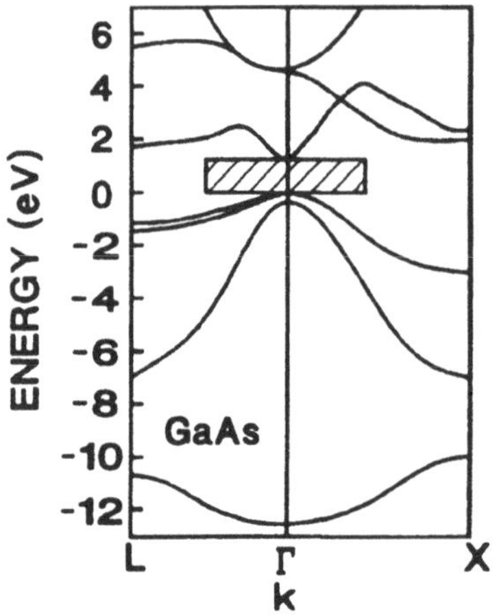


**Fig. 1.5.** Band structure of GaAs. The *hatched region* is the region of interest for optical transitions. $\Gamma$, $X$ and $L$ are high symmetry points in the first Brillouin zone, the $\Gamma$ point is the zone center

the conservation of momentum, the difference in electron momenta must equal the momentum of the photon involved in the transition. The photon momentum is

$$\hbar K = \frac{\hbar \nu n}{c} \ , \tag{1.8}$$

where $\hbar\nu$ is the photon energy, $K$ is the magnitude of the photon wavevector, $n$ is the refractive index of the semiconductor, and $c$ is the speed of light in vacuum. For GaAs, $\hbar\nu \approx 1.4$ eV and $n \approx 3.6$, so that $K \approx 2.54 \times 10^7/$ m, which is negligible on the scale ($\approx 10^{10}/$ m) of the electronic band structure (Brillouin zone). Therefore, we only need to consider a narrow region of the band structure around the bandgap minimum, where optical transitions are most likely to occur. If the region of interest is sufficiently narrow, it is often reasonable to approximate the energy bands in that region as symmetric and parabolic in shape. We discuss deviations from this simple parabolic band approximation in Chaps. 5 to 7.

Another important simplifying factor is that all low lying, completely filled bands do not contribute directly to the optical transitions in the frequency range of interest. Hence, the electronic band structure that we have to consider usually involves only a very small portion of the entire band structure indicated by the hatched area in Fig. 1.5.

Particularly, when we get into the theory of strained quantum-well devices (Chap. 6), it is useful to know roughly why there are three valence bands and one conduction band in Fig. 1.5. The single conduction band in the figure results from a 4$s$-state of the GaAs "molecule", while the three valence bands come from a 4$p$-state. More precisely, there are two spin states for each $\boldsymbol{k}$ in the conduction band. The spin-orbit interaction in the valence band leads to the total angular momentum $j$ with values $j = 3/2$ and $j = 1/2$. The lowest valence band in Fig. 1.5 corresponds to the two spin states of $j = 1/2$, the highest valence band corresponds to the states $m_j = 3/2$, and the middle valence band (degenerate with the highest at $k = 0$) corresponds to $m_j = 1/2$ belonging to $j = 3/2$. The lowest band is variously known as the split-off band or the spin-orbit band. The upper valence bands are called the heavy-hole and light-hole bands corresponding to the reciprocals of their curvatures, as described below.

Most semiconductor lasers may be described by a band structure consisting of one conduction band and several valence bands. Sometimes, even a simple two-band model (one conduction and one valence band) is sufficient to illustrate the physics of semiconductor laser behavior. In this chapter we limit our discussions to such a two-band model. Generalization to the case of multiple valence bands involves introducing a valence band index, something we do in the later chapters.

In the absence of dopant atoms, thermal energy, and pumping processes such as interaction with an optical field, the valence bands of a semiconductor are completely full and the conduction band is empty. As such no states are free for electrons to move to within their respective band, and hence no current can flow. The band structure is calculated for this case of unexcited electrons. As discussed above, if the portion of the band structure of interest is sufficiently small, we may use the effective-mass approximation, where the detailed conduction and valence band structures are approximated by the simple parabolae

$$\varepsilon_{ck} = \frac{\hbar^2 k^2}{2m_c} + \varepsilon_{g0} \ , \tag{1.9}$$

$$\varepsilon_{vk} = \frac{\hbar^2 k^2}{2m_v} \ . \tag{1.10}$$

Here $m_c$ and $m_v$ are the effective masses of the electrons in the conduction and valence bands, respectively, and $\varepsilon_{g0}$ is the bandgap energy in the absence of excited electrons. The band-structure diagram is shown in Fig. 1.6. The effective masses are defined by the reciprocals of the band curvatures, that is, by the second derivatives

$$\frac{1}{m_i} = \frac{1}{\hbar^2} \frac{d^2 \varepsilon_{ik}}{dk^2}\bigg|_{k=0} \qquad i = c, v \tag{1.11}$$

of the conduction- or valence-band energies. As a consequence of the negative valence-band curvature, $m_v$ is negative, while $m_c$ is positive.

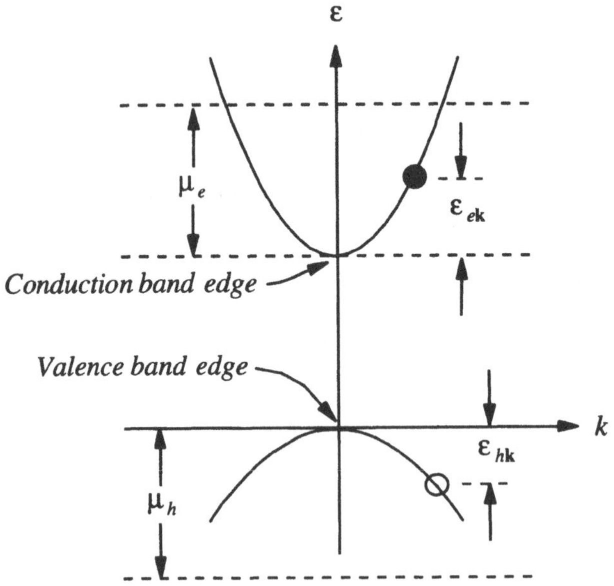


**Fig. 1.6.** Two-band model of a direct bandgap semiconductor. The electron and hole kinetic energies $\varepsilon_{ek}$ and $\varepsilon_{hk}$ are measured from their respective bandedges, with electron (hole) energy increasing upwards (downwards). $\mu_e$ and $\mu_h$ are the quasi-equilibrium chemical potentials described in Sect. 1.6

If an electron in the full valence band absorbs light, it is excited into the empty conduction band leaving behind a missing electron in the valence band. For simplicity, we refer to the conduction electrons simply as electrons and the missing valence-band electrons as holes. Since all states are vacant in the conduction band except the one occupied by the excited electron, a tiny one-electron current can flow in the conduction band. As such the excited electron is called a charge carrier. Similarly, an electron in the valence band can move into the hole, which moves the hole, so that the hole too is a charge carrier with a charge $+e$ opposite that of the electron.

For an optical (vertical) transition, the transition energy at the carrier momentum $\boldsymbol{k}$ is given by

$$\hbar\omega_{\boldsymbol{k}} = \varepsilon_{\mathrm{e}\boldsymbol{k}} + \varepsilon_{\mathrm{h}\boldsymbol{k}} + \varepsilon_{\mathrm{g}0} \quad , \tag{1.12}$$

where the electron and hole energies are

$$\begin{aligned} \varepsilon_{\mathrm{e}\boldsymbol{k}} &= \frac{\hbar^2 k^2}{2m_{\mathrm{e}}} \quad , \\ \varepsilon_{\mathrm{h}\boldsymbol{k}} &= \frac{\hbar^2 k^2}{2m_{\mathrm{h}}} \quad , \end{aligned} \tag{1.13}$$

and $m_{\mathrm{e}}$ and $m_{\mathrm{h}}$ are the effective masses of the electron and hole, respectively. The electron mass $m_{\mathrm{e}}$ equals $m_{\mathrm{c}}$. In this electron-hole description of a semiconductor, the energy of the hole may be thought of as the energy of the completely filled valence band minus the energy of the valence band with a vacant electronic state. In this case, an increase in the hole momentum leads to an increase in the hole energy. Therefore, whereas the effective electron mass in the valence band is negative, the effective hole mass is positive. The relationship between $m_{\mathrm{h}}$ and $m_{\mathrm{v}}$ requires taking into consideration the Coulomb interaction among carriers, which is of course different for a completely filled valence band and for a valence band with a vacancy, see Sect. 3.1 for details.

The resonance energies for the optical transitions can be changed by the Coulomb interaction, which for low densities leads to the creation of *excitons*. Here the Coulomb attraction can bind an excited electron and hole pair into an exciton, which is a hydrogen-like "atom" with a finite lifetime. The exciton life is terminated through electron-hole recombination, which transfers the exciton energy to light (*radiative recombination*), or to the lattice, impurities, etc. (*nonradiative recombination*). By replacing the proton mass by the reduced electron-hole mass, we can use the Bohr hydrogen model to describe an exciton. The radius of the lowest exciton state is given by the exciton Bohr radius (in cgs units)

$$a_0 = \frac{\hbar^2 \epsilon_{\mathrm{b}}}{e^2 m_{\mathrm{r}}} \quad , \tag{1.14}$$

and the energy of the lowest state is given by the exciton Rydberg energy

$$\begin{aligned}\varepsilon_{\mathrm{R}} &= \frac{\hbar^2}{2m_{\mathrm{r}}a_0^2}\\ &= \frac{e^2}{2\epsilon_{\mathrm{b}}a_0}\\ &= \frac{e^4 m_{\mathrm{r}}}{2\epsilon_{\mathrm{b}}^2\hbar^2}\ ,\end{aligned} \tag{1.15}$$

where $\epsilon_{\mathrm{b}}$ is the background dielectric constant and $m_{\mathrm{r}}$ is the reduced electron-hole mass defined by

$$\frac{1}{m_{\mathrm{r}}} = \frac{1}{m_{\mathrm{e}}} + \frac{1}{m_{\mathrm{h}}}\ , \tag{1.16}$$

or from (1.15)

$$m_{\mathrm{r}} = \frac{2\epsilon_{\mathrm{b}}^2\hbar^2\varepsilon_{\mathrm{R}}}{e^4}\ .$$

In GaAs, $a_0 = 124$ Å compared to 0.5 Å in the H atom, and $\varepsilon_{\mathrm{R}} = 4.2$ meV, which is tiny compared with 13.6 eV for the H atom and small compared to room temperature thermal energy $k_{\mathrm{B}}T = 25$ meV. Whether excitons are important in the description of semiconductor behavior depends on $a_0$ compared to the mean distance between electron-hole pairs and the screening length, and $\varepsilon_{\mathrm{R}}$ compared to $k_{\mathrm{B}}T$. The screening length is a measure of the effectiveness of the screening of the Coulomb interaction between two carriers by other carriers. As the carrier density increases (due to an injection current or optical absorption), the Coulomb potential becomes increasingly screened, and for sufficiently high densities the excitons are completely ionized. Similarly, for increasing density the mean particle separation decreases, leading to increasing overlap of the electrons and holes in the excitons. Since electrons and holes are Fermions, each quantum state cannot be occupied by more than one particle (Pauli exclusion principle). Hence, different electrons (holes) compete for the available phase space. Phase-space filling effectively reduces the electron-hole attraction, quite similar to screening. One sees this explicitly in Chaps. 3 and 4, where the many-body effects are treated.

There is some discussion about the values of $m_{\mathrm{e}}$ and $m_{\mathrm{h}}$ to use. For example, for GaAs one may see $m_{\mathrm{e}} = 1.127m_0$, $m_{\mathrm{h}} = 8.82m_0$, and $m_{\mathrm{r}} = 0.05896m_0$, where $m_0$ is the mass of the electron in free space. Alternatively, in this book we usually use $m_{\mathrm{e}} = 1.176m_{\mathrm{r}}$ ($= 0.0665m_0$), $m_{\mathrm{h}} = 6.669m_{\mathrm{r}}$ ($= 0.377m_0$), and $m_{\mathrm{r}} = 0.05653m_0$, which agrees with the Luttinger Hamiltonian discussion in Chap. 5 provided one uses the heavy-hole mass $m_{\mathrm{hh}}$ for $m_{\mathrm{h}}$. The corresponding light-hole mass is $m_{\mathrm{lh}} = 0.09m_0$. Part of the problem arises in attempting to use a two-band theory when three bands (one conduction and two valence bands) participate. For example, the reduced mass $m_{\mathrm{r}}$ given by (1.16) based on measurements of $\varepsilon_{\mathrm{R}}$ and $\epsilon_{\mathrm{b}}$ probably does not exactly agree with that based on using the heavy-hole effective mass from the

Luttinger theory. Crystal strain existing in heterostructures with lattice mismatch can mix valence bands in ways that destroy the accuracy of parabolic band effective mass approximations altogether. Nevertheless, the simplicity of the two-band model merits our careful consideration.

At this point, we need to remind ourselves that the carrier-density independent band structure discussed so far assumes that only one conduction band electron and one valence-band hole are present in the semiconductor. In the presence of more electrons and holes, many-body interactions cause the band structure to change. This band-structure change is a result of changes in the Coulomb repulsion among carriers within the same band, because of screening and exchange interactions which are a consequence of the quantum statistics. One effect of the many-body carrier-carrier interactions is a reduction in the optical transition energy with increasing carrier density. This change is called bandgap renormalization, and it modifies (1.12) to

$$\hbar\omega_{\boldsymbol{k}} = \varepsilon_{\mathrm{e}\boldsymbol{k}} + \varepsilon_{\mathrm{h}\boldsymbol{k}} + \varepsilon_{\mathrm{g}0} + \delta\varepsilon_{\mathrm{g}} \quad , \tag{1.17}$$

where the renormalization energy $\delta\varepsilon_{\mathrm{g}}$ is often assumed to be independent of electron momentum. Bandgap renormalization explains why the laser diode typically oscillates at frequencies just below the zero-density bandgap energy, $\varepsilon_{\mathrm{g}0}$, and has consequences in predictions concerning the laser linewidth and antiguiding. Coulomb attraction between electrons and holes affects semiconductor behavior by reshaping the semiconductor gain spectrum in a way called *Coulomb enhancement.* These many-body effects are discussed in Chaps. 3, 4.

## 1.5 Units

As we can see from (1.14, 15), the question of the choice of units arises early in any discussion on semiconductor lasers. If we deal with laser physics alone, we would encounter no arguments with using MKS units. However semiconductor physicists often use cgs units. Hence, we have the problem of using cgs versus MKS units.

For this book, we propose the following compromise: The exciton Bohr radius often plays the role of a characteristic length scale in semiconductors and the Rydberg energy is the natural energy unit. Expressing results in terms of $a_0$ and $\varepsilon_{\mathrm{R}}$ helps to side step the units problem. This is important because semiconductor lasers are of interest in both solid-state physics and engineering communities. For example, in MKS units, the $e^2$ in (1.14 15) should be replaced by

$$e^2 \to \frac{e^2}{4\pi\epsilon_0} \quad , \tag{1.18}$$

where $\epsilon_0$ is the MKS permittivity of free space.

## 1.6 Fermi-Dirac Distributions

A very important effect of the Coulomb interaction is *carrier-carrier* scattering. This has a counterpart in gas lasers known as velocity-changing collisions, but it is a much stronger effect in semiconductors. In conventional near-infrared semiconductor lasers, the excitons are ionized at the densities high enough to get gain ($> 10^{18}$ cm$^{-3}$). The carrier-carrier scattering drives the electron and hole distributions each into Fermi-Dirac distributions, provided external forces like light fields vary little in the typical carrier-carrier scattering time of a picosecond or less. These distributions are called *quasi-equilibrium* distributions because they result from the equilibration of the carriers within their bands, but not among bands. In true thermodynamic equilibrium, the electrons are described by a single Fermi-Dirac distribution, which for typical temperatures gives filled valence bands and empty conduction bands, that is, a semiconductor in its ground state. Quasi-equilibrium occurs on a time scale long compared to the carrier-carrier scattering time but short compared to interband relaxation times, which are on the order of nanoseconds. It can be maintained in a steady state by pumping the electrons from the valence band to the conduction band with current injection or through optical field pumping.

The rapid carrier equilibration into Fermi-Dirac distributions greatly simplifies the analysis of the semiconductor medium. Instead of having to follow the carrier densities on an individual $\boldsymbol{k}$ basis, we may only need to determine the total carrier density $N$. The individual $\boldsymbol{k}$-dependent carrier population probabilities are then given by

$$f_{\alpha\boldsymbol{k}} = \frac{1}{e^{\beta[\varepsilon_{\alpha\boldsymbol{k}} - \mu_\alpha]} + 1} = f_\alpha(\varepsilon_{\alpha\boldsymbol{k}}) \quad , \tag{1.19}$$

where $\alpha = e(h)$ for electrons (holes), $\beta = 1/k_{\mathrm{B}}T$, and $\mu_\alpha$ is the carrier quasi-chemical potential, which is chosen to yield the total carrier density $N$. We measure the chemical potentials from their respective band edges. From (1.19), we see that independent of temperature, when $\varepsilon_{\alpha\boldsymbol{k}} = \mu_\alpha$, the Fermi-Dirac distribution $f_\alpha(\mu_\alpha) = \frac{1}{2}$. For $\varepsilon_{\alpha\boldsymbol{k}} < \mu_\alpha$, the occupation probability is therefore $> 1/2$ . A negative chemical potential, $\mu_\alpha < 0$, indicates that band $\alpha$ does not contain enough carriers to fill any state with 1/2 probability. At $T = 0$ K, the chemical potential equals the Fermi energy, which is the upper most level filled by carriers. Figure 1.7 is a plot of the Fermi-Dirac distributions for different temperatures. At $T = 0$ K, the Fermi-Dirac distribution is a step function, such that all states with energy below the chemical potential are completely filled and those above are completely empty. As the temperature is increased carriers begin to occupy energetically higher states. Except for high temperatures, the changes in the occupation of the states occur primarily in the energy range $k_{\mathrm{B}}T$ around $\mu_\alpha$.

As mentioned earlier, the chemical potential is determined by the temperature and by the total number of carriers. In the absence of doping, the

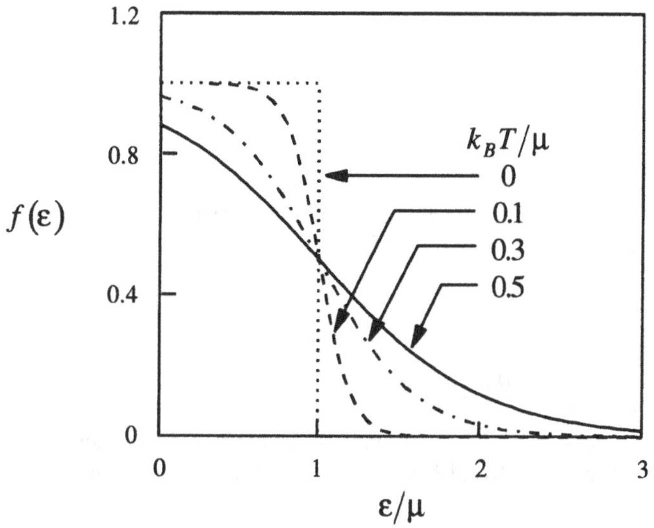


**Fig. 1.7.** Fermi-Dirac distributions versus normalized energy $\varepsilon/\mu$. The *curves* are for the normalized temperatures $k_{\mathrm{B}}T/\mu = 0$ (*dotted*), 0.1 (*dashed*), 0.3 (*dot-dashed*), and 0.5 (*solid*)

total carrier density is the same for electrons and holes. Denoting this by $N$, $\mu_\alpha$ is determined by the condition

$$N = \frac{1}{V} \sum_{\text{states}} f_{\alpha k} \quad , \tag{1.20}$$

where $V$ is the volume of the sample. In a bulk semiconductor, the electron or hole states are specified by the momenta, $k_x$, $k_y$, $k_z$ and spin component $s_z$, so that the summation over states gives

$$N = \frac{1}{V} \sum_{k_x} \sum_{k_y} \sum_{k_z} \sum_{s_z} f_{\alpha k} \quad . \tag{1.21}$$

The $x$-component of the wavefunction, for example, is

$$\psi_{k_x}(x) = \frac{1}{\sqrt{L_x}} \mathrm{e}^{\mathrm{i}k_x x} \quad , \tag{1.22}$$

where $L_x$ is the length of the semiconductor crystal in the $x$ direction. Assuming periodic boundary conditions within the semiconductor, $k_x$ is quantized according to

$$k_x = \frac{2\pi n}{L_x} \quad , \tag{1.23}$$

where $n$ is an integer ranging from $-\infty$ to $\infty$. For $L_x$ sufficiently large, we can assume an essentially continuous range of values for $k_x$. Then we can replace the summation over $k_x$ by an integral

$$\sum_{k_x} \to \int_{-\infty}^{\infty} \mathrm{d}k_x \frac{\mathrm{d}n}{\mathrm{d}k_x} \quad , \tag{1.24}$$

where

$$\frac{\mathrm{d}n}{\mathrm{d}k_x} = \frac{L_x}{2\pi} \quad ,$$

is the number of states within the interval $k_x$ and $k_x + dk_x$. Using an equivalent argument for the other two components of $\boldsymbol{k}$ gives

$$\begin{aligned} N &= \frac{1}{V}\sum_{\boldsymbol{k}} f_{\alpha k} \\ &= \frac{2}{(2\pi)^3}\int_{-\infty}^{\infty} \mathrm{d}k_x \int_{-\infty}^{\infty} \mathrm{d}k_y \int_{-\infty}^{\infty} \mathrm{d}k_z \, f_{\alpha k} \quad , \end{aligned} \tag{1.25}$$

where the factor of 2 comes from the $s_z$ summation and the $1/V$ has been cancelled out by the product $L_x L_y L_z$. If the integrand is spherically symmetric, then

$$\int_{-\infty}^{\infty} \mathrm{d}k_x \int_{-\infty}^{\infty} \mathrm{d}k_y \int_{-\infty}^{\infty} \mathrm{d}k_z \to \int_0^{\infty} \mathrm{d}k \, 4\pi k^2 \quad , \tag{1.26}$$

with $k^2 = k_x^2 + k_y^2 + k_z^2$, so that

$$N = \int_0^{\infty} \mathrm{d}k \, D_\alpha(k) \, f_{\alpha k} \quad , \tag{1.27}$$

where $D_\alpha(k) = (k/\pi)^2$ is the momentum density of states of a bulk semiconductor giving the number of states between $k$ and $k + \mathrm{d}k$.

We can convert the integration variable from $k$ in (1.27) to the energy $\varepsilon_\alpha = \hbar^2 k^2 / 2m_\alpha$, so that

$$N = \int_0^{\infty} \mathrm{d}\varepsilon_\alpha \, D_\alpha(\varepsilon_\alpha) f_\alpha(\varepsilon_\alpha) \tag{1.28}$$

and

$$D_\alpha(\varepsilon_\alpha) = \frac{1}{2\pi^2}\left(\frac{2m_\alpha}{\hbar^2}\right)^{\frac{3}{2}} \sqrt{\varepsilon_\alpha} \tag{1.29}$$

is the energy density of states in a bulk semiconductor giving the number of states between $\varepsilon_\alpha$ and $\varepsilon_\alpha + d\varepsilon_\alpha$. Whenever it is obvious if we are referring to the energy or momentum density of states, the functional dependence is usually omitted. The energy density of states is often more useful than the momentum density of states because transitions involve states that are within a range of energy instead of momentum. Specifically, the initial and final states that contribute strongest to a transition are those with energy separation $\Delta\varepsilon$ within the range $\hbar\nu \pm \hbar\gamma$, where $\nu$ and $\gamma$ are the transition frequency and linewidth, respectively.

We can use (1.28) to check the validity of the two-band model, i.e., the use of the heavy-hole band alone. In quasi-equilibrium, both valence bands share the same chemical potential, $\mu_{\mathrm{h}}$. The total carrier density for the holes is the sum of the light and heavy hole band densities

$$\begin{aligned} N &= N_{\mathrm{lh}} + N_{\mathrm{hh}} \\ &= \frac{\sqrt{2}}{\pi^2 \hbar^3} \left( m_{\mathrm{lh}}^{3/2} + m_{\mathrm{hh}}^{3/2} \right) \int_0^\infty \mathrm{d}\varepsilon \, \frac{\sqrt{\varepsilon}}{\mathrm{e}^{\beta[\varepsilon - \mu_{\mathrm{h}}]} + 1} \ , \end{aligned}$$

giving

$$\frac{N_{\mathrm{lh}}}{N_{\mathrm{hh}}} = \left( \frac{m_{\mathrm{lh}}}{m_{\mathrm{hh}}} \right)^{\frac{3}{2}} \ . \tag{1.30}$$

For the values $m_{\mathrm{hh}} = 0.377 m_0$ and $m_{\mathrm{lh}} = 0.09 m_0$, this gives 12 % of the holes in the light-hole band, which is small enough to neglect for our simpler modeling.

In general, the determination of the chemical potential for a given carrier density has to be done numerically, because there is no analytical solution of the integrals in (1.27, 28). The chemical potential can always be found by iteration, with the necessary integrations done numerically. However there are approximate analytic expressions relating the chemical potential to carrier density for some cases. One such case occurs when only the high energy tail of the Fermi-Dirac distribution is within the band. In other words, the chemical potential lies sufficiently far inside the bandgap (is sufficiently negative) that

$$\varepsilon_{\alpha \boldsymbol{k}} - \mu_\alpha \gg k_{\mathrm{B}} T \ . \tag{1.31}$$

Then the exponential term may dominate the 1 in the denominator of (1.19) and

$$f_{\alpha \boldsymbol{k}} \simeq \mathrm{e}^{\beta \mu_\alpha} \, \mathrm{e}^{-\beta \varepsilon_{\alpha \boldsymbol{k}}} \tag{1.32}$$

which is a Maxwell-Boltzmann distribution. Equation (1.28) then can be readily evaluated as (setting $q = \beta \hbar^2 / 2 m_\alpha$ for typographical simplicity)

$$\begin{aligned} N &\simeq \frac{1}{\pi^2} \mathrm{e}^{\beta \mu_\alpha} \int_0^\infty \mathrm{d}k \, k^2 \, \mathrm{e}^{-q k^2} \\ &= -\frac{1}{\pi^2} \mathrm{e}^{\beta \mu_\alpha} \frac{\partial}{\partial q} \int_0^\infty \mathrm{d}k \, \mathrm{e}^{-q k^2} \\ &= -\frac{1}{\pi^2} \frac{\sqrt{\pi}}{2} \mathrm{e}^{\beta \mu_\alpha} \frac{\partial}{\partial q} \frac{1}{\sqrt{q}} \\ &= \frac{1}{4} \left( \frac{2 m_\alpha k_{\mathrm{B}} T}{\pi \hbar^2} \right)^{\frac{3}{2}} \mathrm{e}^{\beta \mu_\alpha} \ . \end{aligned} \tag{1.33}$$

Hence

$$\mathrm{e}^{\beta \mu_\alpha} = 4N \left( \frac{\pi \hbar^2}{2 m_\alpha k_{\mathrm{B}} T} \right)^{\frac{3}{2}} \equiv \overline{N} \ , \tag{1.34}$$

or

$$\beta \mu_\alpha = \ln(\overline{N}) \ , \tag{1.35}$$

where we introduce the normalized total carrier density $\overline{N}$.

The Maxwell-Boltzmann distribution is sometimes a good approximation for the hole distribution because the high effective mass of the heavy hole band gives rise to a high density of states. This in turn makes it difficult to generate a signifcant hole population in the valence band. Figure 1.8 compares the Fermi-Dirac and Maxwell-Boltzmann distributions for electrons and holes in GaAs. We see that the hole distribution agrees rather well with the Maxwell-Boltzmann approximation, whereas for these parameters the electrons clearly exhibit near degenerate Fermi-Dirac statistics, which cannot be reproduced by a Boltzmann distribution. To force a fit would yield values of $f_e > 1$, a truely unphysical result!

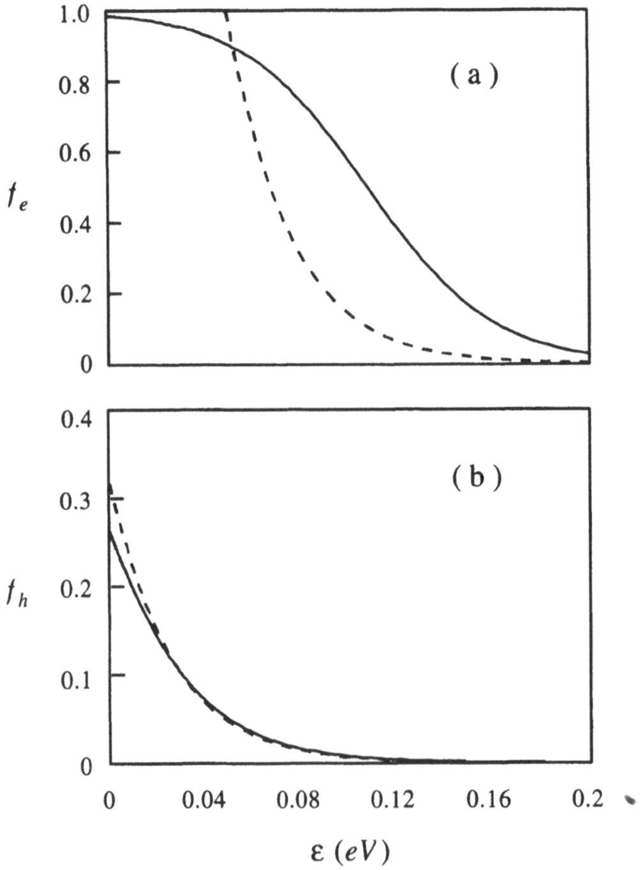


**Fig. 1.8a, b.** Fermi-Dirac (*solid lines*) and Maxwell-Boltzmann (*dashed lines*) distributions as functions of carrier energy $\varepsilon$, for $N = 3 \times 10^{-18}$ cm$^{-3}$ and $T = 300$ K. (**a**) shows the electron population, while (**b**) shows the hole population. We assume bulk GaAs parameters with an electron effective mass $m_c = 0.0665 m_e$, and a hole effective mass $m_h = 0.52 m_e$

For $T \to 0$, $\beta = 1/k_B T \to \infty$, for which the Fermi-Dirac distribution becomes a step function, truncating the integral (1.28) to

$$N = \frac{(2m_\alpha/\hbar^2)^{3/2}}{2\pi^2} \int_0^{\mu_\alpha} d\varepsilon_\alpha \sqrt{\varepsilon_\alpha}$$
$$= \frac{1}{3\pi^2}\left(\frac{2m_\alpha \mu_\alpha}{\hbar^2}\right)^{\frac{3}{2}} .$$

Inverting this, we have the chemical potential

$$\mu_\alpha = \frac{\hbar^2}{2m_\alpha}(3\pi^2 N)^{\frac{2}{3}} . \tag{1.36}$$

## 1.7 Quantum Confinement

GaAs does not occur in nature and as such can be considered a "designer material". Thanks to modern crystal growth techniques, one can not only determine the composition of semiconductors with remarkable precision, but can also determine their shape virtually on an atomic scale. In particular, it is possible to fabricate microstructures so small that their electronic and optical properties deviate substantially from those of bulk materials. The onset of pronounced quantum confinement effects occur when one or more dimensions of a structure become comparable to the characteristic length scale of the elementary excitations. Quantum confinement may be in one spatial dimension, as in *quantum wells*, in two spatial dimensions as in *quantum wires*, or in all three spatial dimensions as in *quantum dots*. The confinement modifies the allowed energy states of the crystal electrons and changes the density of states. In this section, we introduce the basic properties of quantum-well structures that we use in the next chapters on laser gain. In Chap. 6, we discuss the finer but still important modifications to the quantum-well band structure.

We begin our discussion with quantum wells, which are the most developed of the quantum-confined structures. Quantum-well lasers are commercially available, while quantum-wire and quantum-dot lasers are still in the research stages. Some references for these systems are listed at the end of this chapter. An understanding of the basic effects is best obtained by considering *ideal quantum confinement* conditions, for which the elementary excitations are completely confined inside the microstructure and the electronic wavefunctions vanish beyond the surfaces. For this idealized situation, we can write the confinement potential as

$$V_{\rm con}(z) = \begin{cases} 0 & |z| < w/2 \\ \infty & |z| > w/2 \end{cases} . \tag{1.37}$$

In the $xy$ plane there is no quantum confinement and the carriers can move freely. The electron eigenfunction (actually the envelope of the eigenfunction as discussed in Chap. 6) can be separated as

$$\psi_{n,\boldsymbol{k}_\perp}(\boldsymbol{r}) = \phi_{\boldsymbol{k}_\perp}(\boldsymbol{r}_\perp)\zeta_n(z) \quad , \tag{1.38}$$

where the $z$ and transverse components $\boldsymbol{r}_\perp(x,y)$ obey the Schrödinger equations

$$\left[-\frac{\hbar^2}{2m_z}\frac{\mathrm{d}^2}{\mathrm{d}z^2} + V_{\mathrm{con}}(z)\right]\zeta_n(z) = \varepsilon_n\zeta_n(z) \ , \tag{1.39}$$

and

$$-\frac{\hbar^2}{2m_\perp}\nabla_\perp^2\phi_{\boldsymbol{k}_\perp}(\boldsymbol{r}_\perp) = \varepsilon_{\boldsymbol{k}_\perp}\phi_{\boldsymbol{k}_\perp}(\boldsymbol{r}_\perp) \quad , \tag{1.40}$$

respectively. For simplicity, we assume that the bulk-material band structure can be described by parabolic energy bands that are characterized by the effective masses $m_z$ and $m_\perp$. As shown in Chap. 6, $m_z$ and $m_\perp$ are equal for the conduction bands. They differ for the valence bands, which leads to the interesting property of mass reversal. Equation (1.40) describes a two-dimensional free particle (i.e., no external potential and not interacting with other particles) with eigenfunctions

$$\phi_{\boldsymbol{k}\perp}(\boldsymbol{r}_\perp) = \frac{1}{\sqrt{A}}\,\mathrm{e}^{\pm \mathrm{i}\boldsymbol{k}\cdot\boldsymbol{r}_\perp} \tag{1.41}$$

and eigenvalue

$$\varepsilon_{\boldsymbol{k}_\perp} = \frac{\hbar^2 k_\perp^2}{2m_\perp} \quad , \tag{1.42}$$

where $A$ is the area of the quantum-well. Because of the infinite confinement potential, we have the boundary conditions

$$\zeta_n\left(\frac{w}{2}\right) = \zeta_n\left(-\frac{w}{2}\right) = 0 \quad , \tag{1.43}$$

which lead to the even and odd solutions of (1.39)

$$\zeta_n(z) = \sqrt{\frac{2}{w}}\cos(k_n z) \quad , \quad n \text{ even} \quad , \tag{1.44}$$

$$\zeta_n(z) = \sqrt{\frac{2}{w}}\sin(k_n z) \quad , \quad n \text{ odd} \quad , \tag{1.45}$$

where the wavenumbers $k_n$ are given by

$$k_n = \frac{n\pi}{w} \quad , \tag{1.46}$$

and the bound state energies $\varepsilon_n$ are

$$\varepsilon_n = \frac{\hbar^2 k_n^2}{2m_z} = \frac{\pi^2\hbar^2 n^2}{2m_z w^2} \quad . \tag{1.47}$$

Adding the energies of the motion in the $xy$ plane and in the $z$-direction, we find the total energy of the electron subjected to one-dimensional quantum confinement to be

$$\varepsilon = \frac{\pi^2 \hbar^2 n^2}{2m_z w^2} + \frac{\hbar^2 k_\perp^2}{2m_\perp} \ , \tag{1.48}$$

where $n = 1, 2, 3, \ldots$, indicating a succession of energy subbands, i.e., energy parabola $\hbar^2 k_\perp^2 / 2m_\perp$ separated by $\pi^2\hbar^2/2m_z w^2$. The different subbands are labeled by the quantum numbers $n$. Figure 1.9 depicts the energy eigenstates.

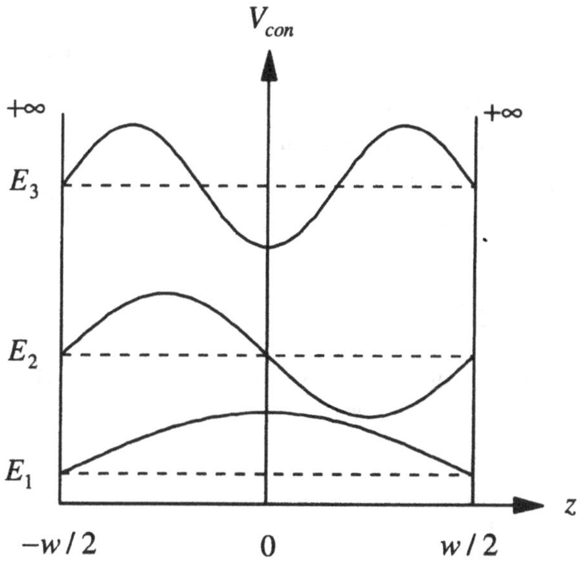


**Fig. 1.9.** Energy eigenfunctions and eigenvalues of the three energetically lowest states for an infinite one-dimensional square potential well

Realistically, we can only fabricate finite confinement potentials, so that

$$V_{\mathrm{con}}(z) = \begin{cases} 0 & |z| < w/2 \\ V_c & |z| > w/2 \end{cases} \ . \tag{1.49}$$

The analysis follows closely the treatment of the infinite potential, with the Schrödinger equation for the $x$-$y$ motion being unchanged. However, solutions in the $z$-direction can no longer be determined analytically. Equation (1.39) now has to be solved separately in the regions, *i)* $|z| < w/2$, *ii)* $z > w/2$, and *iii)* $z < -w/2$. In region *i)*, the solutions are given by (1.44, 45), while in regions *ii)* and *iii)* they are

$$\zeta(z) = C_+ \, \mathrm{e}^{K_z z} + C_- \, \mathrm{e}^{-k_z z} \ , \tag{1.50}$$

where

$$K_z^2 = \frac{2m_z}{\hbar^2}(V_c - \varepsilon_z) \ . \tag{1.51}$$

Since the wavefunction has to be normalizable we have to pick the decaying solutions in (1.50). Also, we have to match the solutions and their derivatives at the interfaces $\pm w/2$. This yields the even states

$$\zeta_{2n}(z) = \begin{cases} B\cos(k_z z) & |z| \leq w/2 \\ C\,\mathrm{e}^{-K_z|z|} & |z| > w/2 \end{cases} \tag{1.52}$$

with the condition

$$\sqrt{\varepsilon_z}\tan\left(\frac{\sqrt{m_z\varepsilon_z}}{2\hbar^2}w\right) = \sqrt{V_c - \varepsilon_z} \ , \tag{1.53}$$

whose solution gives the energy of the even states.

Similarly, the odd-states wavefunctions are given by

$$\zeta_{2n-1}(z) = \begin{cases} A\sin(k_z z) & |z| \leq w/2 \\ -C\,\mathrm{e}^{-K_z|z|} & |z| > w/2 \end{cases} \tag{1.54}$$

with the condition

$$-\sqrt{\varepsilon_z}\cot\left(\frac{\sqrt{m_z\varepsilon_z}}{2\hbar^2}w\right) = \sqrt{V_c - \varepsilon_z} \ , \tag{1.55}$$

whose solution gives the energy of the odd states. Figure 1.10 depicts the solutions of a finite one-dimensional square well.

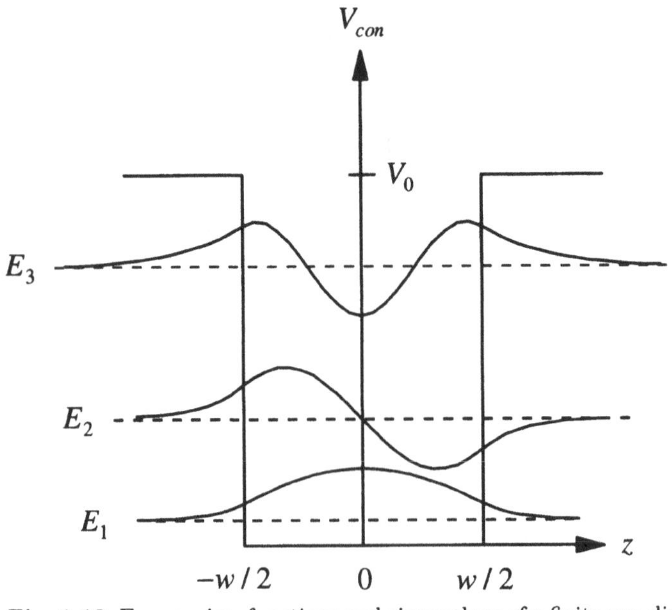


**Fig. 1.10.** Energy eigenfunctions and eigenvalues of a finite one-dimensional square well

An important difference between bulk and quantum-confined structures is in the density of states. Following the steps taken for the bulk material in the previous section, the sum over states in the quantum well may be approximated by the integral

$$\sum_{\text{states}} \to \sum_n \frac{2A}{(2\pi)^2} \int_{-\infty}^{\infty} \mathrm{d}k_x \int_{-\infty}^{\infty} \mathrm{d}k_y \ , \tag{1.56}$$

where the factor of 2 comes from the spin summation and the volume of the material is $V = wA$. If the band structure is symmetric in the $xy$ plane, then

$$\int_{-\infty}^{\infty} \mathrm{d}k_x \int_{-\infty}^{\infty} \mathrm{d}k_y = \int_0^{\infty} \mathrm{d}k_\perp \, 2\pi k_\perp \ , \tag{1.57}$$

where $k_\perp^2 = k_x^2 + k_y^2$, so that

$$\frac{1}{V} \sum_{\text{states}} \to \sum_n \int_0^{\infty} \mathrm{d}k_\perp \, D(k_\perp) \ . \tag{1.58}$$

Here $D(k_\perp) = k/(\pi w)$, is the 2-dimensional momentum density of states giving the number of states between $k_\perp$ and $k_\perp + \mathrm{d}k_\perp$. Using (1.42), we can convert the integration variable into energy, so that

$$\frac{1}{V} \sum_{\text{states}} \to \sum_n \int_0^{\infty} \mathrm{d}\varepsilon \, D(\varepsilon) \ , \tag{1.59}$$

where the constant

$$D(\varepsilon) = \frac{m_z}{\pi w \hbar^2} \tag{1.60}$$

gives the number of states per unit volume between $\varepsilon$ and $\varepsilon + \mathrm{d}\varepsilon$.

The two-dimensional carrier density integral can be evaluated analytically [*Haug* and *Koch* (1994)], yielding the chemical potential

$$\beta\mu_\alpha = \ln\left[\exp\!\left(\frac{\pi\hbar^2 N_{2\mathrm{d}}}{m_\alpha k_\mathrm{B} T}\right) - 1\right] \ , \tag{1.61}$$

where $N_{2\mathrm{d}} = N/w$ is the 2-dimensional carrier density. For $T \to 0$, this reduces to

$$\mu_\alpha = \frac{\pi\hbar^2 N_{2\mathrm{d}}}{m_\alpha} \ . \tag{1.62}$$

## 1.8 Slowly-Varying Maxwell Equations

In this section, we consider how a semiconductor gain medium interacts with a laser field. Most theoretical problems involving lasers may be treated using the semiclassical approximation, where one describes the laser field classically and the gain medium quantum mechanically. Figure 1.11 summarizes

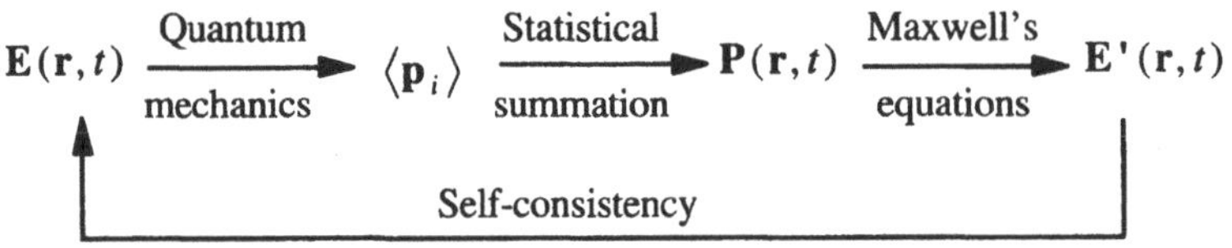


**Fig. 1.11.** Self-consistent semiclassical theory of a laser field $\boldsymbol{E}(\boldsymbol{r}, t)$ interacting with a quantum-mechanical medium. The assumed electric field $\boldsymbol{E}(\boldsymbol{r}, t)$ induces a polarization of the medium that, in turn, drives the field self consistently

the steps involved in the application of semiclassical laser theory. One begins by calculating the microscopic electric-dipole moments induced in the gain medium by a given laser field, $\boldsymbol{E}(\boldsymbol{r}, t)$. These dipoles are summed to yield a macroscopic polarization $\boldsymbol{P}(\boldsymbol{r}, t)$. This polarization then drives the laser field, $\boldsymbol{E}'(\boldsymbol{r}, t)$ according to Maxwell's equations. Self-consistency is imposed by the condition, $\boldsymbol{E}(\boldsymbol{r}, t) = \boldsymbol{E}'(\boldsymbol{r}, t)$. In this section we complete the Maxwell-equation part of this scheme by finding out how a slowly-varying electric field amplitude is affected by the polarization of the medium. The next section outlines the quantum mechanical method for deriving the other part of the scheme, namely how the field induces the polarization of the medium according to the quantum mechanics of the semiconductor.

We describe the laser field by Maxwell's equations (in MKS units)

$$\boldsymbol{\nabla} \cdot \boldsymbol{B} = 0 \ , \tag{1.63}$$

$$\boldsymbol{\nabla} \cdot \boldsymbol{D} = 0 \ , \tag{1.64}$$

$$\boldsymbol{\nabla} \times \boldsymbol{E} = -\frac{\partial \boldsymbol{B}}{\partial t} \ , \tag{1.65}$$

$$\boldsymbol{\nabla} \times \boldsymbol{H} = \frac{\partial \boldsymbol{D}}{\partial t} \ . \tag{1.66}$$

In these equations, the magnetic flux, $\boldsymbol{B}$, and the magnetic field, $\boldsymbol{H}$ are related by the constitutive relation

$$\boldsymbol{B} = \mu \boldsymbol{H} \ , \tag{1.67}$$

where $\mu$ is the permeability of the host medium. The displacement electric field, $\boldsymbol{D}$, is given by

$$\boldsymbol{D} = \epsilon \boldsymbol{E} = \epsilon_\mathrm{b} \boldsymbol{E} + \boldsymbol{P} \ , \tag{1.68}$$

where $\epsilon_\mathrm{b}$ is the permittivity of the host medium and $\boldsymbol{P}$ is the induced polarization (dipoles per unit volume). For a semiconductor laser, the host medium is the lattice. The permeability, $\mu \simeq \mu_0$, where $\mu_0$ is the permeability of the vacuum. The semiconductor lattice typically has a background index of refraction, $n \simeq 3.5$, which is included in $\epsilon_\mathrm{b}$ via $\epsilon_\mathrm{b} = n^2 \epsilon_0$, where $\epsilon_0$ is the permittivity in vaccuum. The polarization, $\boldsymbol{P}$, gives the gain and carrier-induced refractive index, and is induced by the laser field interacting with the electrons in the conduction and valence bands.

Combining the curl of (1.65) with (1.66), gives

$$\nabla \times \nabla \times \boldsymbol{E} = \nabla(\nabla \cdot \boldsymbol{E}) - \nabla^2 \boldsymbol{E} = \mu_0 \frac{\partial^2 \boldsymbol{D}}{\partial t^2} \quad . \tag{1.69}$$

Since most light field vectors vary little along the directions in which they point, $\nabla \cdot \boldsymbol{E} \simeq 0$. For example, a plane-wave field is constant along the direction it points, causing its $\nabla \cdot \boldsymbol{E}$ to vanish identically. Using (1.68) for $\boldsymbol{D}$, we get the wave equation

$$-\nabla^2 \boldsymbol{E} + \left(\frac{n}{c}\right)^2 \frac{\partial^2 \boldsymbol{E}}{\partial t^2} = -\mu_0 \frac{\partial^2 \boldsymbol{P}}{\partial t^2} \quad , \tag{1.70}$$

where $c$ is the speed of light in vacuum and $\mu_0 \epsilon_b = (n/c)^2$.

For the purposes of calculating the gain and index of the medium, we consider a laser field of the simple plane-wave form

$$\boldsymbol{E}(z,t) = \frac{1}{2} \hat{\imath} E(z)\, \mathrm{e}^{\mathrm{i}[Kz - \nu t - \phi(z)]} + \text{c.c.} \quad , \tag{1.71}$$

where $\hat{\imath}$ is the unit vector in the $x$ direction, $E(z)$ and $\phi(z)$ are the real field amplitude and phase shift that vary little in an optical wavelength, $\nu$ is the field frequency in radians/second and $\exp(\mathrm{i}Kz)$ accounts for most of the spatial variation in the laser field. We choose a monochromatic plane travelling wave because it allows us to illustrate the necessary physics of the semiconductor gain medium with the minimum of algebra. We use a plane wave in calculating the local properties of a gain medium, where the volume element considered can always be made sufficiently small compared to the transverse variations in the laser field. The local gain and refractive index are needed for beam propagation and wave optical studies. A limitation to using a monochromatic field is that we cannot deal with the coherent response of a gain medium to multimode fields.

The laser field induces a polarization in the medium,

$$\boldsymbol{P}(z,t) = \frac{1}{2} \hat{\imath} P(z)\, \mathrm{e}^{\mathrm{i}[Kz - \nu t - \phi(z)]} + \text{c.c.} \quad , \tag{1.72}$$

where $P(z)$ is a complex polarization amplitude that varies little in a wavelength. It is related to the complex susceptibility of the medium by

$$P(z) = \epsilon_b \chi(z) E(z) \quad . \tag{1.73}$$

Substituting (1.71) and (1.72) into the wave equation (1.70), we find

$$-\frac{\mathrm{d}^2 E}{\mathrm{d}z^2} - 2\mathrm{i}\left(K - \frac{\mathrm{d}\phi}{\mathrm{d}z}\right)\frac{\mathrm{d}E}{\mathrm{d}z} + \left[\left(K - \frac{\mathrm{d}\phi}{\mathrm{d}z}\right)^2 + \mathrm{i}\frac{\mathrm{d}^2\phi}{\mathrm{d}t^2} - \left(\frac{n\nu}{c}\right)^2\right] E = \mu_0 \nu^2 P \quad . \tag{1.74}$$

This equation can be simplified considerably if one assumes that $E$ and $\mathrm{d}\phi/\mathrm{d}z$ vary little in a wavelength, so that terms containing $\mathrm{d}^2E/\mathrm{d}z^2$, $\mathrm{d}^2\phi/\mathrm{d}z^2$ and $(\mathrm{d}E/\mathrm{d}z)(\mathrm{d}\phi/\mathrm{d}z)$ may be neglected. This *slowly varying envelope approximation* (SVEA) gives

$$-2\mathrm{i}K\frac{\mathrm{d}E}{\mathrm{d}z}+K^2E-2KE\frac{\mathrm{d}\phi}{\mathrm{d}z}-\left(\frac{n\nu}{c}\right)^2E=\mu_0\nu^2P \ ,$$

i.e.,

$$\frac{\mathrm{d}E}{\mathrm{d}z}-\mathrm{i}E\frac{\mathrm{d}\phi}{\mathrm{d}z}=\frac{\mathrm{i}\mu_0\nu^2}{2K}P=\frac{\mathrm{i}K}{2}\chi E \ , \tag{1.75}$$

where we use $K=\nu n/c$. Equating the real and imaginary parts, we find the *self-consistency equations*

$$\frac{\mathrm{d}E(z)}{\mathrm{d}z} = -\frac{\nu}{2\epsilon_0 nc}\,\mathrm{Im}\{P(z)\}=-\frac{K}{2}\chi''(z)E(z) \ , \tag{1.76}$$

$$\frac{\mathrm{d}\phi(z)}{\mathrm{d}z} = -\frac{\nu}{2\epsilon_0 ncE(z)}\,\mathrm{Re}\{P(z)\}=-\frac{K}{2}\chi'(z) \ , \tag{1.77}$$

where $\chi=\chi'+\mathrm{i}\chi''$. Self-consistency refers to the requirement that the field parameters ultimately appearing in the formulas for $P(z)$ are taken to be the very same as those computed from (1.76) and (1.77) for a given $P(z)$.

Two useful parameters for characterizing a laser medium are the gain and the carrier-induced refractive index change. The amplitude gain is defined as

$$\frac{\mathrm{d}E}{\mathrm{d}z}=gE \ , \tag{1.78}$$

where in general, $g$ is a function of $\varepsilon$ and equals one half the intensity gain $G$. The gain has units of inverse length. Comparing (1.76) to (1.78), we find the local gain to be

$$g=-\frac{K}{2}\chi'' \ . \tag{1.79}$$

To find the carrier-induced refractive index $\delta n$, note that the wavenumber of the laser field given by (1.71) is

$$K-\frac{\mathrm{d}\phi}{\mathrm{d}z}=(n+\delta n)K_0 \ , \tag{1.80}$$

where $n$ is the refractive index of the lattice, $\delta n$ is the index change due to the carriers, and $K_0$ is the wavenumber in vacuum. Since $K=nK_0$, we have

$$\frac{\mathrm{d}\phi}{\mathrm{d}z}=-K_0\delta n \ . \tag{1.81}$$

Combining this with (1.77), we have the relative index change

$$\frac{\delta n}{n}=\frac{\chi'}{2} \ . \tag{1.82}$$

## 1.9 Quantum Mechanics of the Semiconductor Medium

In this section, we introduce the second quantized (or Fock) representation to treat the semiconductor gain medium. There are two reasons for going beyond elementary quantum mechanics, which treats the wavefunction as a simple complex function. First, we have to account for the fact that electrons are indistinguishable as depicted by the Feymann diagram for the scattering of two electrons in Fig. 1.12. Similarly the holes are indistinguishable particles. The calculation of the scattering cross section of two electrons has to involve the correct average of the two experimentally indistinguishable events, direct and exchange scattering. We can do that with the properly antisymmetrized wavefunctions, but the second quantized representation takes care of the book keeping more conveniently.

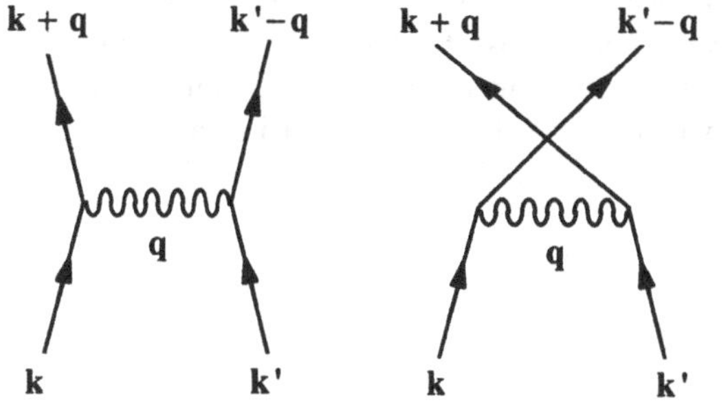


**Fig. 1.12.** Direct and exchange electron-electron scattering. The facts that these two events are experimentally indistinguishable and that their scattering amplitudes have opposite signs are automatically taken into account in the second quantized representation

The second reason for going beyond elementary quantum mechanics is because, for a semiconductor interacting with a light field, the number of electron-hole pairs is not conserved: we can create or annihilate electron-hole pairs. This is very inconvenient to describe at the level of elementary quantum mechanics. On the other hand, it is straightforward with the use of second quantization.

First, we introduce the electron field operator

$$\hat{\psi}(\boldsymbol{r},t) = \sum_{\lambda}\sum_{\boldsymbol{k}}\sum_{s_z} a_{\lambda \boldsymbol{k} s_z}(t)\phi_{\lambda \boldsymbol{k} s_z}(\boldsymbol{r}) \quad , \tag{1.83}$$

where $\phi_{\lambda \boldsymbol{k} s_z}(\boldsymbol{r})$ is the single-particle eigenfunction for an electron in the semiconductor and $a_{\lambda \boldsymbol{k} s_z}(t)$ is the *annihilation operator* for the electron in that state, which we specify by the band index $\lambda = c$ or $v$, momentum $\boldsymbol{k}$, and $z$-component of spin $s_z$. Its Hermitean adjoint $a^\dagger_{\lambda \boldsymbol{k} s_z}$ *creates* an electron in the same state.

For the sake of clarity, we write the operators with all their indices appearing explicitly. Starting with Chap. 2, we usually incorporate the spin variable into $\boldsymbol{k}$ for typographical simplicity. In that case, the subscript $\boldsymbol{k}$ represents the three-dimensional momentum vector $\boldsymbol{k}$, and two possible spin directions $s_z = \pm\frac{1}{2}$. The summation over $\boldsymbol{k}$ then involves summations over $k_x$, $k_y$, $k_z$, and $s_z$.

As Fermion operators, the electron creation and annihilation operators obey anticommutation relations. These relations are a consequence of the Pauli exclusion principle, which states that at most one Fermion can occupy any given state. The anticommutation relations for the electron creation and annihilation operators are

$$\left[a_{\lambda \boldsymbol{k} s_z}, a_{\lambda' \boldsymbol{k}' s_z'}\right]_+ = \left[a^\dagger_{\lambda \boldsymbol{k} s_z}, a^\dagger_{\lambda' \boldsymbol{k}' s_z'}\right]_+ = 0 \tag{1.84}$$

$$\left[a_{\lambda \boldsymbol{k} s_z}, a^\dagger_{\lambda' \boldsymbol{k}' s_z'}\right]_+ = \delta_{\lambda,\lambda'} \delta_{\boldsymbol{k},\boldsymbol{k}'} \delta_{s_z s_z'} \quad , \tag{1.85}$$

where for two operators, $A$ and $B$, the anticommutator is defined by

$$[A, B]_+ = AB + BA \quad .$$

The combination $a^\dagger_{\lambda \boldsymbol{k} s_z} a_{\lambda \boldsymbol{k} s_z}$ is the number operator for an electron in band $\lambda$ with momentum $\boldsymbol{k}$ and $s_z$. The eigenstates for $a^\dagger_{\lambda \boldsymbol{k} s_z} a_{\lambda \boldsymbol{k} s_z}$ are $|0_{\lambda \boldsymbol{k} s_z}\rangle$ and $|1_{\lambda \boldsymbol{k} s_z}\rangle$, which are the states containing no electron and one electron, respectively. These eigenstates when operated on by the creation, annihilation and number operators give

$$\begin{aligned}
& a_{\lambda \boldsymbol{k} s_z} |0_{\lambda \boldsymbol{k} s_z}\rangle = a^\dagger_{\lambda \boldsymbol{k} s_z} |1_{\lambda \boldsymbol{k} s_z}\rangle = 0 \quad , \\
& a_{\lambda \boldsymbol{k} s_z} |1_{\lambda \boldsymbol{k} s_z}\rangle = |0_{\lambda \boldsymbol{k} s_z}\rangle \quad , \\
& a^\dagger_{\lambda \boldsymbol{k} s_z} |0_{\lambda \boldsymbol{k} s_z}\rangle = |1_{\lambda \boldsymbol{k} s_z}\rangle \quad , \\
& a^\dagger_{\lambda \boldsymbol{k} s_z} a_{\lambda \boldsymbol{k} s_z} |0_{\lambda \boldsymbol{k} s_z}\rangle = 0 \quad , \\
& a^\dagger_{\lambda \boldsymbol{k} s_z} a_{\lambda \boldsymbol{k} s_z} |1_{\lambda \boldsymbol{k} s_z}\rangle = |1_{\lambda \boldsymbol{k} s_z}\rangle \quad ,
\end{aligned} \tag{1.86}$$

where the first equation expresses the fact that there is no electron to annihilate in an empty state, and it is impossible to create an electron in an already filled state.

In the electron-hole representation for a two-band model, we define the hole creation operator

$$b^\dagger_{-\boldsymbol{k},-s_z} = a_{v \boldsymbol{k} s_z} \quad . \tag{1.87}$$

This equation indicates that the annihilation of a valence-band electron with a given momentum and $z$-component of spin corresponds to the creation of a hole with the opposite momentum and $z$-component of spin. Note that for clarity in (1.87) we use a comma between the $-\boldsymbol{k}$ and $-s_z$ subscripts, although it is probably clear without the comma since it does not make sense to subtract a spin quantum number from a wavevector $\boldsymbol{k}$. Similarly the hole annihilation operator is given by

$$b_{-\boldsymbol{k},-s_z} = a^\dagger_{v \boldsymbol{k} s_z} \quad . \tag{1.88}$$

The hole operators also obey anticommutation relationships, so that the probability of finding a particular valence-band electron becomes

$$\left\langle a^\dagger_{v \boldsymbol{k} s_z} a_{v \boldsymbol{k} s_z} \right\rangle = 1 - \left\langle b^\dagger_{-\boldsymbol{k},-s_z} b_{-\boldsymbol{k},-s_z} \right\rangle \quad , \tag{1.89}$$

where the brackets $\langle \cdots \rangle$ are used to indicate an expectation (or quantum mechanically averaged) value. As expected, the probability of finding a valence electron is one *minus* the probability of finding a hole. In the electron-hole representation, we use the term electrons to refer to conduction-band electrons and holes to refer to valence-band holes. The electron annihilation operator is

$$a_{ks_z} = a_{cks_z} \tag{1.90}$$

and the electron creation operator is

$$a^\dagger_{ks_z} = a^\dagger_{cks_z} \quad . \tag{1.91}$$

In the second quantized representation, the Hamiltonian for $N$ noninteracting electrons is

$$\begin{aligned} H_{\text{kin}} &= \int \mathrm{d}^3 r\, \hat{\psi}^\dagger(\boldsymbol{r}) \left(-\frac{\hbar^2\nabla^2}{2m_\lambda}\right) \hat{\psi}(\boldsymbol{r}) \\ &= \sum_\lambda \sum_k \sum_{s_z} \varepsilon_{\lambda k} a^\dagger_{\lambda k s_z} a_{\lambda k s_z} \quad , \end{aligned} \tag{1.92}$$

where the index *kin* indicates that this is the kinetic energy part of the interacting case, given by (1.9, 10). Evaluating the band summation and restricting ourselves again to the two-band approximation gives

$$H_{\text{kin}} = \sum_k \sum_{s_z} \left(\varepsilon_{ck} a^\dagger_{cks_z} a_{cks_z} + \varepsilon_{vk} a^\dagger_{vks_z} a_{vks_z}\right) \quad . \tag{1.93}$$

In the electron-hole representation, (1.93) becomes

$$H_{\text{kin}} = \sum_k \sum_{s_z} \left[\varepsilon_{ck} a^\dagger_{ks_z} a_{ks_z} + \varepsilon_{vk}\left(1 - b^\dagger_{-k,-s_z} b_{-k,-s_z}\right)\right] \quad . \tag{1.94}$$

Since the origin of energy is arbitrary, the constant term, $\sum_{ks_z} \varepsilon_{vk}$ is usually left out. Then

$$H_{\text{kin}} = \sum_k \sum_{s_z} (\varepsilon_{\mathrm{g}0} + \varepsilon_{\mathrm{e}k}) a^\dagger_{ks_z} a_{ks_z} + \sum_k \sum_{s_z} \varepsilon_{\mathrm{h}k} b^\dagger_{ks_z} b_{ks_z} \quad , \tag{1.95}$$

where $\varepsilon_{\mathrm{e}k}$ and $\varepsilon_{\mathrm{h}k}$ are given by (1.13). In going from (1.93) to (1.95) we set $m_\mathrm{h} = -m_\mathrm{v}$, where $m_\mathrm{v}$ is the valence electron effective mass and $m_\mathrm{h}$ is the hole effective mass. This is only true for noninteracting electrons. When the Coulomb interactions among electrons are taken into account, the relationship between $m_\mathrm{v}$ and $m_\mathrm{h}$ is more complicated (Sect. 3.1). However the kinetic energy part of the total Hamiltonian can still be put in the form of (1.95).

An example of a physical quantity that is represented by an operator is the density distribution

$$\begin{aligned} n(\boldsymbol{r}) &= \hat{\psi}^\dagger(\boldsymbol{r})\hat{\psi}(\boldsymbol{r}) \\ &= \frac{1}{V} \sum_{k,k',s_z} \mathrm{e}^{\mathrm{i}(k-k')\cdot r} a^\dagger_{k',s_z} a_{k,s_z} \\ &= \sum_q n_q \,\mathrm{e}^{\mathrm{i}q\cdot r} \quad , \end{aligned} \tag{1.96}$$

where

$$n_q = \frac{1}{V} \sum_{k,s_z} a^\dagger_{k-q,s_z} a_{k,s_z} \tag{1.97}$$

is the Fourier amplitude of the density distribution operator. As discussed in the following chapters, the determination of the density distribution involves the solution of the equation of motion for $n_q$.

To compute expectation values for an $N$-particle system, we need to choose a basis set. For our problem a convenient basis is the one made up of the eigenstates of $a^\dagger_{ks_z} a_{ks_z}$ and $b^\dagger_{ks_z} b_{ks_z}$. These are the products

$$\begin{aligned} |\{n_i\}\rangle &= |n_{\mathrm{e}k_1 s_{z1}}\rangle \, |n_{\mathrm{e}k_2 s_{z2}}\rangle \cdots |n_{\mathrm{h}k_1 s_{z1}}\rangle \, |n_{\mathrm{h}k_2 s_{z2}}\rangle \cdots \\ &= |n_{\mathrm{e}k_1 s_{z1}} n_{\mathrm{e}k_2 s_{z2}} \cdots n_{\mathrm{h}k_1 s_{z1}} n_{\mathrm{h}k_2 s_{z2}} \cdots\rangle \ . \end{aligned} \tag{1.98}$$

These products give the occupancy of every state in the portion of the band structure of interest. For example, (1.98) represents the eigenstate with $n_{\mathrm{e}k_1 s_{z1}}$ electrons in conduction band state 1, $n_{\mathrm{e}k_2 s_{z2}}$ electrons in conduction band state 2, and so on. Because of the Pauli exclusion principle, the $n's$ are either 0 or 1.

Any state of the system can then be written as

$$|\psi\rangle = \sum_{\{n_i\}} c_{\{n_i\}} |\{n_i\}\rangle \ , \tag{1.99}$$

where the summation is over all permutations of $n_i's$ and $c_{\{n_i\}} = \langle \{n_i\} | \psi \rangle$ is the probability amplitude of finding the semiconductor in the eigenstate. If we know the state vector for the semiconductor, then the expectation value of an operator $O$ is

$$\langle O \rangle = \langle \psi | O | \psi \rangle \ . \tag{1.100}$$

On the other hand if the system is not in a single state, then the expectation value is

$$\langle O \rangle = \sum_j P_j \langle \psi_j | O | \psi_j \rangle \ , \tag{1.101}$$

where $P_j$ is the probability that the semiconductor is described by the state vector $|\psi_j\rangle$. One should not make the mistake of associating $P_j$ with the quantum mechanical uncertainty given by $c_{\{n_i\}}$. The lack of knowledge that led to $P_j$ is usually classical in origin and in most cases is due to the lack of information on the initial state vector. Inserting the identity operator, $\sum_{\{ni\}} |\{n_i\}\rangle\langle\{n_i\}|$, between $O$ and $|\psi_j\rangle$, we find

$$\begin{aligned} \langle O \rangle &= \sum_{\{n_i\}} \sum_j P_j \, \langle \psi_j | O | \{n_i\} \rangle \, \langle \{n_i\} | \psi_j \rangle \\ &= \sum_{\{n_i\}} \sum_j \langle \{n_i\} | \psi_j \rangle \, P_j \, \langle \psi_j | O | \{n_i\} \rangle \\ &= \mathrm{tr}(\rho O) = \mathrm{tr}(O \rho) \ , \end{aligned} \tag{1.102}$$

where $\rho$ is the statistical operator (often referred to as density operator in quantum optics)

$$\rho = \sum_j P_j \, |\psi_j\rangle\langle\psi_j| \quad . \tag{1.103}$$

This statistical operator is useful for describing lasers because collisions, recombination, and randomness (incoherence) in the excitation processes creating the inversion prohibit a precise knowledge of the state vector of the system. The diagonal elements of the statistical operator

$$\begin{aligned} \langle\{n_i\}|\rho|\{n_i\}\rangle &= \sum_j P_j \, \langle\{n_i\}|\psi_j\rangle \, \langle\psi_j|\{n_i\}\rangle \\ &= \sum_j P_j \, |c_{\{n_i\}}|^2 \end{aligned} \tag{1.104}$$

give the probability of finding the system in the eigenstate $|\{n_i\}\rangle$. The off-diagonal elements of the statistical operator

$$\begin{aligned} \langle\{n_i\}|\rho|\{n_i\}\rangle &= \sum_l P_l \, \langle\{n_i\}|\psi_l\rangle \, \langle\psi_l|\{n_j\}\rangle \\ &= \sum_l P_l \, c^*_{\{n_j\}} c_{\{n_i\}} \quad , \end{aligned} \tag{1.105}$$

contain information concerning the relative phases (coherence) between probability amplitudes.

The statistical average over possible state vectors tends to destroy the coherence in the system, so that whenever collisions or pump effects dominate, we are likely to have a diagonal statistical operator. In a semiconductor laser, the rapid intraband collisions usually dominate the dynamics within each band. These collisions tend to drive the carrier distributions into quasi-equilibrium distributions. As a result, the electron and hole statistical operators are often to a very good approximation

$$\rho_e = \frac{1}{Z_e} \exp\left[-\beta \sum_k \sum_{s_z} (\varepsilon_{ek} - \mu_e) a^\dagger_{ks_z} a_{ks_z}\right] \quad , \tag{1.106}$$

$$\rho_h = \frac{1}{Z_h} \exp\left[-\beta \sum_k \sum_{s_z} (\varepsilon_{hk} - \mu_h) b^\dagger_{ks_z} b_{ks_z}\right] \quad , \tag{1.107}$$

respectively. Here, the partition function for the conduction electrons is

$$\begin{aligned} Z_e &= \mathrm{tr}\left\{\exp\left[-\beta \sum_k \sum_{s_z} (\varepsilon_{ek} - \mu_e) a^\dagger_{ks_z} a_{ks_z}\right]\right\} \\ &= \prod_k \prod_{s_z} \Big[\langle 0_{eks_z}| \, \mathrm{e}^{-\beta(\varepsilon_{ek}-\mu_e) a^\dagger_{ks_z} a_{ks_z}} \, |0_{eks_z}\rangle \\ &\qquad + \langle 1_{eks_z}| \, \mathrm{e}^{-\beta(\varepsilon_{ek}-\mu_e) a^\dagger_{ks_z} a_{ks_z}} \, |1_{eks_z}\rangle\Big] \quad , \end{aligned}$$

which yields

$$Z_{\mathrm{e}} = \prod_{k}\prod_{s_z}\{1+\exp[-\beta(\varepsilon_{\mathrm{e}k}-\mu_{\mathrm{e}})]\} \quad . \tag{1.108}$$

Similarly,

$$\begin{aligned} Z_{\mathrm{h}} &= \mathrm{tr}\left\{\exp\left[-\beta\sum_{k}\sum_{s_z}(\varepsilon_{\mathrm{h}k}-\mu_{\mathrm{h}})b^{\dagger}_{ks_z}b_{ks_z}\right]\right\} \\ &= \prod_{k}\prod_{s_z}\{1+\exp[-\beta(\varepsilon_{\mathrm{h}k}-\mu_{\mathrm{h}})]\} \end{aligned} \tag{1.109}$$

is the partition function for the holes. These statistical operators give the Fermi-Dirac distributions for the expectation value of the particle number operator. For example, the probability of finding an electron with momentum $\boldsymbol{k}$ and $s_z$ is

$$\begin{aligned} \mathrm{tr}\,\rho_{\mathrm{e}}\left(a^{\dagger}_{ks_z}a_{ks_z}\right) &= \frac{1}{Z_{\mathrm{e}}}\exp[-\beta(\varepsilon_{\mathrm{e}k}-\mu_{\mathrm{e}})] \\ &\quad\times\prod_{k'}\prod_{s'_z}\{1+\exp[-\beta\left(\varepsilon_{\mathrm{e}k'}-\mu_{\mathrm{e}}\right)]\} \\ &= \frac{1}{\exp[\beta(\varepsilon_{\mathrm{e}k}-\mu_{\mathrm{e}})]+1} \quad . \end{aligned} \tag{1.110}$$

Note, that (1.110) also gives the diagonal element

$$\sum_{\{n_{k',s'_z}\}}\left\langle\left\{n_{k',s'_z}\right\}1_{k,s_z}\left|\rho_{\mathrm{e}}\right|\left\{n_{k',s'_z}\right\}1_{k,s_z}\right\rangle \quad ,$$

where $\{n_{k',s'_z}\}$ represents all possible permutations of the eigenstates except the one with momentum $\boldsymbol{k}$ and $s_z$.

The dynamics for the expectation value given by (1.102) may reside in the operator $O$ or in the statistical operator $\rho$. The former corresponds to the Heisenberg picture of quantum mechanics, and the latter corresponds to the Schrödinger picture of quantum mechanics. They are equivalent in the sense that both pictures give the same expectation values. In this book we choose to work mostly with the Heisenberg picture because it turns out to be a more convenient approach in the many-body treatment presented in the later chapters. In the Heisenberg picture, the operator $O$ obeys the equation of motion

$$\mathrm{i}\hbar\frac{\mathrm{d}O}{\mathrm{d}t} = [O,H] \quad , \tag{1.111}$$

where the commutator

$$[A,B]\equiv[A,B]_{-}\equiv AB-BA \quad ,$$

and $H$ is the total Hamiltonian. In the Schrödinger picture, the statistical operator obeys the equation

$$i\hbar\frac{\mathrm{d}\rho}{\mathrm{d}t} = [H, \rho] \quad . \tag{1.112}$$

A calculation in the framework of semiclassical laser theory then involves solving the medium equations of motion. These equations are derived using the Heisenberg equation of motion and they are likely to consist of coupled differential equations. The derivation and solution of the medium equations is often complicated. Fortunately, approximations may be made. The next chapters discuss these approximations, and show that different levels of sophistication exist on how we treat the semiconductor gain medium.

# 2. Free-Carrier Theory

As discussed in the previous chapter, the laser field and the semiconductor gain medium are coupled by the gain and carrier-induced refractive index, or equivalently, by the induced complex susceptibility. To determine these quantities, we need to solve the quantum mechanical gain medium equations of motion for the polarization. In principle, these dynamic equations should be derived using the full system Hamiltonian, which contains contributions from the kinetic energies, the many-body Coulomb interactions, the electric-dipole interaction between the carriers and the laser field, as well as, the interactions between the carriers and phonons. The effects of injection current pumping should also be included. Such a complete theory will be very complicated. Therefore, one often makes approximations that allow one to begin with a tractable treatment that is reasonably accurate and hopefully contains the most important effects. By gradually eliminating the approximations, one works toward increasingly rigorous treatments. In this book, we take such an approach.

As a starting point, we assume that the charged particle interactions are sufficiently fast compared to the field transients. We then treat them as reservoir interactions that establish intraband thermodynamic quasi-equilibrium Fermi-Dirac carrier distributions. Furthermore, in this chapter we neglect many-body effects due to Coulomb interactions between the carriers that renormalize the bandgap energy and the electric-dipole interaction energy. As such, we treat the carriers as ideal Fermi gases, which labels our present theory as a "free-carrier" theory. Many-body and band-structure effects are added in the following chapters. On the crudest level, it is the strong reservoir interactions (carrier-carrier scattering) that gives what appears to be a markedly inhomogeneously broadened transition its homogeneously-broadened saturation behavior. On the other hand, the wide tuning characteristics reveal aspects of the underlying inhomogeneously broadened transition.

In this chapter, we derive the equation for the polarization of a semiconductor medium that can explain this dual homogeneously and inhomogeneously broadened nature and at the same time track the medium's response to temperature and carrier density variations. The free-carrier model is a reasonable approximation for describing the bandfilling aspects of the semiconductor laser under most normal operating conditions. We also use the two-band approximation. For the bulk GaAs semiconductor, the contributions of the heavy-hole band dominates that of the light-hole band since it has a much larger density of states. This situation changes in strained quantum-well systems for which the light-hole band can be lifted above the heavy-hole band.

Section 2.1 describes the free-carrier Hamiltonian and derives the free-carrier equations of motion in the Heisenberg picture. Section 2.2 introduces the quasi-equilibrium approximation, which enables us to solve for the gain and carrier-induced refractive index in terms of Fermi-Dirac carrier distributions. In this approximation the $\boldsymbol{k}$-dependent carrier populations and polarization amplitudes are assumed to adiabatically follow temporal variations of the total carrier density and the electric field envelope. Section 2.3 uses the free-carrier gain equation to predict gain spectra in bulk semiconductors and in quantum wells. The width of the gain spectrum is shown to depend strongly on the total chemical potential. Section 2.4 gives predictions and explanations of the dependence of the gain on temperature. The dependence on the applied light intensity (saturation) is analyzed in Sect. 2.5. The chapter closes with Sects. 2.6, 7 on free-carrier predictions of the carrier-induced refractive index and linewidth enhancement or antiguiding factor, respectively. This subject is treated only briefly because an accurate description requires the more complete analysis of the carrier-carrier Coulomb interactions.

## 2.1 Free-Carrier Equations of Motion

The free-carrier theory assumes that the primary effect of the Coulomb interaction among carriers is to relax the carrier distributions within the conduction and the valence bands to quasi-equilibrium distributions. We can phenomenologically account for this effect by treating the Coulomb interaction as a reservoir interaction instead of a dynamical interaction. This leaves us with the free-carrier Hamiltonian

$$\begin{aligned} H &= H_{\text{kin}} + H_{\text{c-f}} \\ &= \sum_{\boldsymbol{k}} \left(\varepsilon_{c\boldsymbol{k}} a^{\dagger}_{c\boldsymbol{k}} a_{c\boldsymbol{k}} + \varepsilon_{v\boldsymbol{k}} a^{\dagger}_{v\boldsymbol{k}} a_{v\boldsymbol{k}}\right) \\ &\quad - \sum_{\boldsymbol{k}} \left(\mu_{\boldsymbol{k}} a^{\dagger}_{c\boldsymbol{k}} a_{v\boldsymbol{k}} + \mu^{*}_{\boldsymbol{k}} a^{\dagger}_{v\boldsymbol{k}} a_{c\boldsymbol{k}}\right) E(z,t) \quad , \end{aligned} \tag{2.1}$$

where $\mu_{\boldsymbol{k}}$ is the dipole matrix element between the valence and conduction band. We have assumed a dipole interaction between the laser field and the carriers

$$H_{\text{c-f}} = -V\boldsymbol{P}\cdot\boldsymbol{E} \quad , \tag{2.2}$$

where $V$ is the volume of the active region, and the active medium polarization is given by the operator

$$P = \frac{1}{V}\sum_{\boldsymbol{k}} \left(\mu_{\boldsymbol{k}} a^{\dagger}_{c\boldsymbol{k}} a_{v\boldsymbol{k}} + \mu^{*}_{\boldsymbol{k}} a^{\dagger}_{v\boldsymbol{k}} a_{c\boldsymbol{k}}\right) \quad . \tag{2.3}$$

In (2.1–3), we used the two-band approximation, and absorbed the spin index into $\boldsymbol{k}$, so that $\sum_{\boldsymbol{k}}$ is actually $\sum_{\boldsymbol{k}s_z}$, and $a_{\alpha\boldsymbol{k}}$ and $a^\dagger_{\alpha\boldsymbol{k}}$ are actually $a_{\alpha\boldsymbol{k}s_z}$ and $a^\dagger_{\alpha\boldsymbol{k}s_z}$, respectively. In the electron-hole representation,

$$P = \frac{1}{V}\sum_{\boldsymbol{k}} \left(\mu_{\boldsymbol{k}} a^\dagger_{\boldsymbol{k}} b^\dagger_{-\boldsymbol{k}} + \mu^*_{\boldsymbol{k}} b_{-\boldsymbol{k}} a_{\boldsymbol{k}}\right) \quad , \tag{2.4}$$

and

$$\begin{aligned} H &= \sum_{\boldsymbol{k}} \left[\left(\varepsilon_{\mathrm{g}0} + \frac{\hbar^2 k^2}{2m_{\mathrm{e}}}\right) a^\dagger_{\boldsymbol{k}} a_{\boldsymbol{k}} + \frac{\hbar^2 k^2}{2m_{\mathrm{h}}} b^\dagger_{-\boldsymbol{k}} b_{-\boldsymbol{k}}\right] \\ &\quad - \sum_{\boldsymbol{k}} \left(\mu_{\boldsymbol{k}} a^\dagger_{\boldsymbol{k}} b^\dagger_{-\boldsymbol{k}} + \mu^*_{\boldsymbol{k}} b_{-\boldsymbol{k}} a_{\boldsymbol{k}}\right) E(z,t) \quad , \end{aligned} \tag{2.5}$$

where we used (1.87, 88, 90, 91). Also, we denote the unrenormalized bandgap energy $\varepsilon_{\mathrm{g}0}$; and reserve $\varepsilon_{\mathrm{g}}$ for the renormalized value obtained from many-body theory.

The link between the classical laser field and the quantum mechanical semiconductor gain medium is obtained via the polarization

$$P(z,t) = \langle P \rangle = \mathrm{tr}\{P\rho\} \quad . \tag{2.6}$$

Using (1.72), (2.4), the polarization amplitude appearing in the slowly varying amplitude and phase equations (1.76, 77) is

$$\mathcal{P}(z) = 2\,\mathrm{e}^{-\mathrm{i}[Kz-\nu t-\phi(z)]} \frac{1}{V} \sum_{\boldsymbol{k}} \mu^*_{\boldsymbol{k}} p_{\boldsymbol{k}} \quad , \tag{2.7}$$

$$p_{\boldsymbol{k}} = \langle b_{-\boldsymbol{k}} a_{\boldsymbol{k}} \rangle \quad . \tag{2.8}$$

The gain medium variables are $p_{\boldsymbol{k}}$ together with the electron and hole occupation numbers

$$n_{\mathrm{e}\boldsymbol{k}} = \langle a^\dagger_{\boldsymbol{k}} a_{\boldsymbol{k}} \rangle \quad , \tag{2.9}$$

$$n_{\mathrm{h}\boldsymbol{k}} = \langle b^\dagger_{-\boldsymbol{k}} b_{-\boldsymbol{k}} \rangle \quad . \tag{2.10}$$

In the Heisenberg picture, the derivation of the equations of motion for the bilinear operators in (2.8–10) involves the evaluation of the commutators appearing in the Heisenberg equation of motion (1.111). Doing the explicit calculations we first note that any bilinear product of Fermion operators for $\boldsymbol{k}$ commutes with any bilinear product of Fermion operators for $\boldsymbol{k}'$, where $\boldsymbol{k} \neq \boldsymbol{k}'$. This follows immediately because four anticommuting exchanges are involved and $(-1)^4 = 1$. Second, we note that the free-carrier Hamiltonian may be written in the from

$$H = \sum_{\boldsymbol{k}} H_{\boldsymbol{k}} \quad , \tag{2.11}$$

where the individual $\boldsymbol{k}$-dependent parts are given by

$$H_k = \left(\varepsilon_{g0} + \frac{\hbar^2 k^2}{2m_e}\right) a_k^\dagger a_k + \frac{\hbar^2 k^2}{2m_h} b_{-k}^\dagger b_{-k} - \left(\mu_k a_k^\dagger b_{-k}^\dagger + \mu_k^* b_{-k} a_k\right) E(z,t) \ . \tag{2.12}$$

Hence, the Heisenberg equation of motion for $O_k$ equal to any of the bilinear operators appearing in (2.8–10) simplifies to

$$\frac{\mathrm{d}O_k}{\mathrm{d}t} = \frac{\mathrm{i}}{\hbar}[H, O_k] = \frac{\mathrm{i}}{\hbar}[H_k, O_k] \ . \tag{2.13}$$

This simplification is no longer possible when the Coulomb interactions are included in the Hamiltonian.

As can be seen from (2.13) the derivation of the equation of motion for $b_{-k}a_k$ involves evaluating the commutator $\left(b_{-k}a_k, a_k^\dagger b_{-k}^\dagger\right)$. Noting that

$$a_k^\dagger b_{-k}^\dagger b_{-k} a_k = a_k^\dagger a_k b_{-k}^\dagger b_{-k} \tag{2.14}$$

$$\begin{aligned} b_{-k} a_k a_k^\dagger b_{-k}^\dagger &= b_{-k} b_{-k}^\dagger a_k a_k^\dagger \\ &= \left(1 - b_{-k}^\dagger b_{-k}\right)\left(1 - a_k^\dagger a_k\right) \ , \end{aligned} \tag{2.15}$$

we have

$$\begin{aligned} \left[b_{-k}a_k, a_k^\dagger b_{-k}^\dagger\right] &= \left(1 - b_{-k}^\dagger b_{-k}\right)\left(1 - a_k^\dagger a_k\right) - a_k^\dagger a_k b_{-k}^\dagger b_{-k} \\ &= 1 - b_{-k}^\dagger b_{-k} - a_k^\dagger a_k \ . \end{aligned} \tag{2.16}$$

The other commutators may be evaluated similarly to give the Heisenberg equations of motion

$$\frac{\mathrm{d}}{\mathrm{d}t} b_{-k} a_k = -\mathrm{i}\omega_k b_{-k} a_k - \frac{\mathrm{i}}{\hbar}\mu_k E(z,t)\left(a_k^\dagger a_k + b_{-k}^\dagger b_{-k} - 1\right) \tag{2.17}$$

$$\frac{\mathrm{d}}{\mathrm{d}t} a_k^\dagger a_k = \frac{\mathrm{i}}{\hbar}\left(\mu_k a_k^\dagger b_{-k}^\dagger - \mu_k^* b_{-k} a_k\right) E(z,t) \tag{2.18}$$

$$= \frac{\mathrm{d}}{\mathrm{d}t} b_{-k}^\dagger b_{-k} \ . \tag{2.19}$$

Here, the transition energy is

$$\begin{aligned} \hbar\omega_k &= \varepsilon_{g0} + \varepsilon_{ek} + \varepsilon_{hk} \\ &= \varepsilon_{g0} + \frac{\hbar^2 k^2}{2m_e} + \frac{\hbar^2 k^2}{2m_h} \\ &= \varepsilon_{g0} + \frac{\hbar^2 k^2}{2m_r} \ , \end{aligned} \tag{2.20}$$

and $m_r$ is the reduced mass given by (1.16). Note that the adjoint has to appear in (2.18, 19) since the number operator is Hermitian but the dipole operator $a_k^\dagger b_{-k}^\dagger$ is not. Furthermore, we expect the electron and hole number operators to be affected by radiative transitions identically, since these transitions either create both an electron and a hole, or they annihilate one of each. Taking the expectation values given in (2.8–10), we have

$$\frac{\mathrm{d}p_{\boldsymbol{k}}}{\mathrm{d}t} = -\mathrm{i}\omega_{\boldsymbol{k}} p_{\boldsymbol{k}} - \frac{\mathrm{i}}{\hbar}\mu_{\boldsymbol{k}} E(z,t)\,(n_{\mathrm{e}\boldsymbol{k}} + n_{\mathrm{h}\boldsymbol{k}} - 1) \quad , \tag{2.21}$$

$$\frac{\mathrm{d}n_{\mathrm{e}\boldsymbol{k}}}{\mathrm{d}t} = \frac{\mathrm{d}n_{\mathrm{h}\boldsymbol{k}}}{\mathrm{d}t} = \frac{\mathrm{i}}{\hbar} E(z,t)\,(\mu_{\boldsymbol{k}} p_{\boldsymbol{k}}^* - \mu_{\boldsymbol{k}}^* p_{\boldsymbol{k}}) \quad . \tag{2.22}$$

If we want to use these equations of motion in a laser model, we must supply the missing terms describing pumping and relaxation processes. While the effects of current injection may be readily incorporated at a phenomenological level, we have to think harder about collisions. The important collisions involve carrier-carrier, carrier-phonon, and carrier-impurity (lattice imperfection) scattering, which are not explicitly present in the free-carrier Hamiltonian (2.5). Carrier recombination via spontaneous emission is also not taken into account since we did not quantize the radiation field. For now, we include all these effects phenomenologically so that (2.21, 22) become

$$\frac{\mathrm{d}p_{\boldsymbol{k}}}{\mathrm{d}t} = -\mathrm{i}\omega_{\boldsymbol{k}} p_{\boldsymbol{k}} - \frac{\mathrm{i}}{\hbar}\mu_{\boldsymbol{k}} E(z,t)\,(n_{\mathrm{e}\boldsymbol{k}} + n_{\mathrm{h}\boldsymbol{k}} - 1) + \left.\frac{\partial p_{\boldsymbol{k}}}{\partial t}\right|_{\mathrm{col}} \quad , \tag{2.23}$$

$$\begin{aligned} \frac{\mathrm{d}n_{\alpha\boldsymbol{k}}}{\mathrm{d}t} &= \frac{\mathrm{i}}{\hbar} E(z,t)\,(\mu_{\boldsymbol{k}} p_{\boldsymbol{k}}^* - \mu_{\boldsymbol{k}}^* p_{\boldsymbol{k}}) + \Lambda_{\alpha\boldsymbol{k}} - B_{\boldsymbol{k}} n_{\mathrm{e}\boldsymbol{k}} n_{\mathrm{h}\boldsymbol{k}} \\ &\quad - \gamma_{\mathrm{nr}} n_{\alpha\boldsymbol{k}} + \left.\frac{\partial n_{\alpha\boldsymbol{k}}}{\partial t}\right|_{\mathrm{col}} \quad . \end{aligned} \tag{2.24}$$

Here $\alpha = \mathrm{e}$ for electrons, $\alpha = \mathrm{h}$ for holes, $\Lambda_{\alpha\boldsymbol{k}}$, is the pump rate due to an injection current, $\gamma_{\mathrm{nr}}$ is the nonradiative decay constant due to capture by defects in the semiconductor, $B_{\boldsymbol{k}}$ is the radiative recombination (spontaneous emission) rate constant, and $\partial p_{\boldsymbol{k}}/\partial t|_{\mathrm{col}}$ and $\partial n_{\alpha\boldsymbol{k}}/\partial t|_{\mathrm{col}}$ are the collision contributions. These contributions include carrier-carrier and carrier phonon scattering, which drive the distribution $n_{\alpha\boldsymbol{k}}$ toward the Fermi-Dirac distribution of (1.19). In the simplest approximation the collision contribution $\partial p_{\boldsymbol{k}}/\partial t|_{\mathrm{col}}$ in the polarization equation describes polarization decay (dephasing) according to

$$\left.\frac{\partial p_{\boldsymbol{k}}}{\partial t}\right|_{\mathrm{col}} \simeq -\gamma p_{\boldsymbol{k}} \quad . \tag{2.25}$$

While rapidly suppressing deviations from the Fermi-Dirac distribution, the scattering does not change the total carrier density $N$ of (1.20). Hence, we can write (1.20) more generally as

$$\begin{aligned} N &= \frac{1}{V}\sum_{\boldsymbol{k}} n_{\mathrm{e}\boldsymbol{k}} = \frac{1}{V}\sum_{\boldsymbol{k}} n_{\mathrm{h}\boldsymbol{k}} \\ &= \frac{1}{V}\sum_{\boldsymbol{k}} f_{\mathrm{e}\boldsymbol{k}} = \frac{1}{V}\sum_{\boldsymbol{k}} f_{\mathrm{h}\boldsymbol{k}} \quad , \end{aligned} \tag{2.26}$$

where we should remember that the spin summation is included in the $\boldsymbol{k}$-summation. Accordingly summing (2.24), we find the equation of motion

$$\frac{\mathrm{d}N}{\mathrm{d}t} = \Lambda - \gamma_{\mathrm{nr}} N - \frac{1}{V}\sum_{k} B_{k} n_{\mathrm{e}k} n_{\mathrm{h}k}$$
$$-\frac{\mathrm{i}}{\hbar V} E(z,t) \sum_{k} \left(\mu_{k} p_{k}^{*} - \mu_{k}^{*} p_{k}\right) \quad . \tag{2.27}$$

Here, the injection current pump $\Lambda$ is given by

$$\Lambda = \frac{\eta J}{ed} \quad , \tag{2.28}$$

$\eta$ is the total quantum efficiency that the injected carriers contribute to the inversion, $J$ is the current density, $e$ is the charge of an electron, and $d$ is the thickness of the active region. For many practical situations, one can assume that by the time the injected carriers reach the active region, they collide often enough to be in equilibrium with one another. Therefore, we have

$$\Lambda_{\alpha k} = \frac{\eta_{\mathrm{tr}} J}{edN_0} f_{\alpha k0} \left(1 - n_{\alpha k}\right) \quad , \tag{2.29}$$

where $\eta_{\mathrm{tr}}$ is the "transport" part of the quantum efficiency, giving the efficiency that the injected carriers reach the active region. Furthermore, $N_0$ and $f_{\alpha k0}$ are the values of $N$ and $f_{\alpha k}$, respectively, in the absence of an electromagnetic field $[E(z,t) = 0]$. The term in the bracket in (2.29) accounts for the fact that the presence of carriers inside the active region, with the distribution $n_{\alpha k}$, reduces the efficiency of the pumping since each quantum state can be occupied only by one carrier. An example of this pump blocking is shown in Fig. 2.1. We see that in the presence of carriers only the high energy part of the additional carriers generated by the pump source can actually enter the active region.

The pump blocking effects can be included in a definition of the total quantum efficiency,

$$\eta = \frac{\eta_{\mathrm{tr}}}{N_0} \sum_{k} f_{\alpha k0} \left(1 - n_{\alpha k}\right) \quad . \tag{2.30}$$

For fixed pump rate, i.e., fixed $f_{\alpha k0}$, the quantum efficiency decreases with increasing carrier density in the active region, as shown in Fig. 2.2.

Chapter 4 will show a more accurate account of intraband scattering that results in Boltzmann equations for the carrier populations, where scattering terms couple the different $\boldsymbol{k}$ states. One then has an infinite set of coupled nonlinear differential equations for the gain medium. The solution of these equations is nontrivial. Fortunately, the problem may sometimes be simplified by noting that in terms of the carrier populations, the net effect of intraband scattering is to return the electron and hole distributions to equilibrium. Then, one way to approximate the intraband scattering terms is

$$\left.\frac{\partial n_{\alpha k}}{\partial t}\right|_{\mathrm{col}} = -\gamma_{\alpha} \left(n_{\alpha k} - f_{\alpha k}\right) \quad , \tag{2.31}$$

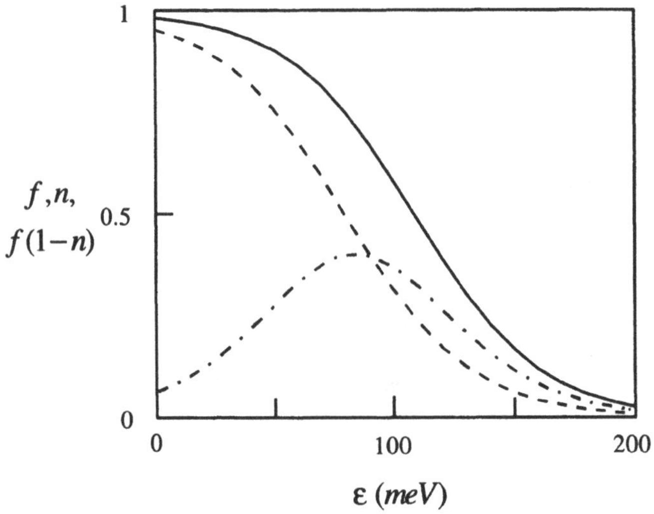


**Fig. 2.1.** Example of pump blocking. The pump generated distribution $f_{k0}$ is shown as *solid line*, the actual carrier distribution $n_k$ inside the active region is shown as *dashed line*, and the effective pump rate $f_{k0}(1-n_k)$ is plotted as a *dot-dashed line*. The $x$-axis is given in units of the energy $\varepsilon = \hbar^2 k^2/2m_r$

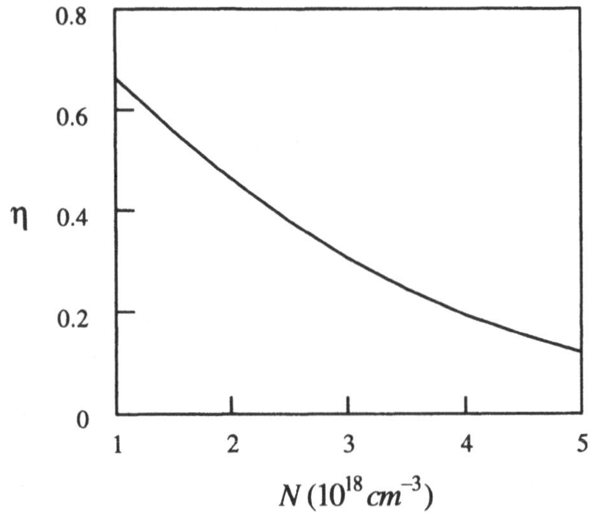


**Fig. 2.2.** Effective quantum efficiency $\eta$ (in units of $\eta_{tr}/N_0$) for fixed pump rate and increasing carrier density $N$ inside the active region

where $f_{\alpha k}$ is the Fermi-Dirac distribution satisfying (2.26) for a saturated total carrier density, i.e., the carrier density in the presence of the laser field. The rates $\gamma_\alpha$ with which a perturbed carrier distribution returns to equilibrium are sometimes referred to as the carrier-relaxation rates.

In semiconductor gain media, the carrier-carrier scattering rates are typically of the order of $10^{13}\ \mathrm{s}^{-1}$. Except for the dynamics on short time scales or for very strong laser fields that induce stationary nonequilibrium distributions, these rates are sufficiently large to dominate any other mechanism that tries to cause the carrier distributions to deviate from quasi-equilibrium. The quasi-equilibrium approximation takes the limit of $\gamma_\alpha \to \infty$, so that

$$n_{ek} \simeq f_{ek} \quad \text{and} \quad n_{\mathrm{h}k} \simeq f_{\mathrm{h}k} \quad . \tag{2.32}$$

Furthermore, the polarization dynamics is usually eliminated adiabatically, as will be discussed in Sect. 2.2. Of course, both the $\gamma_\alpha$ and free-carrier models are approximations.

The total radiative recombination rate is

$$\Gamma_{\mathrm{rr}} = \frac{1}{V}\sum_k B_k f_{ek} f_{\mathrm{h}k} \quad . \tag{2.33}$$

For small $N$, the Fermi-Dirac distributions have negative chemical potentials and can be approximated by Maxwell-Boltzmann distributions. Using (1.32), it is easy to show that these distributions are proportional to $N$, so that (2.33) yields $\Gamma_{\mathrm{rr}} = BN^2$. In fact, using the same approach as in (1.33), we set $q = \beta\hbar^2/2m_{\mathrm{r}}$ for typographical simplicity and find

$$\begin{aligned}
\Gamma_{\mathrm{rr}} &\simeq \frac{\overline{B}}{V}\sum_k f_{ek} f_{\mathrm{h}k} \\
&= \frac{\overline{B}}{\pi^2}\, \mathrm{e}^{\beta\mu_{\mathrm{e}}}\, \mathrm{e}^{\beta\mu_{\mathrm{h}}} \int_0^\infty \mathrm{d}k\, k^2\, \mathrm{e}^{-qk^2} \\
&= \frac{\overline{B}\overline{N}_{\mathrm{e}}\overline{N}_{\mathrm{h}}}{\pi^2} \frac{\partial}{\partial q} \int_0^\infty \mathrm{d}k\, \mathrm{e}^{-qk^2} \\
&= \frac{\overline{B}\overline{N}_{\mathrm{e}}\overline{N}_{\mathrm{h}}}{\pi^2} \frac{\sqrt{\pi}}{2} \frac{\partial}{\partial q} \frac{1}{\sqrt{q}} \\
&= \frac{\overline{B}\overline{N}_{\mathrm{e}}\overline{N}_{\mathrm{h}}}{4(\pi\beta\hbar^2/2m_{\mathrm{r}})^{3/2}} \\
&= \overline{B}N^2 \left(\frac{\pi\beta\hbar^2 m_{\mathrm{r}}}{2m_{\mathrm{e}}m_{\mathrm{h}}}\right)^{\frac{3}{2}} \quad ,
\end{aligned} \tag{2.34}$$

where we define

$$B \equiv \overline{B}\left(\frac{\pi\beta\hbar^2 m_{\mathrm{r}}}{2m_{\mathrm{e}}m_{\mathrm{h}}}\right)^{\frac{3}{2}} \quad ,$$

which decreases as $T^{-3/2}$.

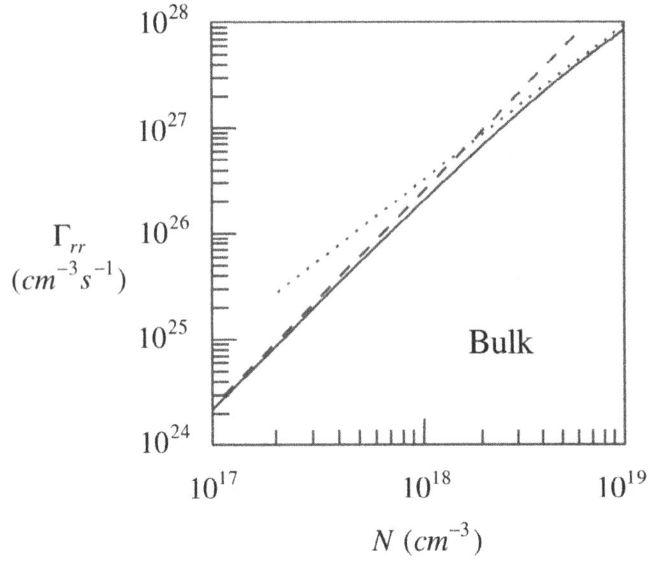


**Fig. 2.3.** Radiative-recombination coefficient $\Gamma_{\mathrm{rr}}$ *(solid curve)* versus total carrier density. The *dashed curve* shows a relatively good fit at low density with $\Gamma_{\mathrm{rr}} = BN^2$ and $B = 2.4 \times 10^{-16}\,\mathrm{cm}^3\,\mathrm{s}^{-1}$. The *dotted curve* shows the fit at high density with $\Gamma_{\mathrm{rr}} = \Gamma N^{3/2}$ and $\Gamma = 3.0 \times 10^{-4}\,\mathrm{cm}^{3/2}\,\mathrm{s}^{-1}$

This is illustrated in Fig. 2.3, where we see that for $N < 10^{18}\,\mathrm{cm}^{-3}$, $\Gamma_{\mathrm{rr}}$ is proportional to $N^2$. However, for larger $N$, the rate of increase is less than $N^2$, due to the fact that the Fermi-Dirac limit $f_{\mathrm{e}\boldsymbol{k}}$, $f_{\mathrm{h}\boldsymbol{k}} \leq 1$ comes into play.

At present, most semiconductor lasers operate with either undoped or lightly doped gain regions. The results so far are for an undoped gain region. To generalize to the lightly doped case, we need to take into account the difference between the total electron density $N_{\mathrm{e}}$ and the total hole density $N_{\mathrm{h}}$. In a $p$-doped gain medium,

$$N_{\mathrm{h}} = N_{\mathrm{e}} + N_{\mathrm{A}} \ , \tag{2.35}$$

where $N_{\mathrm{A}}$ is the acceptor density, and in an $n$-doped gain medium,

$$N_{\mathrm{e}} = N_{\mathrm{h}} + N_{\mathrm{D}} \ , \tag{2.36}$$

where $N_{\mathrm{D}}$ is the donor density. Also, in the absence of an injection current, there is a residual carrier density due to the dopants. For example, the spontaneous emission term is

$$\Gamma_{\mathrm{rr}} = \frac{1}{V} \sum_{\boldsymbol{k}} B_{\boldsymbol{k}} \left(n_{\mathrm{e}\boldsymbol{k}} - f_{\mathrm{D}\boldsymbol{k}}\right) n_{\mathrm{h}\boldsymbol{k}} \ , \tag{2.37}$$

for an $n$-doped gain medium,

$$\Gamma_{\mathrm{rr}} = \frac{1}{V} \sum_{\boldsymbol{k}} B_{\boldsymbol{k}} n_{\mathrm{e}\boldsymbol{k}} \left(n_{\mathrm{h}\boldsymbol{k}} - f_{\mathrm{A}\boldsymbol{k}}\right) \ , \tag{2.38}$$

for a $p$-doped gain medium, where $f_{Dk}$ and $f_{Ak}$ are the Fermi-Dirac distributions for the carriers due to the donor and acceptor populations. For a heavily doped gain region, the band structure is modified by the doping, and laser transitions are possible between free-carrier and $\boldsymbol{k}$-independent bound impurity states. Since heavily doped gain regions are encountered only rarely in modern lasers, we do not discuss this situation further.

## 2.2 Quasi-Equilibrium Approximation

Our goal is to solve the active medium equations of motion for the gain and carrier-induced refractive index in the limit that the electric field envelope and the total carrier density vary little in the dipole lifetime $T_2 \equiv 1/\gamma$. To do that, we multiply (2.23) by the integrating factor $\exp[(\mathrm{i}\omega_{\boldsymbol{k}} + \gamma)t]$ to find

$$\frac{\mathrm{d}}{\mathrm{d}t}\left(p_{\boldsymbol{k}}\,\mathrm{e}^{(\mathrm{i}\omega_{\boldsymbol{k}}+\gamma)t}\right) = -\frac{\mathrm{i}}{\hbar}\mu_{\boldsymbol{k}}E(z,t)\left(n_{\mathrm{e}\boldsymbol{k}} + n_{\mathrm{h}\boldsymbol{k}} - 1\right)\mathrm{e}^{(\mathrm{i}\omega_{\boldsymbol{k}}+\gamma)t} \quad , \tag{2.39}$$

which can be formally integrated to give

$$p_{\boldsymbol{k}}(t) = -\frac{\mathrm{i}\mu_{\boldsymbol{k}}}{\hbar}\int_{-\infty}^{t}\mathrm{d}t'\,E(z,t')\,\mathrm{e}^{(\mathrm{i}\omega_{\boldsymbol{k}}+\gamma)(t'-t)}\left[n_{\mathrm{e}\boldsymbol{k}}(t') + n_{\mathrm{h}\boldsymbol{k}}(t') - 1\right] \; . \tag{2.40}$$

For a simple steady-state or nearly steady-state theory, we assume that the $\boldsymbol{k}$-dependent carrier densities $n_{\alpha\boldsymbol{k}}(t')$, and the field envelope vary little in the time $T_2 = \frac{1}{\gamma}$. Then these quantities are replaced by their values at the time $t$ and removed from the integral. This approximation is called the *rate equation approximation*, because as we see shortly it leads to a rate equation for the total carrier density. For our present purposes, we consider the constant field envelope described by the plane wave field

$$E(z,t) = \frac{1}{2}E(z)\,\mathrm{e}^{\mathrm{i}[Kz-\nu t-\phi(z)]} + \mathrm{c.c.} \quad . \tag{2.41}$$

As we show in Chap. 4, fluctuations due to the carrier-carrier scattering term $\partial n_{\alpha\boldsymbol{k}}/\partial t|_{\mathrm{col}}$ have a time scale on the order of $T_2$. However, since we consider a field envelope that varies little in the time $T_2$ (in fact here one that is constant), the $n_{\alpha\boldsymbol{k}}(t)$ are driven by the scattering into a quasi-equilibrium distribution in which they can adiabatically track the slow time variations of the total carrier density $N$ via the density-dependent quasi-chemical potential $\mu(N(t))$ determined by (2.26). $N$ varies significantly only on relatively long times like those associated with interband processes. With this approximation, the integral in (2.40) can be readily performed, giving

$$\begin{aligned} p_{\boldsymbol{k}}(t) &= -\frac{\mathrm{i}}{\hbar}\mu_{\boldsymbol{k}}E(z)\left[n_{\mathrm{e}\boldsymbol{k}}(t) + n_{\mathrm{h}\boldsymbol{k}}(t) - 1\right] \\ &\quad \times \left(\frac{\mathrm{e}^{\mathrm{i}[Kz-\nu t-\phi(z)]}}{\mathrm{i}\,(\omega_{\boldsymbol{k}} - \nu) + \gamma} + \frac{\mathrm{e}^{-\mathrm{i}[Kz-\nu t-\phi(z)]}}{\mathrm{i}\,(\omega_{\boldsymbol{k}} + \nu) + \gamma}\right) \quad . \end{aligned} \tag{2.42}$$

We can neglect the second term in the $\{\cdots\}$ of (2.42), because its denominator is very large compared to that of the first term at optical frequencies. This is called the *rotating-wave approximation* (RWA) since in the time domain the second term rotates very rapidly in comparison to the first one, and thus more or less averages to zero. Using the RWA reduces (2.42) to

$$\begin{aligned} p_{\boldsymbol{k}}(t) &= -\frac{\mathrm{i}}{2\hbar}\mu_{\boldsymbol{k}}E(z)\,\mathrm{e}^{\mathrm{i}[Kz-\nu t-\phi(z)]} \\ &\quad\times\left[n_{\mathrm{e}\boldsymbol{k}}(t)+n_{\mathrm{h}\boldsymbol{k}}(t)-1\right]\frac{1}{\mathrm{i}\left(\omega_{\boldsymbol{k}}-\nu\right)+\gamma} \quad . \end{aligned} \tag{2.43}$$

By substituting (2.43) into (2.7), we find the complex polarization

$$\begin{aligned} P(z) &= -\frac{\mathrm{i}E(z)}{2\hbar\gamma V}\sum_{\boldsymbol{k}}\left|\mu_{\boldsymbol{k}}\right|^2\left[n_{\mathrm{e}\boldsymbol{k}}(t)+n_{\mathrm{h}\boldsymbol{k}}(t)-1\right] \\ &\quad\times\left(1-\mathrm{i}\frac{\omega_{\boldsymbol{k}}-\nu}{\gamma}\right)L\left(\omega_{\boldsymbol{k}}-\nu\right) \quad , \end{aligned} \tag{2.44}$$

where we again used the rotating-wave approximation. The factor

$$L\left(\omega_{\boldsymbol{k}}-\nu\right)=\frac{\gamma^2}{\gamma^2+\left(\omega_{\boldsymbol{k}}-\nu\right)^2} \tag{2.45}$$

is the Lorentzian lineshape function. Substituting (2.43) into (2.27), we obtain the equation of motion for the total carrier density

$$\begin{aligned} \frac{\mathrm{d}N}{\mathrm{d}t} &= \Lambda-\gamma_{\mathrm{nr}}N-\frac{1}{V}\sum_{\boldsymbol{k}}B_{\boldsymbol{k}}n_{\mathrm{e}\boldsymbol{k}}n_{\mathrm{h}\boldsymbol{k}} \\ &\quad-\frac{E(z)^2}{2\hbar^2\gamma V}\sum_{\boldsymbol{k}}\left|\mu_{\boldsymbol{k}}\right|^2\left[n_{\mathrm{e}\boldsymbol{k}}(t)+n_{\mathrm{h}\boldsymbol{k}}(t)-1\right]L\left(\omega_{\boldsymbol{k}}-\nu\right) \quad . \end{aligned} \tag{2.46}$$

To evaluate the spontaneous emission term in (2.46), we use the Fermi-Dirac quasi-equilibrium distributions $n_{\alpha\boldsymbol{k}}=f_{\alpha\boldsymbol{k}}$, $\alpha=\mathrm{e,h}$, for the electrons and holes. Note, that in addition to the rate-equation requirement that the field envelope vary little in the time $T_2$, the field intensity should not be so strong that it can burn holes in the Fermi-Dirac distributions.

To get a feeling of what might happen for such intense fields, let us suppose that the carrier-carrier scattering can be modeled according to (2.31). Substituting (2.31, 43) into (2.24), and solving in steady state gives

$$n_{\alpha\boldsymbol{k}}\simeq f_{\alpha\boldsymbol{k}}-\frac{\left|\mu_{\boldsymbol{k}}\right|^2E(z)^2}{2\hbar^2\gamma\gamma_\alpha}\left(n_{\mathrm{e}\boldsymbol{k}}+n_{\mathrm{h}\boldsymbol{k}}-1\right)L\left(\omega_{\boldsymbol{k}}-\nu\right) \quad ,$$

where we note that the pump and interband decay rates are small in comparison to the carrrier-carrier scattering rate. The above equation yields

$$n_{\mathrm{e}\boldsymbol{k}}+n_{\mathrm{h}\boldsymbol{k}}-1\simeq\frac{f_{\mathrm{e}\boldsymbol{k}}+f_{\mathrm{h}\boldsymbol{k}}-1}{1+\frac{I}{I_{\mathrm{sc}}}L\left(\omega_{\boldsymbol{k}}-\nu\right)} \quad , \tag{2.47}$$

where

$$I = \frac{1}{2}\epsilon_0 cn|E(z)|^2 \quad , \tag{2.48}$$

and

$$I_{sc} = \frac{\epsilon_0 nc\hbar^2\gamma}{\left(\frac{1}{\gamma_e} + \frac{1}{\gamma_h}\right)|\mu_k|^2} \quad . \tag{2.49}$$

In (2.47), the depletion of the inversion and hence the saturation of gain are described by the denominator $1+L(\omega_k - \nu)\, I/I_{sc}$ and by the saturated carrier distributions $f_{ek}$ and $f_{hk}$. The former is due to spectral hole burning, which is the frequency-dependent saturation of the inversion by stimulated emission or absorption. The latter describes the decrease in the overall inversion due to the filling of the spectral holes by intraband collisions. Spectral hole burning along with other nonlinear effects of the gain medium play an important role in both multimode operation and high-speed modulation of semiconductor lasers. On the other hand, using $|\mu_k| \simeq e \times 3$ Å for the dipole matrix element, we find $I_{sc} \simeq 60\ \mathrm{MW\,cm^{-2}}$. Therefore, $I/I_{sc} \ll 1$ for most semiconductor laser problems, and the quasi-equilibrium condition should be reasonable.

Using the quasi-equilibrium approximation in the complex polarization (2.44), we obtain the free-carrier complex susceptibility

$$\begin{aligned}\chi^{(0)}(z) &= -\frac{\mathrm{i}}{\hbar\gamma\epsilon_b V}\sum_k |\mu_k|^2\left[f_{ek}(t) + f_{hk}(t) - 1\right] \\ &\quad \times L(\omega_k - \nu)\left(1 - \mathrm{i}\frac{\omega_k - \nu}{\gamma}\right) \quad .\end{aligned} \tag{2.50}$$

Similarly, the equation of motion (2.46) for the total carrier density becomes

$$\begin{aligned}\frac{\mathrm{d}N}{\mathrm{d}t} &= \Lambda - \gamma_{nr}N - \frac{1}{V}\sum_k B_k f_{ek} f_{hk} \\ &\quad - \frac{E(z)^2}{2\hbar^2\gamma V}\sum_k |\mu_k|^2\left[f_{ek}(t) + f_{hk}(t) - 1\right] L(\omega_k - \nu) \quad ,\end{aligned} \tag{2.51}$$

where (2.26) relates the Fermi-Dirac distributions to the total carrier density $N$. Substituting $\chi^{(0)}(z)$ into (1.76, 77), and using (1.78, 81), we find the amplitude gain, $g = G/2$, and carrier-induced refractive index

$$g = \frac{\nu}{2\epsilon_0 nc\hbar\gamma V}\sum_k |\mu_k|^2\left[f_{ek}(t) + f_{hk}(t) - 1\right] L(\omega_k - \nu) \quad , \tag{2.52}$$

$$\delta n = \frac{1}{2\epsilon_0 n\hbar\gamma V}\sum_k |\mu_k|^2\left[f_{ek}(t) + f_{hk}(t) - 1\right] L(\omega_k - \nu)\frac{\omega_k - \nu}{\gamma} \quad . \tag{2.53}$$

The equation of motion (2.51) for $N$ can be simplified by substituting the gain $g$ of (2.52),

$$\frac{\mathrm{d}N}{\mathrm{d}t} = \Lambda - \gamma_{nr}N - \frac{1}{V}\sum_k B_k f_{ek} f_{hk} - \frac{2gI}{\hbar\nu} \quad . \tag{2.54}$$

As we see in Chap. 4, this formula is also valid for the quasi-equilibrium many-body case, provided g is the many-body gain. Equation (2.54) is sometimes further simplified by replacing the spontaneous emission term simply by $-BN^2$ as in (2.34). This is a reasonable thing to do when using the linear-density gain model of (1.3), since then we do not have to calculate Fermi-Dirac distributions. However, if we use a scheme where we deal with these distributions, the more accurate spontaneous emission value in (2.54) does not increase the calculation time significantly and allows for the fact that in gain media the rate is somewhere in between an $N^2$ and an $N$ dependence as illustrated in Fig. 2.3.

Hence, given that the carrier density is related to the carrier distributions by the Fermi-Dirac distributions, the simultaneous solution of (2.52, 54) gives us the saturated gain as a function of laser intensity and injection current. The quasi-chemical potential obtained in the process can then be used in (2.53) to calculate the carrier-induced refractive index.

We can simplify the summations over $\boldsymbol{k}$ by assuming a symmetric band structure with an essentially continuous distribution of states. Then similar to the density of states discussion of Sect. 1.6, we find for a bulk gain medium

$$\begin{aligned}\frac{1}{V}\sum_{\boldsymbol{k}} f(\boldsymbol{k}) &= \frac{2}{(2\pi)^3}\int_0^\infty \mathrm{d}k\, k^2 \int_0^{2\pi} \mathrm{d}\phi \int_0^\pi \mathrm{d}\theta \sin\theta\, f(k,\theta,\phi) \\ &\rightarrow \frac{2}{(2\pi)^3}\int_0^\infty dk\; 4\pi k^2\; f(k) \quad , \qquad (2.55)\end{aligned}$$

where the factor of 2 comes from the spin summation and the second line can be used if the function $f(\boldsymbol{k})$ is spherically symmetric in $k$, such as the quasi-equilibrium Fermi-Dirac distributions.

It is often more convenient to integrate over the reduced-mass energy

$$\varepsilon = \frac{\hbar^2 k^2}{2m_\mathrm{r}} \quad , \qquad (2.56)$$

which enters into the transition energy (2.20). Similar to the derivation for the carrier density of states, we find a joint density of states, $D(\varepsilon)$, defined by the equations (for spherically symmetric functions)

$$\frac{1}{V}\sum_{\boldsymbol{k}} \rightarrow \int_0^\infty \mathrm{d}\varepsilon\, D(\varepsilon) \quad , \qquad (2.57)$$

where the *reduced* or *joint density of states*

$$D(\varepsilon) = \frac{1}{2\pi^2}\left(\frac{2m_\mathrm{r}}{\hbar^2}\right)^{\frac{3}{2}} \sqrt{\varepsilon} \quad . \qquad (2.58)$$

Under these conditions, the total carrier density of (2.26) is given by

$$N = \frac{1}{2\pi^2}\left(\frac{2m_\mathrm{r}}{\hbar^2}\right)^{\frac{3}{2}} \int_0^\infty \mathrm{d}\varepsilon \frac{\sqrt{\varepsilon}}{\exp\left[\beta\left(\frac{m_\mathrm{r}}{m_\alpha}\varepsilon - \mu_\alpha\right)\right] + 1} \quad . \qquad (2.59)$$

In terms of $\varepsilon$, we have the energy detuning

$$\hbar\,(\omega_{\boldsymbol{k}} - \nu) = (\hbar\omega_{\boldsymbol{k}} - \varepsilon_{g0}) - (\hbar\nu - \varepsilon_{g0}) = \varepsilon - \hbar\delta \ , \tag{2.60}$$

where the field detuning relative to the bandgap is given by

$$\hbar\delta = \hbar\nu - \varepsilon_{g0} \ . \tag{2.61}$$

To carry out integrations over carrier energy, we express all frequencies in meV. To convert $\gamma$ to meV, we take advantage of the fact that $\gamma$ is usually given in terms of its inverse, the dephasing time $T_2$. For example, a dephasing time of $T_2 = 10^{-13}\,$s gives $\hbar\gamma = 6.58\,$meV. Similarly we evaluate the frequency difference $\hbar\delta$ in meV. For room temperature, $1/\beta \simeq 25\,$meV. The optical gain coefficient of (2.52) becomes

$$g = \frac{\nu}{4\pi^2\epsilon_0 nc\hbar\gamma}\left(\frac{2m_\mathrm{r}}{\hbar^2}\right)^{\frac{3}{2}}\int_0^\infty \mathrm{d}\varepsilon\,\sqrt{\varepsilon}\,|\mu_{\boldsymbol{k}}|^2\,\frac{f_\mathrm{e}(\varepsilon)+f_\mathrm{h}(\varepsilon)-1}{1+\left(\frac{\varepsilon-\hbar\delta}{\hbar\gamma}\right)^2}\ , \tag{2.62}$$

where the energy-dependent Fermi-Dirac distribution is given by

$$f_\alpha(\varepsilon) = \frac{1}{\exp\left[\beta\left(\varepsilon\frac{m_\mathrm{r}}{m_\alpha}-\mu_\alpha\right)\right]+1}\ . \tag{2.63}$$

In an ideal 2D quantum-well laser, the electrons are free to move only in the plane of the active layer, so that the summation over $\boldsymbol{k}$ is restricted to two dimensions according to

$$\frac{1}{A}\sum_{\boldsymbol{k}} \rightarrow \frac{2}{(2\pi)^2}\int_0^\infty \mathrm{d}k\,2\pi k\ , \tag{2.64}$$

where A is the area of the quantum well, and we assumed that the function summed over is cylindrically symmetric. When written in the form given by (2.57), we have a two-dimensional joint density of states

$$D = \frac{m_\mathrm{r}}{\pi\hbar^2} \tag{2.65}$$

which unlike the three dimensional value (2.58) does not depend on the reduced-mass energy $\varepsilon$.

## 2.3 Semiconductor Gain

Armed with the free-carrier gain equation (2.62), we can investigate at this level of approximation the dependence of semiconductor gain on the total carrier density and temperature. We begin by plotting a gain spectrum for several values of the homogeneous linewidth factor $\gamma$. Figure 2.4 shows the free-carrier gain spectrum of a bulk GaAs gain medium. The different curves are for different values of $\gamma$. The solid curve, which is for $\gamma \rightarrow 0$, gives the

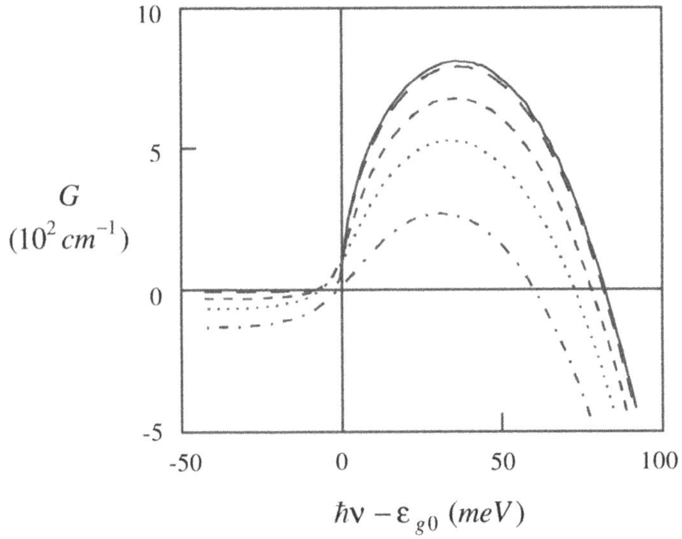


**Fig. 2.4.** Intensity gain $G = 2g$ versus detuning $\hbar\nu - \varepsilon_{g0}$ as predicted by the free-carrier model (2.62). The curves are for carrier density $N = 3 \times 10^{18}\,\mathrm{cm}^{-3}$, temperature $T = 300\,\mathrm{K}$, and $\gamma = 0$ (*solid line*), $10^{12}\,\mathrm{s}^{-1}$ (*long-dashed*), $5 \times 10^{12}\,\mathrm{s}^{-1}$ (*short-dashed*), $10^{13}\,\mathrm{s}^{-1}$ (*dotted*) and $2\times 10^{13}\,\mathrm{s}^{-1}$ (*dot-dashed*). The material parameters are $m_e = 0.067 m_0$, $m_h = 0.52 m_0$, $n = 3.6$ and $\mu_0 = e \times 4.73$ Å

inhomogeneously broadened limit, where the gain spectrum is the product of the density of states and the inversion, i.e.,

$$g = g_0 \sqrt{\varepsilon}\,[f_e(\varepsilon) + f_h(\varepsilon) - 1] \quad , \tag{2.66}$$

where the reduced mass energy $\varepsilon = \hbar\delta = \hbar\nu - \varepsilon_{g\phi}$ and (2.62) gives

$$g_0 = \frac{\nu}{4\pi\epsilon_0 n}\,|\mu(\varepsilon)|^2 \left(\frac{2m_r}{\hbar^2}\right)^{\frac{3}{2}} \quad . \tag{2.67}$$

Note that $\hbar\gamma = 10^{12}\,\mathrm{s}^{-1}$ is very close to the inhomogeneously broadened limit given by $\gamma = 0$. The more typical value $\gamma \approx 10^{13}\,\mathrm{s}^{-1}$ yields a significantly different curve from that at the inhomogeneously broadened limit. In general, the peak gain decreases with increasing $\gamma$.

The use of a Lorentzian lineshape function overestimates the effects of homogeneous broadening because of its slowly decaying tails. This leads to some absorption at photon frequencies below the bandgap. The correct lineshape function follows from the microscopic treatment of the collision contributions in (2.23). This is discussed in Chap. 4. To have a simple empirical expression which removes some of the artifacts of the Lorzentian lineshape one may use a sech lineshape

$$\frac{\gamma^2}{\gamma^2 + (\omega_k - \nu)^2} \to \mathrm{sech}\left(\frac{\omega_k - \nu}{\gamma}\right) \quad . \tag{2.68}$$

Note that the area under both of these functions is $\pi\gamma$. Figure 2.5 illustrates that the sech function decays faster (exponentially) than the Lorentzian. A comparison of the gain spectra computed with the different lineshapes is shown in Fig. 2.6. We can see that the use of the sech lineshape function removes the problem of absorption below the bandgap energy. Furthermore, it predicts that transparency occurs at lower carrier densities than for the Lorentzian lineshape approximation. This is quite apparent in the plots of peak gain versus carrier density in Fig. 2.7. The different curves correspond to different dephasing rates. One would not expect to use the same linewidth for the different lineshape functions, and Fig. 2.7 shows that in order to obtain similar predictions, $\gamma_{\rm sech}$ needs to be slightly larger than $\gamma_{\rm Lor}$.

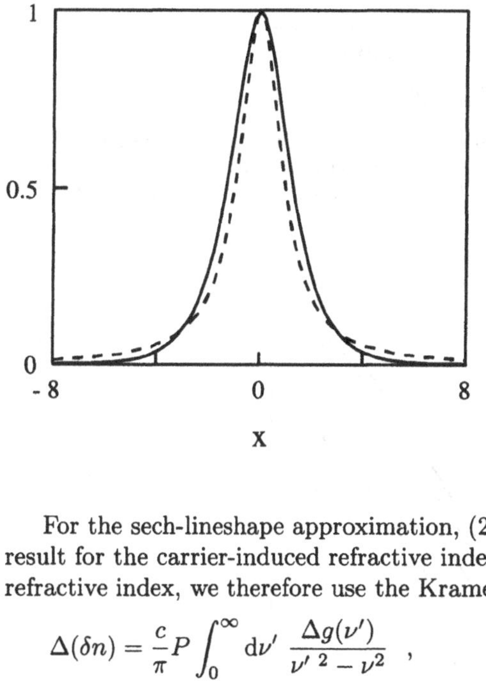


**Fig. 2.5.** The *solid curve* is a plot of sech($x$), while the *dashed curve* is a plot of $1/(1+x^2)$. The areas under both curves equal $\pi\gamma$

For the sech-lineshape approximation, (2.53) no longer yields the correct result for the carrier-induced refractive index. To obtain the carrier-induced refractive index, we therefore use the Kramers-Kronig transformation,

$$\Delta(\delta n) = \frac{c}{\pi} P \int_0^\infty \mathrm{d}\nu' \, \frac{\Delta g(\nu')}{\nu'^{\,2} - \nu^2} \quad , \tag{2.69}$$

where $\Delta(\delta n)$ and $\Delta g$ are the differences in refractive index and gain at two different carrier densities, and $P$ denotes the principal-value integral.

In order to have a rough approximation for the spectral width of the gain region we use the bracketed part of the $\delta$-function lineshape formula (2.66) written in terms of the valence-band electron distribution as

$$\begin{aligned} f_{\rm e}(\varepsilon) + f_{\rm h}(\varepsilon) - 1 &= f_{\rm e}(\varepsilon) - [1 - f_{\rm h}(\varepsilon)] \\ &= f_{\rm e}(\varepsilon) - f_{\rm v}(\varepsilon) \quad . \end{aligned} \tag{2.70}$$

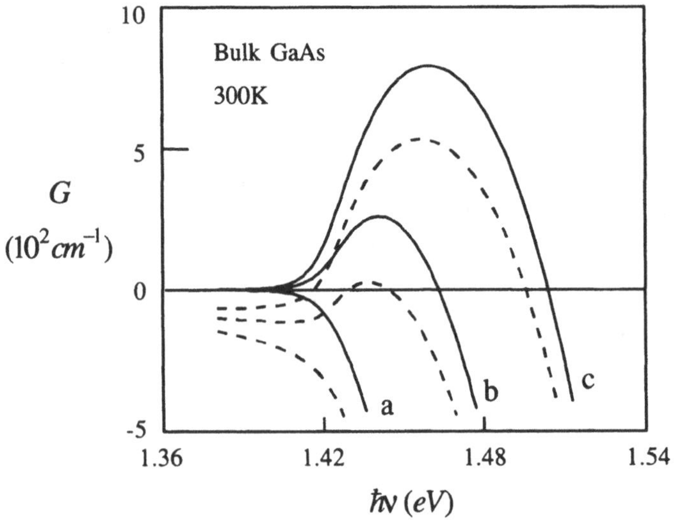


**Fig. 2.6.** The *solid curves* are the gain spectra calculated with a sech lineshape function, while the *dashed curves* are those calculated with a Lorentzian lineshape function. The carrier densities are ($a$) 1×, ($b$) 2× and ($c$) $3 \times 10^{18}\,\mathrm{cm}^{-3}$ and $\gamma = 10^{13}\,\mathrm{s}^{-1}$

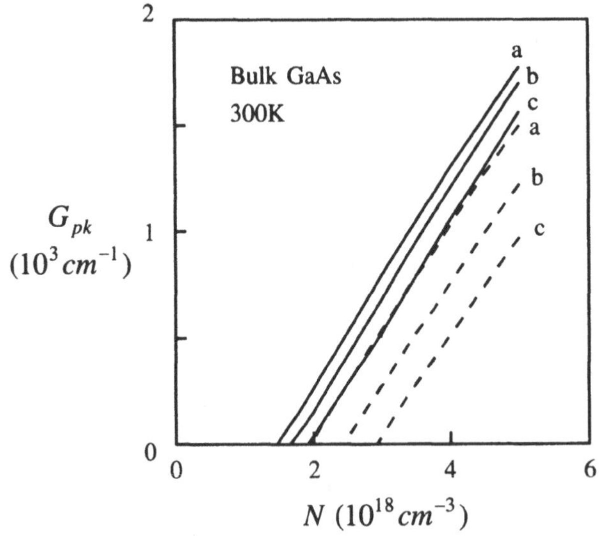


**Fig. 2.7.** Peak value of intensity gain versus carrier density for sech (*solid curves*) and Lorentzian (*dashed curves*) lineshape functions. The different curves corresponds to different dephasing rates $\gamma$ = ($a$) $10^{13}\,\mathrm{s}^{-1}$, ($b$) $2 \times 10^{13}\,\mathrm{s}^{-1}$ and ($c$) $3 \times 10^{13}\,\mathrm{s}^{-1}$

Here the valence-band electron distribution is given by

$$\begin{aligned} f_{\mathrm{v}}(\varepsilon) &= 1 - \frac{1}{\exp\left[\beta\left(\varepsilon \frac{m_{\mathrm{r}}}{m_{\mathrm{h}}} - \mu_{\mathrm{h}}\right)\right] + 1} \\ &= \frac{1}{1 + \exp\left[-\beta\left(\varepsilon \frac{m_{\mathrm{r}}}{m_{\mathrm{h}}} - \mu_h\right)\right]} \\ &= \frac{1}{\exp\left[\beta\left(\varepsilon \frac{m_{\mathrm{r}}}{mv} + \mu_{\mathrm{h}}\right)\right] + 1} \quad . \end{aligned} \tag{2.71}$$

The probability difference (2.70) is positive if $f_{\mathrm{e}}(\varepsilon) > f_{\mathrm{v}}(\varepsilon)$, i.e., if the conduction-band probability at energy $\varepsilon$ is greater than the valence-band probability. The transparency point (crossover from gain to absorption) is given by the equal probability condition $f_{\mathrm{e}}(\varepsilon) = f_{\mathrm{v}}(\varepsilon)$, which occurs when the arguments of the two Fermi-Dirac exponential equal one another,

$$\frac{m_{\mathrm{r}}}{m_{\mathrm{e}}}\varepsilon - \mu_{\mathrm{e}} = \frac{m_{\mathrm{r}}}{m_{\mathrm{h}}}\varepsilon + \mu_{\mathrm{h}} = 0 \quad ,$$

that is,

$$\varepsilon = \mu_{\mathrm{e}} + \mu_{\mathrm{h}} \quad . \tag{2.72}$$

The *total chemical potential*

$$\mu = \mu_{\mathrm{e}} + \mu_{\mathrm{h}} \tag{2.73}$$

is an important parameter in semiconductor laser theory since it defines the upper limit of the gain region with respect to the bandgap energy $\varepsilon_{\mathrm{g0}}$, i.e., gain occurs in the spectral region

$$\varepsilon_{\mathrm{g0}} < \hbar\nu < \varepsilon_{\mathrm{g0}} + \mu \quad . \tag{2.74}$$

For the many-body case, this inequality has to be replaced by

$$\varepsilon_{\mathrm{g0}} \rightarrow \varepsilon_{\mathrm{g}} = \varepsilon_{\mathrm{g0}} + \delta\varepsilon_{\mathrm{g}} \quad ,$$

where $\delta\varepsilon_{\mathrm{g}}$ is the electronic bandgap renormalization discussed in Chap. 3. The dominant effect of this bandgap renormalization is a net frequency shift of the gain spectrum with carrier density.

Although the free-carrier model does not predict a bandgap shift, it is useful in helping us to understand the bandfilling effects of the semiconductor gain medium. For example, for bulk materials it gives a reasonably accurate description of the change in peak gain with carrier density, as long as we do not care about the absolute peak gain value and the energy at which that peak gain occurs. Figure 2.8 plots the peak gain as a function of carrier density for bulk and quantum-well GaAs gain media and various temperatures. These curves show an essentially linear dependence of the bulk peak gain on carrier density, which agrees reasonably well with the phenomenological gain expression, (1.3),

$$G = A_{\mathrm{g}}(N - N_{\mathrm{g}}) \quad . \tag{2.75}$$

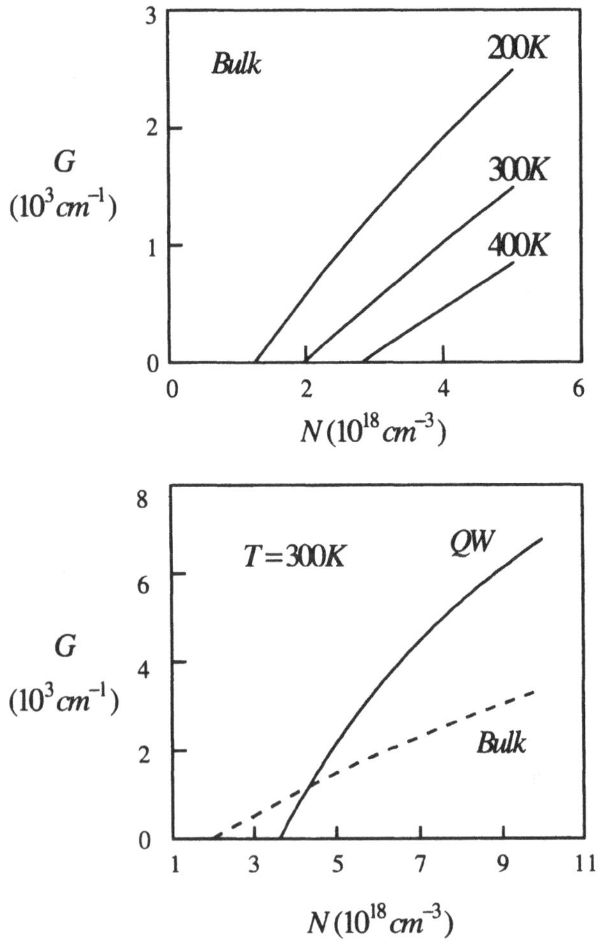


**Fig. 2.8.** (**top**) Peak value of intensity gain (in $10^3\,\mathrm{cm}^{-1}$) versus carrier density for a bulk GaAs gain medium. The different curves are for $T = 200\,\mathrm{K}$, $300\,\mathrm{K}$, and $400\,\mathrm{K}$. (**bottom**) Peak gain versus carrier density for an idealized two-dimensional quantum-well gain medium at $T = 300\,\mathrm{K}$ (*solid line*). The *dashed curve* is for the bulk gain medium. For the quantum well we assume the same effective masses as the bulk and choose a well width of 4 nm

Equation (2.75) is often used together with a similar expression for the carrier-induced refractive index change

$$\delta n = -RA_{\mathrm{g}}N\frac{n}{2K_0} \quad , \tag{2.76}$$

where $R$ is an antiguiding factor. These two formulas for $G$ and $\delta n$ form the phenomenological gain model. We must be careful when using (2.75, 76), because of their limitations. One of them is that $A_{\mathrm{g}}$, $N_{\mathrm{g}}$ and $R$, which are adjustable parameters, have different values for different temperatures and effective masses. The free-carrier model may be used to calculate the peak gain versus carrier density curves as functions of temperature and effective masses. For example, in Fig. 2.8 we use the free-carrier model to show that the gain degrades with increasing temperature. However, we should be aware that the free-carrier model neglects some important physics, namely the Coulomb effects. Therefore, its predictions can be inaccurate under some experimental conditions.

Another important limitation of the phenomenological gain expression (2.75) is that it applies only to a bulk gain medium. It is fortuitous that the product of the $\sqrt{\varepsilon}$ factor in the bulk density of states (2.58) with the probability inversion (2.70) has a nearly linear total-carrier-density dependence. Obviously, the constant $2D$ joint density of states (2.65) cannot give the same result. In fact, the peak gain versus carrier-density curve shown in the lower part of Fig. 2.8 is calculated using (2.65) in the free-carrier gain formula. Notice the rollover in gain at high carrier density.

The difference in behavior between the free-carrier bulk gain and the idealized two-dimensional quantum-well gain is due to the different density of states and may be understood qualitatively by examining the Fermi distributions. In Fig. 2.9 (top), we plot the electron and hole contributions to the integrand in the gain (2.62) versus the reduced-mass energy for three different carrier densities. We see that because of the relatively large hole mass $m_{\mathrm{h}}$ and consequently the large density of hole states (1.29), the hole states are only partially filled; in fact they never reach the inversion value of one half. The addition of carriers then leads to a fairly uniform increase in $f_{\mathrm{h}}(\varepsilon)$ versus $\varepsilon$. On the other hand, because of the relatively small $m_{\mathrm{e}}$ and small density of electron states, the lower energy electron states are almost completely filled, i.e., $f_{\mathrm{e}}(\varepsilon \simeq 0) \simeq 1$. Consequently for $\varepsilon \simeq 0$, there is no more "room at the top" so that the addition of carriers cannot significantly increase the maximum value of $f_{\mathrm{e}}(\varepsilon \simeq 0)$. Instead, the exclusion principle causes the additional electrons to preferentially occupy the vacant higher energy states.

In the middle part of Fig. 2.9 we plot the corresponding curves with the bulk density of states factor $D(\varepsilon) \propto \sqrt{\varepsilon}$. Two factors contribute to the increase in the product $D(\varepsilon)f_{\mathrm{e}}(\varepsilon)$ and $D(\varepsilon)f_{\mathrm{h}}(\varepsilon)$ with carrier density. One is the increase in $f_{\mathrm{e}}$ and $f_{\mathrm{h}}$ and the other is the increase in the relative density of states as the higher energy states become occupied. Both are responsible for the increase in $D(\varepsilon)f_{\mathrm{h}}$ with carrier density, whereas because of the exclusion principle only the latter plays a role in the case of $D(\varepsilon)f_{\mathrm{e}}$.

The changes in the carrier energy distributions lead to changes in the inversion energy distribution as shown in the bottom part of Fig. 2.9. Notice that the peak value varies essentially linearly with carrier density. Therefore, we expect the peak gain also to increase linearly with carrier density. The gain peaks for $\gamma = 10^{13}\,\mathrm{s}^{-1}$ are indicated by the dots.

Figure 2.10 shows plots for a 2D quantum-well medium. Since the 2D density of states is independent of the reduced-mass energy $\varepsilon$, the corresponding

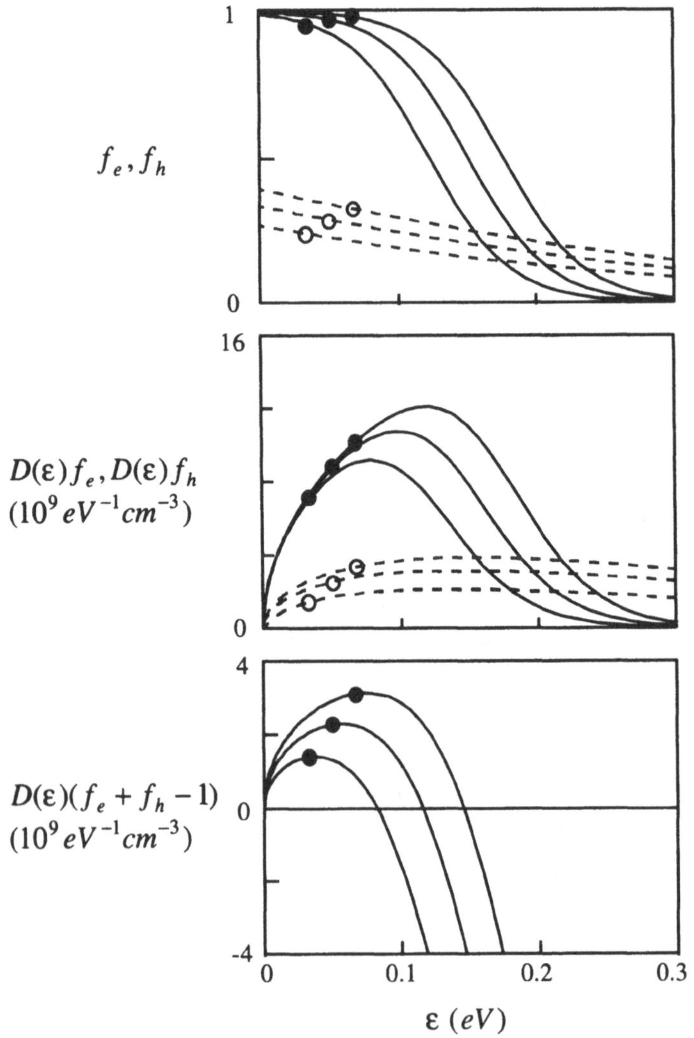


**Fig. 2.9.** (**top**) $f_e(\varepsilon)$ (*solid lines*), $f_h(\varepsilon)$ (*dashed lines*), (**middle**) $D(\varepsilon)f_e(\varepsilon)$ (*solid lines*), $D(\varepsilon)f_h(\varepsilon)$ (*dashed lines*), and (**bottom**) $D(\varepsilon)[f_e(\varepsilon)+ f_h(\varepsilon) - 1]$ versus $\varepsilon \equiv \hbar\omega - \varepsilon_{g0}$, for bulk total carrier densities $N = 3\times$, $4\times$, and $5 \times 10^{18}\,\text{cm}^{-3}$. The *dots* indicate the gain peaks for $\gamma = 10^{13}\,\text{s}^{-1}$

gain contributions are proportional to the inversion. Hence, the peak gain occurs for $\varepsilon \approx 0$ and the occupation of higher energy states populated by increasing $N$ does not increase the peak values for the carrier energy distributions. As a result, saturation effects are evident in Fig. 2.10; the peak values for $f_e(\varepsilon) + f_h(\varepsilon) - 1$ do not increase linearly with carrier density, which causes the peak gain in Fig. 2.8 to roll over. For a nonzero linewidth,

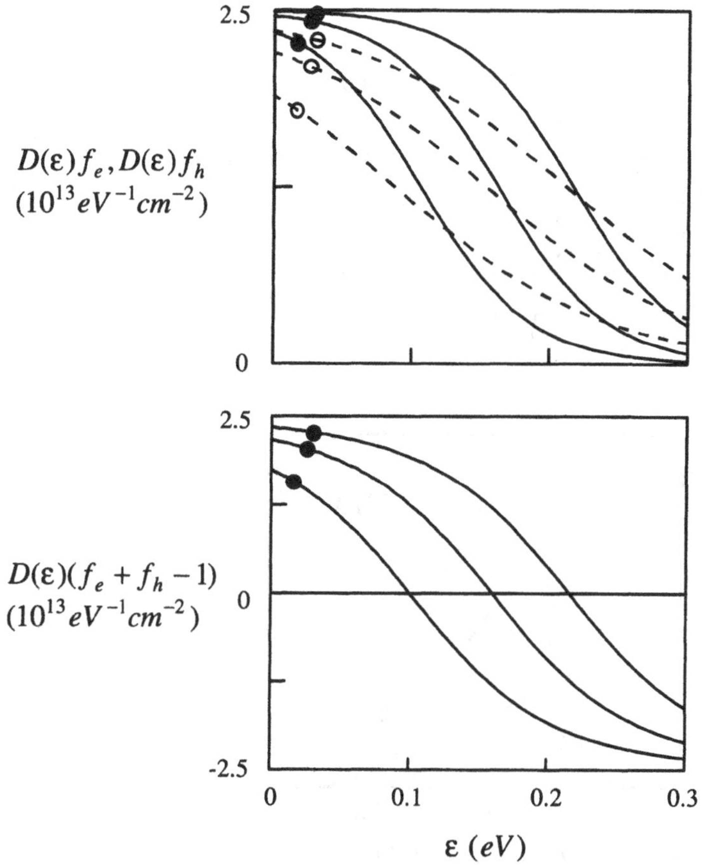


**Fig. 2.10.** (**top**) $D(\varepsilon)f_e(\varepsilon)$ (*solid lines*), $D(\varepsilon)f_h(\varepsilon)$ (*dashed lines*), and (**bottom**) $D(\varepsilon)[f_e(\varepsilon)+f_h(\varepsilon)-1]$ versus $\varepsilon \equiv \hbar\omega - \varepsilon_{g0}$ for a quantum well. The two-dimensional total carrier density $N_{2d} = 2 \times 10^{12}\,\text{cm}^{-2}$ (*lowest curves*), $3 \times 10^{12}\,\text{cm}^{-2}$ (*middle curves*), and $4 \times 10^{12}\,\text{cm}^{-2}$ (*top curves*). The *dots* indicate the gain peaks for $\gamma = 10^{13}\,\text{s}^{-1}$

the peak gain occurs somewhat above the bandgap as indicated by the dots for $\gamma = 10^{13}\,\text{s}^{-1}$.

It is interesting to investigate the individual gain and emission contributions to the gain factor

$$\begin{aligned} f_e(\varepsilon) + f_h(\varepsilon) - 1 &= f_e(\varepsilon)f_h(\varepsilon) - [1 - f_e(\varepsilon)]\,[1 - f_h(\varepsilon)] \\ &= \left(1 - e^{\beta(\varepsilon-\mu)}\right) f_e(\varepsilon)f_h(\varepsilon) \ . \end{aligned} \tag{2.77}$$

Comparison of (2.33, 52) shows that the second line in (2.77) relates the spontaneous emission and gain expression. In Fig. 2.11 we compare the gain (Fig. 2.11a) and the spontaneous emission spectra (Fig. 2.11b) for bulk GaAs

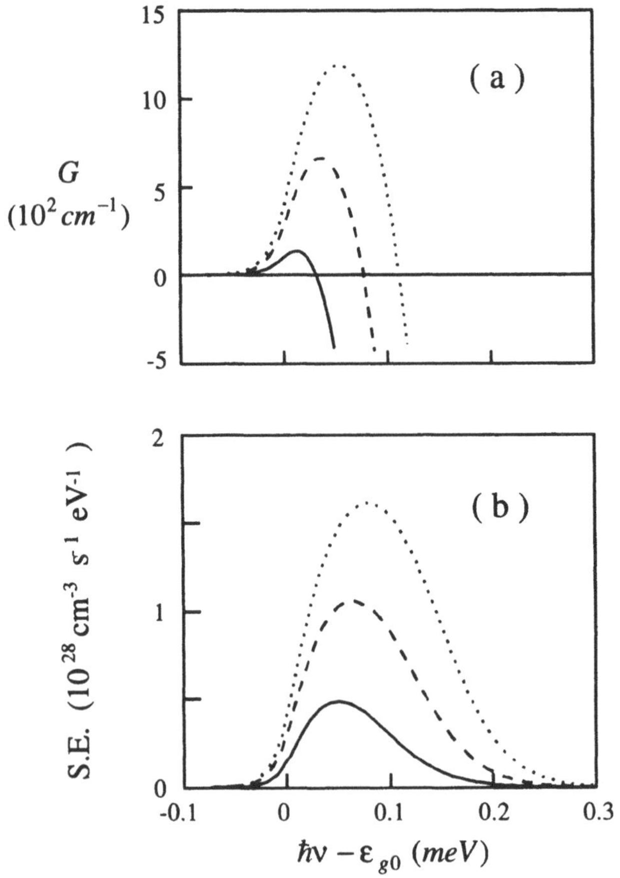


**Fig. 2.11a, b.** (**a**) Intensity gain and (**b**) spontaneous emission spectra for bulk medium and carrier densities $N = 2\times10^{18}$ cm$^{-3}$ (*solid lines*), $3\times10^{18}$ cm$^{-3}$ (*dashed*), $4\times10^{18}$ cm$^{-3}$ (*dotted*). We used a sech lineshape function with $\gamma = 2\times10^{13}$ s$^{-1}$ in the gain calculation

at different carrier densities. The spontaneous emission spectra are considerably broader than the gain region since they are determined by the product of electron and hole population function, which has no upper bound given by the chemical potential as in the case of the gain region.

## 2.4 Temperature Dependence of Gain

We have seen in Fig. 2.8, that the semiconductor gain decreases with increasing temperature. To understand this, we use a set of figures that are very similar to those in the previous section. Figure 2.12 plots the occupational probabilities of electron and hole states multiplied by the bulk relative

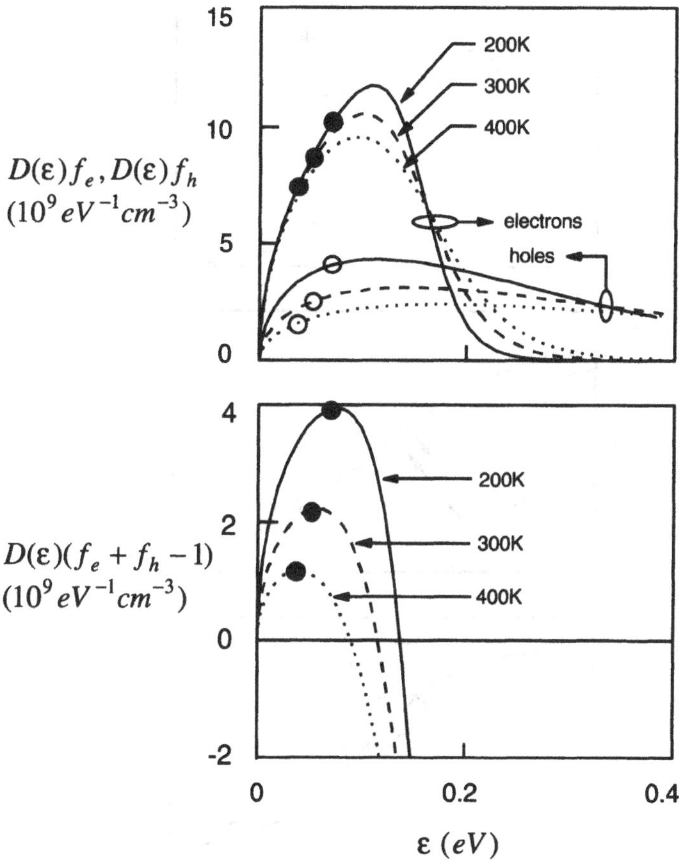


**Fig. 2.12.** **(top)** $D(\varepsilon)f_e(\varepsilon)$ and $D(\varepsilon)f_{he}(\varepsilon)$ versus $\varepsilon \equiv \hbar\omega - \varepsilon_{g0}$ for temperatures $T = 200$ (*solid curve*), 300 (*dashed curve*) and 400 K (*dotted curve*). The carrier density is $N = 4 \times 10^{18}\,\mathrm{cm}^{-3}$. **(bottom)** The corresponding curves for $D(\varepsilon)\,[f_e(\varepsilon) + f_h(\varepsilon) - 1]$. The *dots* mark the peak gain energies

density of states factor. Notice that the peaks of the distributions decrease with increasing temperature, which leads to a decrease in the gain. Note also that some of the hole distributions have negative chemical potentials and look very much like the decaying tails of Maxwell-Boltzmann distributions. We see that both the electron and hole distributions are particularly sensitive to temperature changes in the region of $\varepsilon \simeq \mu/2$, which is where the gain tends to peak due to the $\sqrt{\varepsilon}$ factor. The peak gain positions are marked by the black dots in Fig. 2.12. Bulk semiconductor materials therefore exhibit a substantial temperature dependence of the gain, as contrasted with the idealized $2D$ quantum-well case discussed next.

Figure 2.13 shows similar plots for the two-dimensional idealized quantum-well medium. Since at the gain peak ($\varepsilon \approx 0$), only the hole distributions are

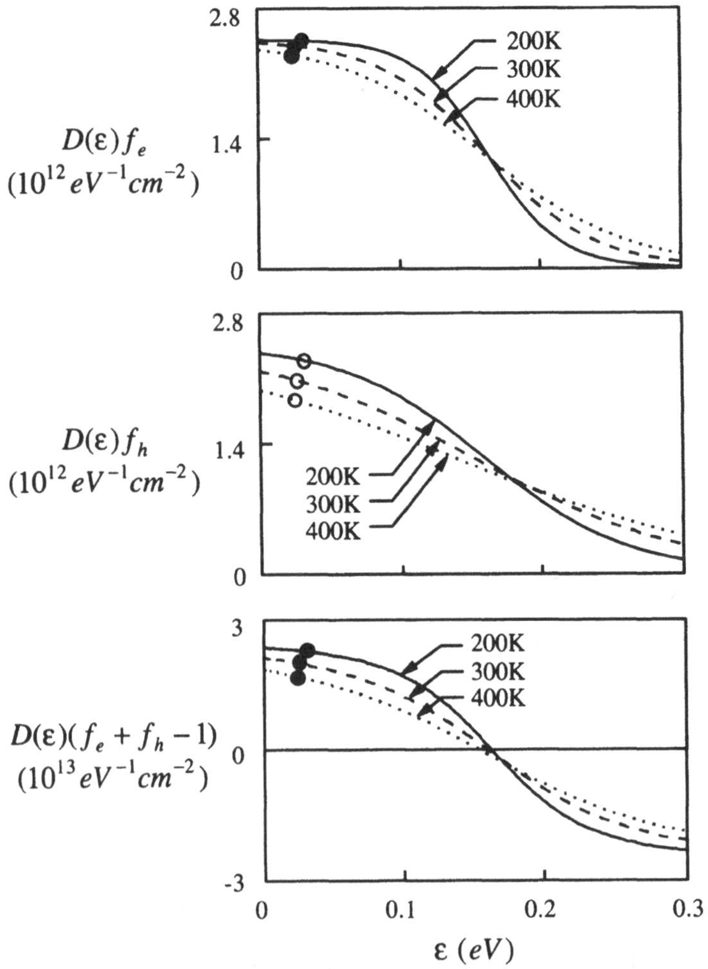


**Fig. 2.13.** (**top**) $D(\varepsilon)f_e(\varepsilon)$ (**middle**) $D(\varepsilon)f_{he}(\varepsilon)$ and (**bottom**) $D(\varepsilon)\,[f_e(\varepsilon)+f_h(\varepsilon)-1]$ versus $\varepsilon \equiv \hbar\omega - \varepsilon_{g0}$ for quantum well and temperatures $T = 200$ (*solid curve*), 300 (*dashed curve*) and 400 K (*dotted curve*). The carrier density is $N_{2d} = 3 \times 10^{12}\,\mathrm{cm}^{-2}$. The *dots* mark the gain peak energies

changed significantly, the inversion is less affected by temperature than in the bulk case. As a result, the degradation of the idealized two-dimensional quantum-well gain with increasing temperature is less than that of the bulk. This is shown by the two sets of curves in Fig. 2.14. These curves are calculated using (2.52, 65).

The lower group of quantum-well curves in Fig. 2.14 suggest that a small density of states is advantageous for temperature insensitivity. A more Fermi-Dirac (or less Boltzmann) like energy distribution, with filled low energy

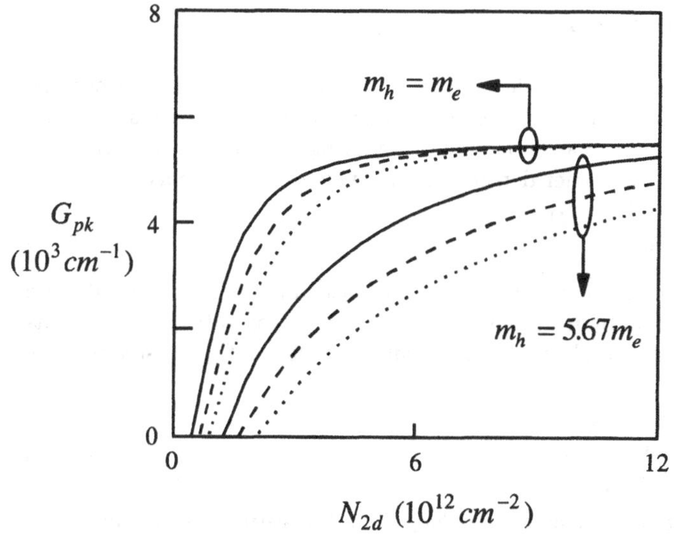


**Fig. 2.14.** Peak values of intensity gain versus carrier density for the idealized quantum-well system for the temperatures $T$ = 200 (*solid curve*), 300 (*dashed curve*) and 400 K (*dotted curve*). The hole masses are $m_h = 5.67 m_e$ (used in previous examples) and $m_e = m_h$

states reduces the temperature dependence. Therefore, if we can alter the valence band curvature so that the hole effective mass, and consequently, the hole density of states is as small as that of the electrons, then gain degradation due the temperature increase should be reduced. To see if this is indeed the case, we compute the gain for an artificial quantum-well medium with equal hole and electron effective masses. The result is shown in the upper group of curves in Fig. 2.14. In practice, a reduction in the hole effective mass may be achieved with some strained-layer quantum wells. Of course, the actual results are neither as good nor as straightforward as shown in Fig. 2.14, because the strain generally deforms the band curvature nonuniformly (nonparabolic bands) and alters the transition matrix elements. In fact, as we see in Chaps. 6 and 7, the strained-layer quantum-well structures involve more than just a change in the effective masses, since the various valence bands are mixed together. Notice that a consequence of a small hole effective mass is an increase in gain rollover. This too can be understood in terms of the exclusion principle embodied in the Fermi-Dirac distributions, since for high $N$ *both* the electron and the hole distributions fail to have room at the top for more carriers.

## 2.5 Gain Saturation

In any oscillator saturation effects play an important role since otherwise the oscillation amplitude would keep building up indefinitely. A simple description of saturation in a semiconductor gain medium starts with the rate equation for the total carrier density, (2.54), which we rewrite as

$$\frac{\mathrm{d}N}{\mathrm{d}t} = \frac{J\eta}{ed} - \gamma_{\text{eff}} N - \frac{GI}{\hbar\nu} \ , \tag{2.78}$$

where we have simplified the description of both radiative and nonradiation carrier recombination by using an effective carrier decay rate $\gamma_{\text{eff}}$. Substituting (2.75) for the gain and solving the resulting equation for the steady-state value of $N$ gives

$$N - N_{\text{g}} = \frac{\frac{J\eta}{ed\gamma_{\text{eff}}} - N_{\text{g}}}{1 + \frac{I}{I_{\text{sat}}}} \ , \tag{2.79}$$

where the saturation intensity $I_{\text{sat}} = \hbar\nu\gamma_{\text{nr}}/A_{\text{g}}$. Substituting (2.79) back into (2.75) gives the saturated gain,

$$G = \frac{G_0}{1 + \frac{I}{I_{\text{sat}}}} \ , \tag{2.80}$$

where the small signal gain

$$G_0 = A_{\text{g}}\left(\frac{J\eta}{ed\gamma_{\text{eff}}} - N_{\text{g}}\right) \ . \tag{2.81}$$

The saturation behavior described by (2.80) is illustrated in Fig. 2.15 for both absorbing (no injection current) and gain cases. We see the $S$-shaped saturation curves familiar in the saturation of homogeneously broadened two-level media.

To get a better understanding of the nature of gain saturation, we consider the $\delta$-function linewidth gain formula (2.66) for fixed detunings $\hbar\delta = \hbar\nu - \varepsilon_{\text{g}0}$ above the bandgap for both gain and absorption media (with and without injected currents, respectively). Equation (2.66) is similar to (1.3) in that both consist of the difference between a term that increases as N increases and a transparency value. However, (2.66) has the advantages that it depends explicitly on the temperature, the carrier effective masses, the electric-dipole matrix element, and the field detuning $\hbar\delta$, and it takes the Fermi-Dirac character of the gain into account. On the other hand, we should not forget that it still neglects many-body effects and spectral hole burning.

Figure 2.16 illustrates the dependence of $g$ on $I$ for a number of detunings $\hbar\delta$. The curves are generated by solving (2.54) for $I$ in steady state ($\dot{N} = 0$) as

$$I = \frac{\hbar\nu}{2g}\left(\Lambda - \frac{N}{T_1}\right) \ , \tag{2.82}$$

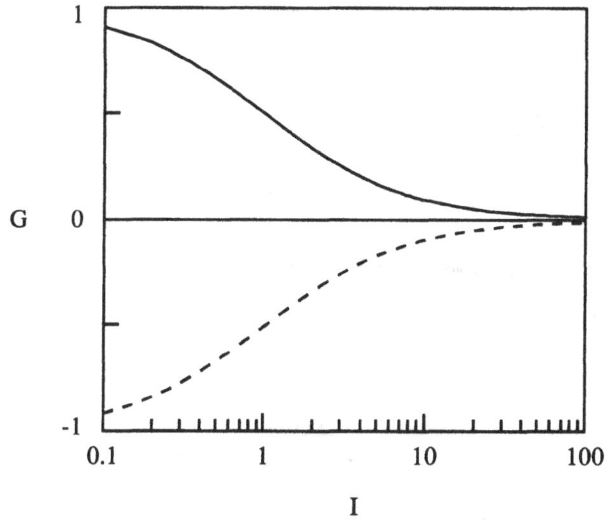


**Fig. 2.15.** Linear-density gain given by (2.80) versus applied light intensity $I$ for gain (*solid line*) and absorbing (*dashed line*) media. The parameters are chosen such that the saturation intensity $\hbar\nu/A_g T_1 = 1$, $A_g(JT_1/ed - N_g) = 1$ for the gain case, and $-A_g N_g = -1$ and $J = 0$ for the absorbing case

and plotting $g$ versus $I$ as $N$ is varied from $10^{17}\,\mathrm{cm}^{-3}$ to $10^{19}\,\mathrm{cm}^{-3}$ ($2\times 10^{11}\,\mathrm{cm}^{-2}$ to $2\times 10^{13}\,\mathrm{cm}^{-2}$ for the 2-dimensional case). For $g < 0$ (below transparency, in the absorption region), $\Lambda T_1$ is set equal to $10^{18}\,\mathrm{cm}^{-3}$, while for $g > 0$, $\Lambda T_1 = 10^{19}\,\mathrm{cm}^{-3}$.

For $I = 0$, the gain is that for $N = \Lambda T_1$, which is chosen to be quite large so that $f_e$ and $f_h$ have the maximum value of unity. As $I$ increases, $N$ decreases according to (2.82), but initially with little effect on $g$ since the Fermi-Dirac distributions remain at their maximum value of 1. Since the holes are heavier, $f_h$ starts to fall off first, later followed by $f_e$, thereby decreasing $g$. The Fermi-Dirac unity limit plays a relatively smaller role in the absorption case ($g < 0$), since $f_h$ is substantially less than one half over the entire intensity range, and $f_e$ is less than a half for all but very large $I$.

We see from the curves in Fig. 2.16 and from (2.82) that generally, increasing $I$ indefinitly cannot make a transition between absorption and gain at the photon energy $\hbar\nu$. Instead, increasing $I$ increases $N$ for the absorption case and decreases N for the gain case, in either case driving $f_e(\hbar\delta) + f_h(\hbar\delta) - 1$ to zero, i.e., pulling the total chemical potential $\mu = \mu_e + \mu_h$ to $\hbar\delta = \hbar\nu - \varepsilon_{g0}$ and creating transparency at $\hbar\nu$. Since gain occurs for energies in the range $0 < \varepsilon < \mu$, this shows that in the absence of an injection current, a sufficiently strong pump wave of frequency $\nu$ can create a gain region below $\nu$. This is due to the fact that carrier-carrier scattering redistributes the electron-hole pairs created by optical pumping above the carrier chemical potentials into the appropriate quasi-equilibrium Fermi-Dirac distributions.

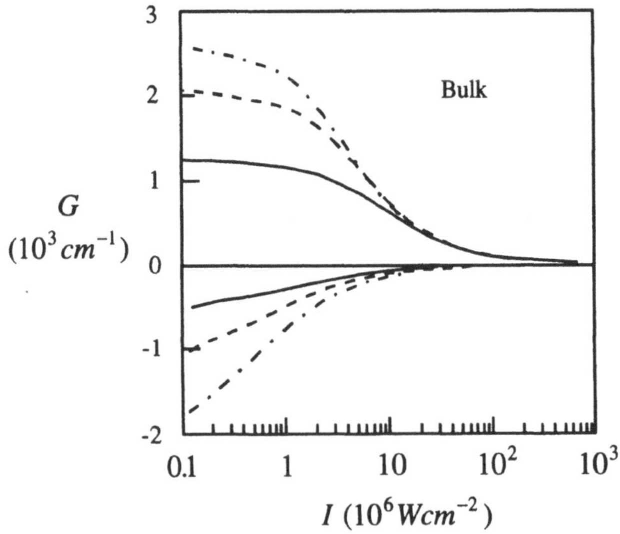


**Fig. 2.16.** Intensity gain $G = 2g$ versus intensity $I$ for the detuning $\hbar\nu - \varepsilon_{g0} = 10\,\text{meV}$ (*solid lines*), 30 meV (*dashed lines*), and 60 meV (*dot-dashed lines*). Positive $G$ values are drawn for $\Lambda T_1 = 10^{19}\,\text{cm}^{-3}$ and the negative $G$ values are shown for $\Lambda T_1 = 10^{18}\,\text{cm}^{-3}$, respectively

We can gain some further insight by expanding the susceptibility $\chi(N)$ in a first-order Taylor series about the zero-field value $N_0$. This $N_0$ can be generated by an injection current, optical pumping above the interaction region, or a combination of both. The first nonlinear term is called $\chi^{(3)}$, which, as we see shortly, is a strong function of $N_0$, field frequency, and temperature. Our derivation is valid also for the quasi-equilibrium many-body theory and for all carrier densities and temperatures as long as the quasi-equilibrium approximations are justified, i.e., as long as Fermi-Dirac distributions of the carriers exist in the laser. The main approximation is that the field intensity must be small enough to be treated by a third-order theory.

We write $N$ as

$$N = N_0 + \Delta N \tag{2.83}$$

and expand the susceptibility $\chi(N)$ in the first-order Taylor series

$$\chi(N) \simeq \chi(N_0) + \left.\frac{\partial \chi(N)}{\partial N}\right|_{N_0} \Delta N \ . \tag{2.84}$$

To find the $\Delta N$ resulting from weak field saturation, we expand the total carrier-density equation of motion to first order in $\Delta N \propto |E|^2$and take steady state ($\dot{N} = 0$). We write the equation of motion (2.54) for $N$ as

$$\frac{\mathrm{d}N}{\mathrm{d}t} = \Lambda - \Gamma(N) + \frac{\epsilon_b}{2\hbar}\chi''(N)|E|^2 \ , \tag{2.85}$$

where $\Lambda$ represents the optical pumping or carrier injection, $\Gamma(N)$ is a decay function including both radiative and nonradiative decay, $\epsilon_b$ is the host susceptibility, and $E$ is the complex electric-field envelope. To lowest order and in steady state, we have

$$\Lambda = \Gamma(N) \ . \tag{2.86}$$

To first order in $\Delta N$ and $|E|^2$, we have

$$\begin{aligned} 0 &= \Lambda - \Gamma(N_0) - \Delta N\, \Gamma_1 + \frac{\epsilon_b}{2\hbar}\chi''(N_0)|E|^2 \\ &= -\Delta N\, \Gamma_1 + \frac{\epsilon_b}{2\hbar}\chi''(N_0)|E|^2 \ , \end{aligned}$$

where the decay-rate coefficient

$$\Gamma_1 = \left.\frac{\mathrm{d}\Gamma(N)}{\mathrm{d}N}\right|_{N=N_0} \ . \tag{2.87}$$

This gives the weak-field intensity-induced carrier-density change

$$\Delta N = \frac{\epsilon_b}{2\hbar\Gamma_1}\chi''(N_0)|E|^2 \ . \tag{2.88}$$

Substituting this change into (2.84), we have the approximate nonlinear susceptibility

$$\chi(N) \simeq \chi(N_0) + \frac{\epsilon_b\chi''(N_0)}{2\hbar\Gamma_1}\left.\frac{\mathrm{d}\chi(N)}{\mathrm{d}N}\right|_{N=N_0} |E|^2 \ . \tag{2.89}$$

This gives the first two terms in the intensity expansion of the susceptibility as

$$\chi^{(1)}(N_0) = \chi(N_0) \ , \tag{2.90}$$

$$\chi^{(3)}(N_0) = \frac{\epsilon_b\chi''(N_0)}{2\hbar\Gamma_1}\left.\frac{\mathrm{d}\chi(N)}{\mathrm{d}N}\right|_{N=N_0} \ . \tag{2.91}$$

## 2.6 Carrier Induced Refractive Index

For the free-carrier theory, the background host variation represented by the $-1$ in (2.53) can be included in the host index $n$, that is, the carrier-induced refractive index reduces to

$$\delta n = \frac{1}{2\epsilon_0 n\hbar\gamma V}\sum_{k}|\mu_k|^2\left[f_{ek}(t) + f_{hk}(t)\right] L\left(\omega_k - \nu\right)\frac{\omega_k - \nu}{\gamma} \ . \tag{2.92}$$

This approach avoids the problem that the $\sum_k$ as given by (2.57) does not converge due to the long unphysical tails of the $L(\omega_k - \nu)(\omega_k - \nu)/\gamma$ factor. Otherwise, convergence has to be achieved e.g. with the $k$-dependence of the dipole matrix element. To see this we investigate the matrix element of the space operator between two bands $\lambda$ and $\lambda'$

$$r_{\lambda\lambda'} = \langle \phi_{\lambda' k'} | \, r \, | \phi_{\lambda k} \rangle \quad , \tag{2.93}$$

where $|\phi_{\lambda k}\rangle$ and $|\phi_{\lambda' k'}\rangle$ are eigenstates of the semiconductor, i.e.,

$$H_0 \, |\phi_{\lambda k}\rangle = \varepsilon_{\lambda k} \, |\lambda k\rangle \quad , \tag{2.94}$$

and $H_0$ is the system Hamiltonian. To evaluate (2.93) we consider only interband transitions, $\lambda \neq \lambda'$. Furthermore, we notice, using (2.94), that

$$\langle \phi_{\lambda' k'} | \, r \, | \phi_{\lambda k} \rangle \left( \varepsilon_{\lambda k} - \varepsilon_{\lambda' k'} \right) = \langle \phi_{\lambda' k'} | \, [r, H_0] \, | \phi_{\lambda k} \rangle \quad . \tag{2.95}$$

Equation (2.95) and the fact that

$$[r, H_0] = \frac{\mathrm{i}\hbar}{m_0} p \quad , \tag{2.96}$$

where $p$ is the momentum operator, give

$$r_{\lambda'\lambda}\left(k', k\right) = \frac{\mathrm{i}\hbar}{m_0 \left( \varepsilon_{\lambda k} - \varepsilon_{\lambda' k'} \right)} \langle \phi_{\lambda' k'} | \, p \, | \phi_{\lambda k} \rangle \quad . \tag{2.97}$$

The momentum operator is diagonal in the coordinate representation

$$\langle r' | \, p \, | r \rangle = \delta \left( r - r' \right) \frac{\hbar}{\mathrm{i}} \nabla \quad , \tag{2.98}$$

so that

$$\langle \phi_{\lambda' k'} | \, p \, | \phi_{\lambda k} \rangle = \int_V \mathrm{d}^3 r \, \phi^*_{\lambda' k'}(r) \, p \, \phi_{\lambda k}(r) \quad . \tag{2.99}$$

To the lowest order, we can approximate

$$\langle r | \phi_{\lambda k} \rangle \simeq \frac{\mathrm{e}^{\mathrm{i} k \cdot r}}{\sqrt{V}} \langle r | \lambda \rangle \quad , \tag{2.100}$$

leading to

$$\langle \phi_{\lambda' k'} | \, p \, | \phi_{\lambda k} \rangle = \frac{1}{V} \int_V \mathrm{d}^3 r \, \mathrm{e}^{-\mathrm{i}(k'-k)\cdot r} \langle \lambda' | r \rangle \left( \hbar k + p \right) \langle r | \lambda \rangle \quad . \tag{2.101}$$

To continue with our evaluation, we split the integral over the crystal volume into the sum over the unit cells and the integral within a unit cell

$$\frac{1}{V} \int_V \mathrm{d}^3 r \to \frac{1}{N} \sum_\nu \frac{1}{v} \int_v \mathrm{d}^3 \rho \quad , \tag{2.102}$$

where $V = Nv$ and the space vector is written as

$$r = R_\nu + \rho$$

so that $\rho$ varies within one unit cell and $R_\nu$ is the position vector of unit cell $\nu$. Inserting these expressions into (2.101), we get

$$\langle \phi_{\lambda' k'} | \, p \, | \phi_{\lambda k} \rangle = \sum_v \frac{\mathrm{e}^{-\mathrm{i}(k'-k)\cdot R_\nu}}{N} \times \int_v \mathrm{d}^3 \rho \, \frac{\mathrm{e}^{-\mathrm{i}(k'-k)\cdot \rho}}{v} \langle \lambda' | \rho \rangle \left( \hbar k + p \right) \langle \rho | \lambda \rangle \quad . \tag{2.103}$$

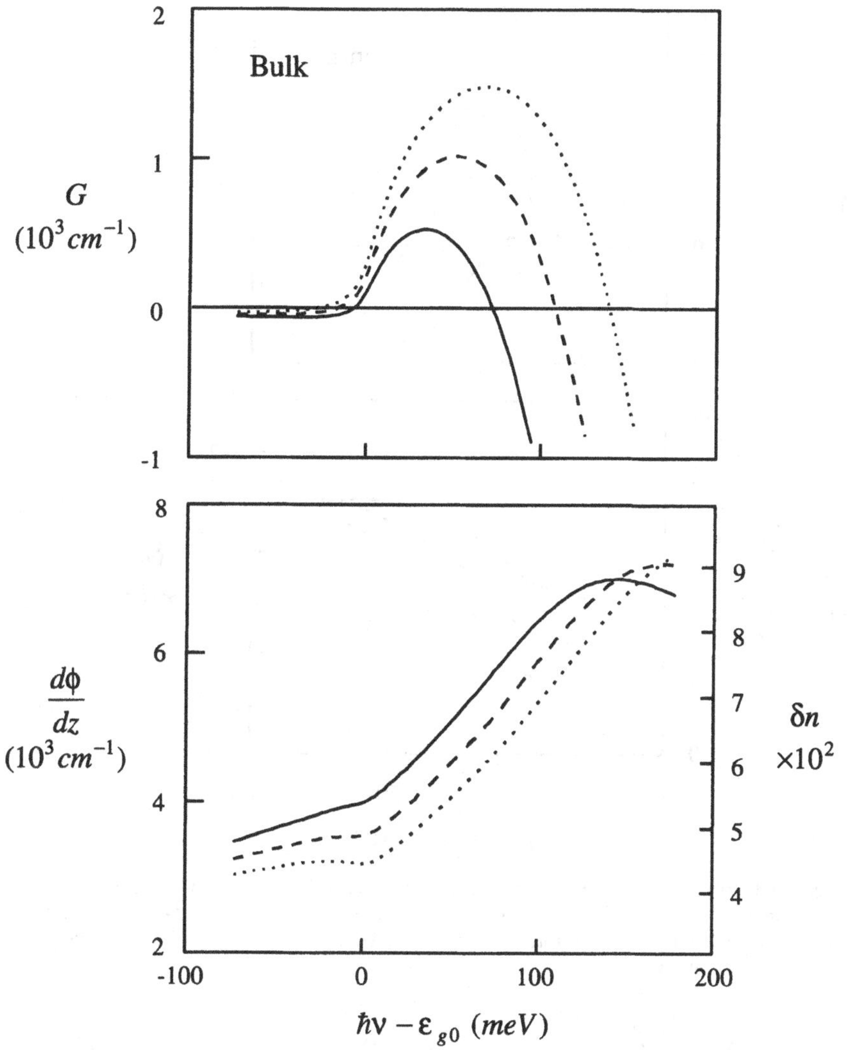


**Fig. 2.17.** Intensity gain and carrier-induced phase shift versus detuning $\hbar\nu - \varepsilon_{g0}$ for bulk medium. $N = 3 \times 10^{18}\,\text{cm}^{-3}$ (*solid lines*), $4 \times 10^{18}\,\text{cm}^{-3}$ (*dashed*), $5 \times 10^{18}\,\text{cm}^{-3}$ (*dotted*). The values for the carrier-induced refractive index are also shown (right scale in bottom part of figure)

The unit-cell integral yields the same result for all unit cells and can be moved out of the summation over the unit cells, which then yields $\delta_{\boldsymbol{k},\boldsymbol{k}'}$ and

$$\langle \phi_{\lambda'\boldsymbol{k}'} | \boldsymbol{p} | \phi_{\lambda \boldsymbol{k}} \rangle = \delta_{\boldsymbol{k},\boldsymbol{k}'} \langle \lambda' | \boldsymbol{p} | \lambda \rangle = \delta_{\boldsymbol{k},\boldsymbol{k}'} \boldsymbol{p}_{\lambda',\lambda}(0) \quad . \tag{2.104}$$

Here, we denote

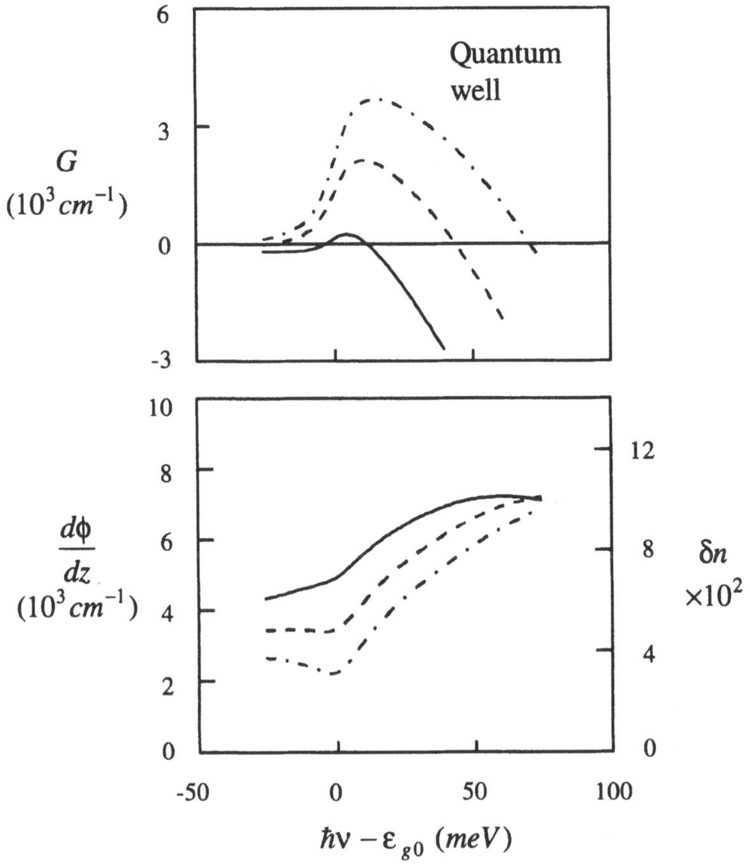


**Fig. 2.18.** Intensity gain and carrier-induced phase shift versus detuning $\hbar\nu - \varepsilon_{g0}$ for quantum-well medium. $N_{2d} = 1.5 \times 10^{12}\,\text{cm}^{-2}$ (*solid lines*), $2 \times 10^{12}\,\text{cm}^{-2}$ (*dashed*), and $2.5 \times 10^{12}\,\text{cm}^{-2}$ (*dash-dotted*). The values for the carrier-induced refractive index are also shown

$$\langle \lambda' | \boldsymbol{p} | \lambda \rangle = \frac{1}{v} \int_v \mathrm{d}^3\rho \, \langle \lambda' | \boldsymbol{\rho} \rangle \, (\hbar \boldsymbol{k} + \boldsymbol{p}) \, \langle \boldsymbol{\rho} | \lambda \rangle \quad . \tag{2.105}$$

The term $\propto \hbar\boldsymbol{k}$ disappears in going from (2.103) to (2.104) because of the orthogonality of the lattice periodic functions and the $\lambda \neq \lambda'$ requirement.

Collecting all contributions to the dipole matrix element, we get

$$e\boldsymbol{r}_{\lambda'\lambda}\left(\boldsymbol{k}', \boldsymbol{k}\right) = \boldsymbol{\mu}_{\lambda'\lambda}\left(\boldsymbol{k}', \boldsymbol{k}\right) = \frac{\mathrm{i}e\hbar \boldsymbol{p}_{\lambda'\lambda}(0)}{m_0\left(\varepsilon_{\lambda'\boldsymbol{k}} - \varepsilon_{\lambda\boldsymbol{k}}\right)} \delta_{\boldsymbol{k},\boldsymbol{k}'} \quad . \tag{2.106}$$

or

$$\boldsymbol{\mu}_{\lambda'\lambda}\left(\boldsymbol{k}', \boldsymbol{k}\right) = \delta_{\boldsymbol{k},\boldsymbol{k}'} \boldsymbol{\mu}_{\lambda'\lambda}(0) \, \frac{\varepsilon_{\lambda'} - \varepsilon_{\lambda}}{\varepsilon_{\lambda'\boldsymbol{k}} - \varepsilon_{\lambda\boldsymbol{k}}} \equiv \delta_{\boldsymbol{k},\boldsymbol{k}'} \boldsymbol{\mu}_{\lambda'\lambda}(\boldsymbol{k}) \quad . \tag{2.107}$$

For the case of two parabolic bands with effective masses $m_\lambda$ and $m_{\lambda'}$, and dispersions

$$\varepsilon_{\lambda' \boldsymbol{k}} = \varepsilon_{g0} + \frac{\hbar^2 k^2}{2m_{\lambda'}} \quad \text{and} \quad \varepsilon_{\lambda \boldsymbol{k}} = -\frac{\hbar^2 k^2}{2m_\lambda} \ , \tag{2.108}$$

we recover the Kane matrix element $\mu_{\boldsymbol{k}}$

$$\mu_{\boldsymbol{k}} = \frac{\mu_0}{1 + \frac{\hbar^2 k^2}{2m_r \varepsilon_{g0}}} = \frac{\mu_0}{1 + \varepsilon/\varepsilon_{g0}} \ , \tag{2.109}$$

where $\mu_0$ is the matrix element at $k = 0$. Even though the above energy dependent denominator leads to convergence in (2.92), there still remains an overestimation of the contributions from the high energy states. The problem stems both from the simple complex-Lorentzian lineshape factor and from the limitations of the effective-mass approximation. In real semiconductors the bands do not remain parabolic for higher $\boldsymbol{k}$-values; they usually flatten out, leading to a finite bandwidth limiting the spectral range of optical transitions. As an illustration we show in Figs. 2.17, 18 the carrier-induced refractive index $\delta n$ and the intensity gain $G = 2g$ for a bulk and a quantum-well medium.

## 2.7 Linewidth Enhancement or Antiguiding Factor

We often do not need to compute the absolute phase shift since only the phase shift changes enter our expressions. For example, the linewidth enhancement factor (or antiguiding parameter)

$$\begin{aligned} \alpha &= \frac{\partial \chi'/\partial N}{\partial \chi''/\partial N} \\ &= -\frac{K}{n_b} \frac{\partial (\delta n)/\partial N}{\partial g/\partial N} \end{aligned} \tag{2.110}$$

is the change in the phase shift with respect to carrier density divided by the corresponding change in the gain. Discussions of the carrier-induced refractive index often involve the $\alpha$ factor since it provides a simple way to model such index contributions. This factor can be written in terms of $\chi^{(3)}$ of (2.91) as

$$\alpha = \frac{\mathrm{Re}\{\chi^{(3)}\}}{\mathrm{Im}\{\chi^{(3)}\}} \ . \tag{2.111}$$

For a linear-density gain value $G = 2g = A_g(N - N_g)$ of (2.75), we have using (1.79) that

$$\frac{\partial \chi''}{\partial N} = -\frac{A_g}{K} \ . \tag{2.112}$$

Similarly, combining the index change $\delta n = -RA_gN$ of (2.76) with (1.82), we have

$$\frac{\partial \chi'}{\partial N} = -\frac{RA_g}{K} \quad . \tag{2.113}$$

Dividing this by (2.112), we find that the antiguiding factor $R$ is just the linewidth enhancement factor

$$\alpha = R \quad . \tag{2.114}$$

If we lump the host index $n$ into the $N_g$ term, we can write the susceptibility as

$$\chi \simeq (\mathrm{i} + \alpha)\chi'' \quad . \tag{2.115}$$

Note that (2.112–115) are only valid as long as the susceptibility is a linear function of the total carrier density, which is typically only true for the gain peak of a bulk-gain medium. Nevertheless, the linewidth enhancement factor is an interesting quantity to study, since many problems, such as the laser linewidth itself, involve small changes in the susceptibility, for which $\alpha$ as a function of $N$, $T$, and tuning is a relevant measure.

Figure 2.19 shows $\alpha$ spectra computed using (2.110). The value of $\alpha$ at the peak gain is a good parameter for comparing the importance of the carrier-induced phase shift in determining the laser performance under different experimental conditions. The larger the value of $\alpha$, the greater is the effect of the carrier-induced phase shift on the linewidth or the filamentation of a laser. Figure 2.20 depicts the dependence of $\alpha(\nu_{pk})$ on carrier density. Note, that

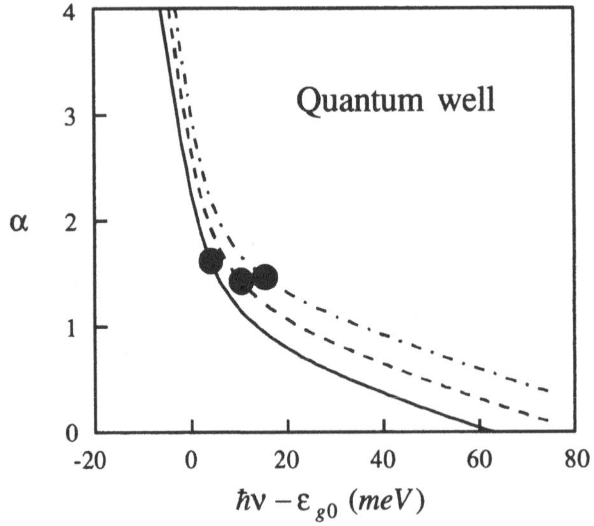


**Fig. 2.19.** Linewidth enhancement or antiguiding parameter versus detuning $\hbar\nu - \varepsilon_{g0}$ for quantum-well medium. $N_{2d} = 1.5 \times 10^{12}\,\mathrm{cm}^{-2}$ (*solid lines*), $2 \times 10^{12}\,\mathrm{cm}^{-2}$ (*dashed*), and $2.5 \times 10^{12}\,\mathrm{cm}^{-2}$ (*dash-dotted*). The points show the location of the gain peak

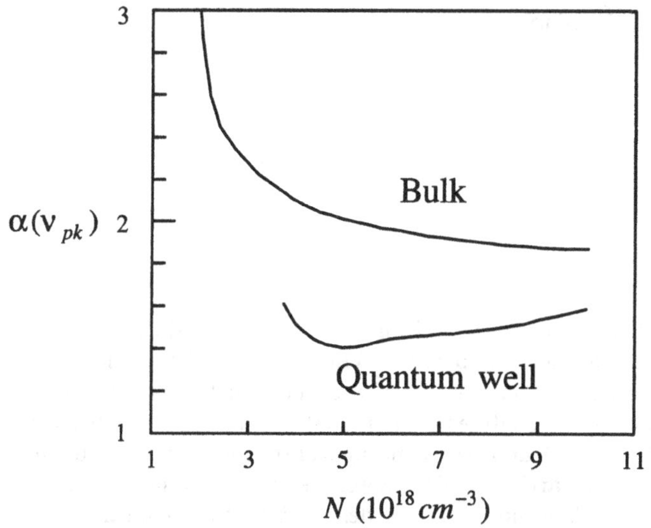


**Fig. 2.20.** $\alpha$ at the peak gain frequency versus carrier density for bulk and an idealized quantum-well system

$\alpha$ changes considerably as a function of carrier density, which contradicts a major assumption of the phenomenological model.

It is tempting to use Fig. 2.20 to come to the conclusion that the quantum-well laser has a smaller $\alpha$. While this statement is often true, the free-carrier results used in arriving at this conclusion do not describe the entire picture. Most importantly, they neglect many-body and band-structure effects, which significantly affect the behavior of the gain medium. We elaborate on these features in the next chapters.

# 3. Coulomb Effects

In Chap. 2, we describe a simple model for semiconductor gain from a free (i.e., noninteracting) electron-hole plasma. While this model provides some useful insight to the elementary physics of a semiconductor gain medium, its inadequacies show up in analyses of high-quality samples and advanced laser structures, where one clearly sees signatures of the more subtle Coulomb interaction effects among carriers. This chapter, as well as the next one, discusses approaches towards a more realistic description of the gain medium, where one includes the Coulomb interaction between charge carriers. The Coulomb potential is attractive between electron and holes (interband attraction) and repulsive for carriers in the same band (intraband repulsion). Since Coulomb interaction processes always involve more than one carrier, the resulting effects are often called *many-body effects*, and quantum mechanical many-body techniques have to be used to analyze these phenomena.

As mentioned in Chap. 2, one of the most important consequences of the Coulomb interaction is the rapid equilibration of the electrons and holes into quasi-equilibrium Fermi-Dirac distributions. Under typical laser conditions the carrier-carrier scattering is also the dominant contributer to optical dephasing, which is the decay of the polarization of the medium. Another important many-body effect is plasma screening, which is the carrier density dependent weakening of the Coulomb interaction potential due to the presence of background charge carriers. These many-body Coulomb interactions significantly modify the gain and refractive index spectra. The spectral positions of the gain and index spectra are shifted through bandgap renormalization, and the shapes of the gain and index spectra are modified through Coulomb correlation effects.

A schematic outline of a many-body theory is given in Fig. 3.1. Beginning with the many-body Hamiltonian for the interacting carriers, we derive equations of motion for the electron probability $n_{e\boldsymbol{k}}$, the hole probability $n_{h\boldsymbol{k}}$, and the interband polarization $p_{\boldsymbol{k}}$. The equations of motion couple these expectation values of products of two particle operators to those of products of four particle operators, which, in turn, are coupled to six operator expectation values, and so on. This process produces an infinite hierarchy of coupled differential equations involving expectation values of products of ever higher numbers of field operators. In order to approximately deal with this many-body hierarchy, we have to use suitable truncation procedures. For example,

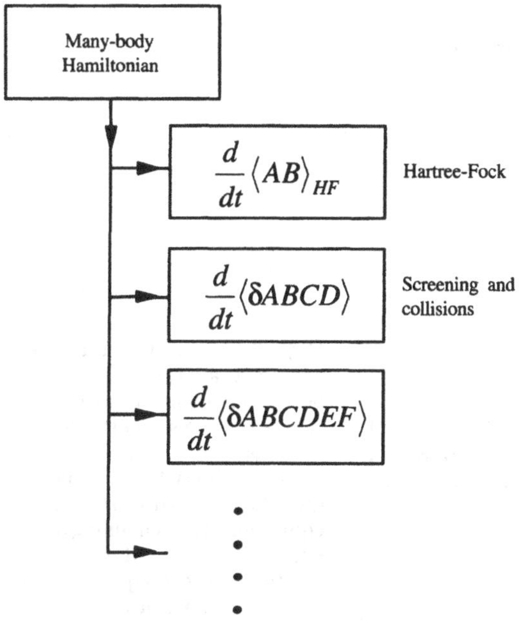


**Fig. 3.1.** A schematic outline of the many-body approach used in this book. $AB$ represents a product of two creation or annihilation operators. The equations are split into the Hartree-Fock (HF) part and the higher-order correlations, $\delta\langle ABCD\rangle$, $\delta\langle ABCDEF\rangle$ and so on. The higher-order correlations describe effects such as plasma screening and carrier collisions

we can truncate the hierarchy by factorizing higher-order expectation values into products of the second-order averages, $n_{ek}$, $n_{hk}$ and $p_k$. Factorizing the equations of motion for $n_{ek}$, $n_{hk}$ and $p_k$ in this way, we obtain the Hartree-Fock equations, so named because Hartree and Fock made a similar kind of approximation in studying the many-electron atom. Important many-body Coulomb effects appear in these Hartree-Fock equations, such as bandgap renormalization and interband Coulomb enhancement. Improvements of the Hartree-Fock equations are obtained by delaying the factorization procedure to the next level of the hierarchy. This yields the collision terms.

The equations for $n_{ek}$, $n_{hk}$ and $p_k$, combined with some form of the collision and screening contributions form a set of equations that play the same role as the optical Bloch equations in two-level systems. Therefore, it has become customary to refer to them as the *semiconductor Bloch equations*. These equations reduce to atomic Bloch equations without decay when we drop the Coulomb interaction potential altogether, that is, in the limit of no carrier-carrier scattering, no plasma screening, no bandgap renormalization, and no Coulomb enhancement.

We derive the semiconductor Bloch equations in Sect. 3.1. Section 3.2 discusses the physical origin of the Coulomb enhancement phenomenon by treating the low carrier-density limit of the semiconductor Bloch equations. For this limiting case, the semiconductor Bloch equations contain the "hydrogen-like" Wannier equation, which describes the fundamental electron-hole pair or excitonic properties of a dielectric medium. At higher densities, the con-

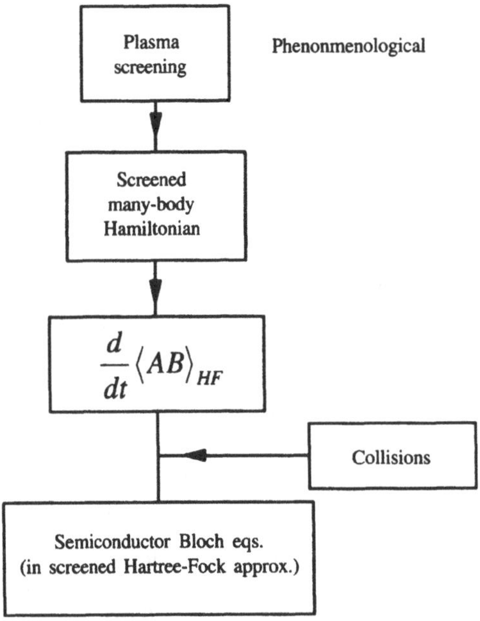


**Fig. 3.2.** Outline of an approach that takes advantage of a phenomenological derivation of plasma screening, which is incorporated directly into the many-body Hamiltonian to give the screened Hartree-Fock equations. The remaining steps needed to arrive at the semiconductor Bloch equations are similar to those in Fig. 3.1

tributions from carrier collisions, which lead to population and polarization relaxation as well as screening, become important. Because the results of a systematic analysis of these contributions are quite complicated, we delay its discussion until Chap. 4, and use the remainder of this chapter to describe a phenomenological alternative. Figure 3.2 depicts this approach, which leads to a more easily usable laser model. Part of the simplifications come from the *screened Hartree-Fock approximation* discussed in Sect. 3.3. This approximation involves replacing the bare Coulomb potential in the system Hamiltonian with a screened Coulomb potential. We note that even though this approach is justified at the dynamical Hartree-Fock level, one should not use such an *effective Hamiltonian* to derive correlations of higher order, since this would amount to double counting of some many-body effects. Section 3.3 also describes the standard approach to obtaining the screened Coulomb potential. Using the results of Sect. 3.3, Sect. 3.4 discusses bandgap renormalization, and shows that in the screened Hartree-Fock approximation the contributions to the renormalization of the bandgap energy comes from a Coulomb-hole self energy and a screened-exchange shift. In Sect. 3.5, we derive a high-density approximation, the so-called Padé approximation, of the screened Hartree-Fock semiconductor Bloch equations. The remaining sections use the resulting gain and phase shift formulas and assume a simple two-band semiconductor to illustrate the many-body effects. Section 3.6 shows the influence of Coulomb effects on gain-medium properties involving carrier-density-dependent changes in gain or index in a bulk semiconductor.

As an important nonlinear quantity, results for the linewidth enhancement factor are discussed. The quantum-well active region is treated in Sect. 3.7, where we demonstrate that the many-body effects are enhanced in two dimensions as a consequence of the generally increasing importance of interaction and correlation effects in systems of reduced dimensionality.

## 3.1 Semiconductor Bloch Equations

For an interacting electron system in a dielectric medium, the system Hamiltonian is

$$H = H_{\mathrm{kin}} + H_{\mathrm{c\text{-}f}} + H_{\mathrm{C}} \quad , \tag{3.1}$$

where the kinetic energy $H_{\mathrm{kin}}$ and the carrier-laser-field interaction energy $H_{\mathrm{c\text{-}f}}$ are those of (2.1). The Coulomb interaction energy among electrons in the various bands is, in cgs units,

$$H_{\mathrm{C}} = \int \mathrm{d}^3 r_1 \int \mathrm{d}^3 r_2 \, \hat{\psi}^\dagger(\boldsymbol{r}_2)\hat{\psi}^\dagger(\boldsymbol{r}_1) \frac{e^2}{\epsilon_{\mathrm{b}}|\boldsymbol{r}_1 - \boldsymbol{r}_2|} \hat{\psi}(\boldsymbol{r}_1)\hat{\psi}(\boldsymbol{r}_2) \quad , \tag{3.2}$$

where $\epsilon_{\mathrm{b}}$ is the background dielectric constant in the medium. Using in (3.2) the expansion (1.83) for the electron field operator $\hat{\psi}(\boldsymbol{r}_n)$, and reordering the terms, one obtains in the two-band approximation

$$\begin{aligned} H_{\mathrm{C}} &= \tfrac{1}{2} \textstyle\sum_{\boldsymbol{k},\boldsymbol{k}'} \sum_{\boldsymbol{q}\neq 0} V_q \big( a^\dagger_{c,\boldsymbol{k}+\boldsymbol{q}} a^\dagger_{c,\boldsymbol{k}'-\boldsymbol{q}} a_{c\boldsymbol{k}'} a_{c\boldsymbol{k}} + a^\dagger_{v,\boldsymbol{k}+\boldsymbol{q}} a^\dagger_{v,\boldsymbol{k}'-\boldsymbol{q}} a_{v\boldsymbol{k}'} a_{v\boldsymbol{k}} \\ &\qquad + 2 a^\dagger_{c,\boldsymbol{k}+\boldsymbol{q}} a^\dagger_{v,\boldsymbol{k}'-\boldsymbol{q}} a_{v\boldsymbol{k}'} a_{c\boldsymbol{k}} \big) \quad . \end{aligned} \tag{3.3}$$

Here, the first two terms describe the intraband carrier interaction, and the last term describes the interband interaction between the electrons in the valence and the conduction bands. As long as we are not in the electron-hole representation, we only deal with electrons in different bands and all Coulomb terms, i.e. intra- and interband interaction are repulsive.

As in the previous chapter, the momentum index $\boldsymbol{k}$ includes the spin index so that summation over $\boldsymbol{k}$ implies also summing over the two possible spin orientations. The subscript includes a comma only when ambiguity might otherwise arise. The Coulomb interaction Hamiltonian (3.3) contains the Fourier transform of the Coulomb potential energy $V_q$, which we obtain for a bulk-material as

$$\begin{aligned} V_q &= \frac{1}{V} \int \mathrm{d}^3 r \, \mathrm{e}^{-\mathrm{i}\boldsymbol{q}\cdot\boldsymbol{r}} V(r) \\ &= \frac{1}{V} \int \mathrm{d}^3 r \, \mathrm{e}^{-\mathrm{i}\boldsymbol{q}\cdot\boldsymbol{r}} \frac{e^2}{\epsilon_{\mathrm{b}} r} \\ &= \frac{4\pi e^2}{\epsilon_{\mathrm{b}} V q^2} \quad , \end{aligned} \tag{3.4}$$

and for an idealized 2D quantum-well system as

$$\begin{aligned} V_q &= \int \mathrm{d}^2 r\, \mathrm{e}^{-\mathrm{i}\boldsymbol{q}\cdot\boldsymbol{r}} \frac{e^2}{\epsilon_\mathrm{b} r} \\ &= \frac{2\pi e^2}{\epsilon_\mathrm{b} q A} \ . \end{aligned} \tag{3.5}$$

In (3.5), $A$ is the quantum-well area. In the derivation of (3.3), we used the fact that the Coulomb scattering does not alter the spin orientation of an electron and that the $\boldsymbol{q} = 0$ contribution is cancelled by the corresponding terms of the electron-ion and ion-ion Coulomb interaction. Furthermore, we omit Coulomb terms that fail to conserve the number of electrons in each band, since such terms involve Coulomb induced interband transitions, which are very unfavorable energetically.

As discussed in Sect. 1.9, we now transform to the electron-hole representation. For this purpose we use the electron and hole operators of (1.90, 91, 87, 88) in all the terms of (3.1) and restore normal ordering of all creation and annihilation operators. This gives the two-band Hamiltonian for interacting electrons and holes

$$\begin{aligned} H &= \sum_{\boldsymbol{k}} \left[ (\varepsilon_{e\boldsymbol{k}} + \varepsilon_{\mathrm{g}0})\, a^\dagger_{\boldsymbol{k}} a_{\boldsymbol{k}} + \varepsilon_{\mathrm{h}\boldsymbol{k}} b^\dagger_{-\boldsymbol{k}} b_{-\boldsymbol{k}} \right] \\ &\quad + \frac{1}{2} \sum_{\boldsymbol{k},\boldsymbol{k}'} \sum_{\boldsymbol{q}\neq 0} V_q \left( a^\dagger_{\boldsymbol{k}+\boldsymbol{q}} a^\dagger_{\boldsymbol{k}'-\boldsymbol{q}} a_{\boldsymbol{k}'} a_{\boldsymbol{k}} + b^\dagger_{\boldsymbol{k}+\boldsymbol{q}} b^\dagger_{\boldsymbol{k}'-\boldsymbol{q}} b_{\boldsymbol{k}'} b_{\boldsymbol{k}} - 2 a^\dagger_{\boldsymbol{k}+\boldsymbol{q}} b^\dagger_{\boldsymbol{k}'-\boldsymbol{q}} b_{\boldsymbol{k}'} a_{\boldsymbol{k}} \right) \\ &\quad - \sum_{\boldsymbol{k}} \left( \mu_{\boldsymbol{k}} a^\dagger_{\boldsymbol{k}} b^\dagger_{-\boldsymbol{k}} + \mu^*_{\boldsymbol{k}} b_{-\boldsymbol{k}} a_{\boldsymbol{k}} \right) E(z,t) \ , \end{aligned} \tag{3.6}$$

where constant terms have been dropped because they only lead to an irrelevant shift of the reference energy. The kinetic energies in (3.6) are

$$\varepsilon_{e\boldsymbol{k}} = \frac{\hbar^2 k^2}{2 m_\mathrm{e}} \ , \tag{3.7}$$

$$\begin{aligned} \varepsilon_{\mathrm{h}\boldsymbol{k}} &= -\varepsilon_{\mathrm{v}\boldsymbol{k}} + \sum_{\boldsymbol{q}\neq 0} V_q \\ &= \frac{\hbar^2 k^2}{2 m_\mathrm{h}} \ , \end{aligned} \tag{3.8}$$

where the term containing $V_q$ in $\varepsilon_{\mathrm{h}\boldsymbol{k}}$ originates from the replacement of valence-band electron operators by hole operators in the interaction term

$$\frac{1}{2} \sum_{\boldsymbol{k},\boldsymbol{k}'} \sum_{\boldsymbol{q}\neq 0} V_q \, a^\dagger_{v,\boldsymbol{k}+\boldsymbol{q}} a^\dagger_{v,\boldsymbol{k}'-\boldsymbol{q}} a_{v\boldsymbol{k}'} a_{v\boldsymbol{k}}$$

of (3.3). Equation (3.8) differs from the free-carrier result [see (1.10, 13)] in that the kinetic energy and therefore the hole effective mass includes the Coulomb energy of the full valence band.

# Model Based Fuzzy Control

Springer
*Berlin*
*Heidelberg*
*New York*
*Barcelona*
*Budapest*
*Hong Kong*
*London*
*Milan*
*Paris*
*Santa Clara*
*Singapore*
*Tokyo*

Rainer Palm   Dimiter Driankov
Hans Hellendoorn

# Model Based Fuzzy Control

## Fuzzy Gain Schedulers and Sliding Mode Fuzzy Controllers

With 86 Figures

Springer

Dr. Rainer Palm
Dr. Hans Hellendoorn
Siemens AG
D-81730 München, Germany

Prof. Dr. Dimiter Driankov
University of Linköping
S-58183 Linköping, Sweden

Library of Congress Cataloging-in-Publication Data

Palm, Rainer.
Model-based fuzzy control : fuzzy gain schedulers and sliding mode fuzzy control / Rainer Palm, Dimiter Driankov, Hans Hellendoorn.
p. cm.
Includes bibliographical references and index.

1. Automatic control. 2. Fuzzy systems. I. Driankov, Dimiter. II. Hellendoorn, Hans. III. Title.
TJ213.P222 1996
629.8'9--dc21 96-47571
CIP

ISBN 978-3-642-08262-7

Springer-Verlag Berlin New York Heidelberg



Cover design: Künkel + Lopka, Ilvesheim

# Foreword

Despite the excitement about its capabilities, and its success in some challenging applications, fuzzy control is still young and hence some important questions arise: In what situations, or for what applications, is fuzzy control superior to conventional control methods? Is it possible to combine some of the best ideas from the conventional methods with ideas from fuzzy control to provide more effective control solutions? How do we measure the success or failure of the methods? For many applications, simulations and experimental evaluations are not sufficient to verify the behavior of a control system, and in such situations there is a need for mathematical analysis of closed-loop system properties such as stability, performance, and robustness to determine whether a fuzzy control system is successful. Verification of such properties may be especially important for "safety-critical applications" (e.g., an aircraft or nuclear power plant) where the engineer must gain as much confidence in the closed-loop system as is possible before implementation.

Conventional control focuses on the use of models to construct controllers for dynamical systems. Fuzzy control focuses on the use of heuristics in the construction of a controller. While each of these seemingly disjoint approaches uses some ideas from the other, there has been relatively little work on how to more closely combine them to exploit the best characteristics of each. This book helps to remedy this problem by providing schemes that allow for the incorporation of heuristics and mathematical models. It also provides important ideas on how to marry some very successful conventional control ideas (e.g., sliding mode control and gain scheduling) with ideas from fuzzy control. While the past lack of focus on the use of mathematical models in fuzzy control systems development resulted in researchers ignoring mathematical analysis of stability, the basic approach used allows the authors to confront this important problem.

Overall, this book makes important steps toward bridging the apparent gap between fuzzy and conventional control. I would expect the techniques studied here to be quite useful for a wide range of challenging applications and would expect these ideas to form a foundation on which more research in this promising area will be based.

Columbus, Ohio, USA, 1996 *Kevin M. Passino*

# Preface

During the past few years two principally different approaches to the design of fuzzy logic controllers (FLC) have emerged: heuristics based design and model based design.

The main motivation for the heuristics based design is given by the fact that many industrial processes are still controlled in one of the following two ways:

- The process is controlled manually by an experienced operator.
- The process is controlled by an automatic control system which needs additional manual on-line "trimming" from an experienced operator.

In both cases it is enough to translate the operator's manual control algorithm in terms of a set of fuzzy if-then rules in order to obtain an equally good, or an even better, wholy automatic control system incorporating an FLC. This implies that the design of an FLC can only be done *after* a "control algorithm" already exists.

In the first case, the existing control algorithm may consist of sequential and/or parallel manual control actions performed by the operator upon a process whose mathematical model is either impossible to derive or of negligible utility for cost related reasons. In this case the FLC simply makes explicit the existing manual control knowledge, and consequently automates the use of this knowledge thus becoming a part of the closed loop system. In the second case, the existing control algorithm is a conventional control algorithm in need of additional manual "trimming". An FLC is then again used to automate the manual "trimming" algorithm employed by the operator and thus acts as a supervisor to the conventional closed loop system already in place.

It is admitted in the literature on fuzzy control that the heuristics based design is very difficult to apply to multiple-input/multiple-output control problems, which represent the largest part of challenging industrial process control applications. Furthermore, the heuristics based design lacks systematic and formally verifiable tuning techniques, and studies of stability, performance, and robustness can only be done via extensive simulations. Last but not least, there is a lack of systematic and easily verifiable knowledge acquisition techniques via which the qualitative knowledge about the process and/or available manual control algorithm can be extracted.

The above difficulties faced by the heuristics based design explain the recent surge of interest in the derivation of *black box* fuzzy models of the plant under control, in terms of the identification of a set of fuzzy if-then rules, by the use of conventional identification techniques, neural networks, genetic algorithms, or a mixture of these techniques.

This interest in the identification of fuzzy models is accompanied by a similar surge of interest in the model based design of fuzzy controllers. Model based fuzzy control uses a given conventional or fuzzy open loop model of the plant under control in order to derive the set of fuzzy if-then rules constituting the corresponding FLC. Interest then centers on the stability, performance, and robustness analysis of the resulting closed loop system involving a conventional model and an FLC, or a fuzzy model and an FLC. The major objective of model based fuzzy control is to use existing conventional linear and nonlinear design and analysis methods for the design of such FLCs that have better stability, performance, and robustness properties than the corresponding non-fuzzy controllers designed by the use of these same techniques. How to achieve this objective in terms of the design and analysis of sliding mode fuzzy controllers and fuzzy gain schedulers is the subject of this book.

In **Chapter 1** we introduce the basic notions and concepts in fuzzy control, the basic types of FLCs treated in the book, the major types of nonlinear control problems, and the existing methods for model based design and analysis of nonlinear control systems relevant for model based fuzzy control. We finally discuss informally the motivation for the design of fuzzy gain schedulers and fuzzy sliding mode controllers.

In Section 1.1 of this introductory chapter we present the basic fuzzy control related concepts used throughout the book involving the notions of a fuzzy state, fuzzy input, fuzzy output variables, and fuzzy state space.

In Section 1.2 we present the basic types of FLCs whose model based design and analysis is our subject. These include different types of Takagi–Sugeno FLCs (TSFLC) and the sliding mode FLC (SMFLC). We describe the open loop models used for the design of the different types of FLCs and the form of the fuzzy rules constituting these FLCs. We also present the control schemes incorporating an FLC that are relevant for model based fuzzy control.

In Section 1.3 we first describe the two basic types of nonlinear control problems considered in the book, namely nonlinear regulation and nonlinear tracking. Then, we discuss the specifications of the desired behavior of nonlinear closed loop systems in terms of stability, robustness, accuracy, and response speed.

In Section 1.4 we present the major existing methods for the model based design and analysis of nonlinear control systems and identify those of them whose fuzzy counterparts (with appropriate modifications) concern us.

In Section 1.5 we introduce and discuss informally the motivation for two basic types of FLCs, namely the Takagi–Sugeno FLC and the sliding mode FLC.

In **Chapter 2** we describe computation with an FLC and its formal description as a static nonlinear transfer element and thus provide the background knowledge needed to understand control with an FLC. We show the relationship between conventional and rule-based transfer elements and establish the compatibility between these two conceptually different, in terms of representation, types of transfer elements. We also introduce the basic stability concepts used in the model based design and analysis of FLCs.

In Section 2.1 we describe the computational structure of an FLC involving the computational steps of input scaling, fuzzification, rule firing, defuzzification, and output scaling.

In Section 2.2 we present the sources of nonlinearity in the computational structure of an FLC by relating them to particular computational steps.

In Section 2.3 we describe the relationship between conventional transfer elements and rule-based transfer elements. We show the gradual transition from a conventional transfer element to a crisp rule-based transfer element, and finally to a fuzzy rule-based transfer element. We also describe the computational structure of Takagi–Sugeno FLCs.

In Section 2.4 we present the stability concepts used in the model based design and analysis of an FLC including Lyapunov-linearization and the Lyapunov direct method for autonomous and nonautonomous systems.

In **Chapter 3** we make use of the similarity between the so-called diagonal form FLC and a sliding mode controller (SMC) to redefine a diagonal form FLC in terms of an SMC with boundary layer (BL).

In Section 3.1 we describe the control law of an SMC for an $n$-th-order SISO nonlinear nonautonomous system and its design for a tracking control problem with and without an integrator term.

In Section 3.2 we describe in detail the diagonal form FLC for a second order SISO nonlinear autonomous system and derive the similarities between the control law of a diagonal form sliding mode FLC and the control law of an SMC with BL.

In Section 3.3 we describe the design of the control law of a sliding mode FLC for an $n$-th-order SISO system for the tracking control problem, with and without integrator term.

In Section 3.4 we discuss the tuning of input scaling factors of a sliding mode FLC. In Section 3.5 we give an example of a force adapting manipulator arm for the design of a sliding mode FLC. In Section 3.6 we extend the sliding mode FLC design method to MIMO systems.

In **Chapter 4** we present the design methods for each of the different types of Takagi–Sugeno FLCs outlined in Chap. 1.

In Section 4.1 we present the Takagi–Sugeno FLC-1. We confine ourselves only to the presentation of the form of the open loop system, the form of the Takagi–Sugeno FLC-1, the form of the closed loop system and its stability properties, and an outline of a trial-and-error type of design method.

In Section 4.2 and 4.3 we present in detail the design of the Takagi–Sugeno FLC-2 and its use in local stabilization and tracking of a nonlinear autonomous system. With respect to local stabilization, a Takagi–Sugeno FLC-2 is able to stabilize a nonlinear autonomous system around *any* operating point without the need to change its gains. With respect to tracking, the FLC-2 performs gain scheduling on *any* reference state trajectory under the restriction that the reference state trajectories are slowly time varying.

In **Chapter 5** we illustrate in detail the design of fuzzy sliding mode controllers and fuzzy gain schedulers on a MIMO control problem concerning the control of a two-link robot arm.

Munich, 1996

*Rainer Palm*
*Dimiter Driankov*
*Hans Hellendoorn*

# Table of Contents

Proceeding as in Sect. 2.1, we derive coupled equations of motion for the electron and hole populations $n_{ek}$ and $n_{hk}$, (2.9, 10), respectively, as well as for the interband polarization $p_k$, (2.8). The derivation requires simple but lengthy operator rearrangements to reduce the commutators in the Heisenberg equations to

$$\begin{aligned}\frac{\mathrm{d}p_k}{\mathrm{d}t} &= -\mathrm{i}\omega'_k p_k - \mathrm{i}\hbar^{-1}\mu_k E(z,t)\,(n_{ek}+n_{hk}-1) + \frac{\mathrm{i}}{\hbar}\sum_{k',q\neq 0} V_q \\ &\quad\times \Big(\langle a^\dagger_{k'+q} b_{-k} a_{k'} a_{k+q}\rangle + \langle b^\dagger_{k'-q} b_{k'} a_k b_{-k-q}\rangle - \langle a^\dagger_{k'+q} b_{-k+q} a_{k'} a_k\rangle \\ &\quad - \langle b^\dagger_{k'-q} b_{-k} b_{k'} a_{k-q}\rangle + \langle b_{-k+q} a_{k-q}\rangle \delta_{k,k'}\Big) \quad , \end{aligned} \tag{3.9}$$

$$\begin{aligned}\frac{\mathrm{d}n_{ek}}{\mathrm{d}t} &= \frac{\mathrm{i}}{\hbar} E(z,t)\,(\mu_k p^*_k - \mu^*_k p_k) \\ &\quad + \frac{\mathrm{i}}{\hbar}\sum_{k',q\neq 0} V_q \Big(\langle a^\dagger_k a^\dagger_{k'-q} a_{k-q} a_{k'}\rangle - \langle a^\dagger_{k+q} a^\dagger_{k'-q} a_k a_{k'}\rangle \\ &\quad + \langle a^\dagger_k a_{k-q} b^\dagger_{k'-q} b_{k'}\rangle - \langle a^\dagger_{k+q} a_k b_{k'-q} b_{k'}\rangle\Big) \quad , \end{aligned} \tag{3.10}$$

and

$$\begin{aligned}\frac{\mathrm{d}n_{hk}}{\mathrm{d}t} &= \frac{\mathrm{i}}{\hbar} E(z,t)\,(\mu_k p^*_k - \mu^*_k p_k) \\ &\quad + \frac{\mathrm{i}}{\hbar}\sum_{k',q\neq 0} V_q \Big(\langle b^\dagger_{-k} b^\dagger_{k'-q} b_{-k-q} b_{k'}\rangle - \langle b^\dagger_{-k+q} b^\dagger_{k'-q} b_{-k} b_{k'}\rangle \\ &\quad + \langle a^\dagger_{k'+q} a_{k'} b^\dagger_{-k} b_{-k+q}\rangle - \langle a^\dagger_{k'+q} a_{k'} b^\dagger_{-k-q} b_k\rangle\Big) \quad . \end{aligned} \tag{3.11}$$

Here we denoted the transition energy $\hbar\omega_k$ of (2.20) by

$$\hbar\omega'_k = \varepsilon_{ek} + \varepsilon_{hk} + \varepsilon_{g0} \quad , \tag{3.12}$$

where the prime has been introduced in order to distinguish the unrenormalizated energy $\hbar\omega'_k$ from the renormalized one $\hbar\omega_k$, which appears later in (3.20). Equations (3.9–11) show that the Coulomb interaction couples the two-operator dynamics to four-operator terms. One way to proceed is to factorize these terms into products of two-operator terms, yielding the Hartree-Fock limit of the equations. To obtain a systematic hierarchy of equations, we separate out the Hartree-Fock contributions. For example, we write for a two-operator combination $AB$,

$$\begin{aligned}\frac{\mathrm{d}}{\mathrm{d}t}\langle AB\rangle &= \frac{\mathrm{d}}{\mathrm{d}t}\langle AB\rangle_{\mathrm{HF}} + \left(\frac{\mathrm{d}}{\mathrm{d}t}\langle AB\rangle - \frac{\mathrm{d}}{\mathrm{d}t}\langle AB\rangle_{\mathrm{HF}}\right) \\ &\equiv \frac{\mathrm{d}}{\mathrm{d}t}\langle AB\rangle_{\mathrm{HF}} + \frac{\mathrm{d}}{\mathrm{d}t}\langle AB\rangle_{\mathrm{col}} \quad . \end{aligned} \tag{3.13}$$

Here, HF indicates the Hartree-Fock contribution. The quantity inside the square bracket then contains both two and four-operator products, which we represent in general by $\langle ABCD\rangle$. These contributions beyond the Hartree-Fock approximation are often called collision (subscript col) or correlation contributions. They will be discussed in detail in the following Chap. 4. Here, we only mention that with the full many-body Hamiltonian (3.1), the Heisenberg equation of motion gives the equation of motion for $\langle ABCD \rangle$ as

$$\frac{\mathrm{d}}{\mathrm{d}t}\langle ABCD\rangle = \frac{\mathrm{d}}{\mathrm{d}t}\langle ABCD\rangle_{\mathrm{F}} + \left(\frac{\mathrm{d}}{\mathrm{d}t}\langle ABCD\rangle - \frac{\mathrm{d}}{\mathrm{d}t}\langle ABCD\rangle_{\mathrm{F}}\right) \quad , \tag{3.14}$$

where $\mathrm{d}\langle ABCD\rangle/\mathrm{d}t$ contains expectation values of products of up to six operators. The label F is used to indicate the result from a Hartree-Fock factorization of the four and six operator expectation values. We can continue by deriving the equation of motion for

$$\langle ABCDEF\rangle \equiv \left(\frac{\mathrm{d}}{\mathrm{d}t}\langle ABCD\rangle - \frac{\mathrm{d}}{\mathrm{d}t}\langle ABCD\rangle_{\mathrm{F}}\right)$$

and so on. The result is a hierarchy of equations, where each succeeding equation describes a correlation contribution that is of higher order than the one before. In practice, we truncate the hierarchy at some point.

Returning to (3.9–11), we first evaluate the Hartree-Fock contributions. To do so, we factorize all the expectation values of four-operator products into *all possible* operator combinations leading to products of densities and/or polarizations. For example, for $\langle a_{\boldsymbol{k}}^\dagger a_{\boldsymbol{k}'}^\dagger a_l\, a_{l'}\rangle$, we can have the two-operator combinations $\langle a_{\boldsymbol{k}}^\dagger a_{\boldsymbol{k}'}^\dagger\rangle\langle a_l\, a_{l'}\rangle$, $\langle a_{\boldsymbol{k}}^\dagger a_l\rangle\langle a_{\boldsymbol{k}'}^\dagger a_{l'}\rangle$, and $\langle a_{\boldsymbol{k}}^\dagger a_{l'}\rangle\langle a_{\boldsymbol{k}'}^\dagger a_l\rangle$. Taking the anticommutation relations into account to get the proper signs between these combinations we find

$$\begin{aligned}\langle a_{\boldsymbol{k}}^\dagger a_{\boldsymbol{k}'}^\dagger a_l\, a_{l'}\rangle_{\mathrm{F}} &= \langle a_{\boldsymbol{k}}^\dagger a_{\boldsymbol{k}'}^\dagger\rangle\langle a_l\, a_{l'}\rangle - \langle a_{\boldsymbol{k}}^\dagger a_l\rangle\langle a_{\boldsymbol{k}'}^\dagger a_{l'}\rangle + \langle a_{\boldsymbol{k}}^\dagger a_{l'}\rangle\langle a_{\boldsymbol{k}'}^\dagger a_l\rangle \\ &= 0 + (-\delta_{\boldsymbol{k},l}\delta_{\boldsymbol{k}',l'} + \delta_{\boldsymbol{k},l'}\delta_{\boldsymbol{k}',l})n_{\mathrm{e}\boldsymbol{k}}n_{\mathrm{e}\boldsymbol{k}'} \quad .\end{aligned} \tag{3.15}$$

Another example is

$$\langle a_{\boldsymbol{k}'+\boldsymbol{q}}^\dagger b_{\boldsymbol{q}-\boldsymbol{k}} a_{\boldsymbol{k}'} a_{\boldsymbol{k}}\rangle_{\mathrm{F}} = \delta_{\boldsymbol{k}',\boldsymbol{k}-\boldsymbol{q}} n_{\mathrm{e}\boldsymbol{k}} p_{\boldsymbol{k}'} \quad . \tag{3.16}$$

Factorizing all the other four-operator products in this way and formally adding the contributions beyond Hartree-Fock, we find the *semiconductor Bloch equations*

$$\frac{\mathrm{d}p_{\boldsymbol{k}}}{\mathrm{d}t} = -\mathrm{i}\omega_{\boldsymbol{k}} p_{\boldsymbol{k}} - \mathrm{i}\Omega_{\boldsymbol{k}}(z,t)\,(n_{\mathrm{e}\boldsymbol{k}} + n_{\mathrm{h}\boldsymbol{k}} - 1) + \left.\frac{\partial p_{\boldsymbol{k}}}{\partial t}\right|_{\mathrm{col}} \tag{3.17}$$

$$\frac{\mathrm{d}n_{\mathrm{e}\boldsymbol{k}}}{\mathrm{d}t} = \mathrm{i}\,[\Omega_{\boldsymbol{k}}(z,t)p_{\boldsymbol{k}}^* - \Omega_{\boldsymbol{k}}^*(z,t)p_{\boldsymbol{k}}] + \left.\frac{\partial n_{\mathrm{e}\boldsymbol{k}}}{\partial t}\right|_{\mathrm{col}} \tag{3.18}$$

$$\frac{\mathrm{d}n_{\mathrm{h}\boldsymbol{k}}}{\mathrm{d}t} = \mathrm{i}\,[\Omega_{\boldsymbol{k}}(z,t)p_{\boldsymbol{k}}^* - \Omega_{\boldsymbol{k}}^*(z,t)p_{\boldsymbol{k}}] + \left.\frac{\partial n_{\mathrm{h}\boldsymbol{k}}}{\partial t}\right|_{\mathrm{col}} \quad . \tag{3.19}$$

We have written the terms containing the Hartree-Fock contributions explicitly, while those due to higher order correlations (collisions) are denoted formally by the partial derivatives $\partial/\partial t|_{\mathrm{col}}$. The Hartree-Fock contributions in (3.17–19) contain two important many-body effects, namely a density dependent contribution to the transition energy, and a renormalization of the electric-dipole interaction energy. Specifically, $\hbar\omega'_{\boldsymbol{k}}$ of (3.12) is replaced by the renormalized transition energy

$$\hbar\omega_{\boldsymbol{k}} = \hbar\omega'_{\boldsymbol{k}} - \sum_{\boldsymbol{k}'\neq\boldsymbol{k}} V_{|\boldsymbol{k}-\boldsymbol{k}'|}\left(n_{\mathrm{e}\boldsymbol{k}'} + n_{\mathrm{h}\boldsymbol{k}'}\right) \quad , \tag{3.20}$$

and the Rabi frequency $\mu_{\boldsymbol{k}}E(z,t)/\hbar$ is renormalized as

$$\Omega_{\boldsymbol{k}}(z,t) = \frac{\mu_{\boldsymbol{k}}E(z,t)}{\hbar} + \frac{1}{\hbar}\sum_{\boldsymbol{k}'\neq\boldsymbol{k}} V_{|\boldsymbol{k}-\boldsymbol{k}'|}p_{\boldsymbol{k}'} \quad , \tag{3.21}$$

where the Coulomb terms ($\propto V_{|\boldsymbol{k}-\boldsymbol{k}'|}$) in (3.20, 21) are called the *exchange shift* and the *Coulomb field renormalization*, respectively.

The semiconductor Bloch equations look like the two-level Bloch equations, with the exceptions that the transition energy and the electric-dipole interaction are renormalized, and the carrier probabilities $n_{\alpha\boldsymbol{k}}$ enter instead of the probability difference between upper and lower levels. The renormalizations are due to the many-body Coulomb interactions, and they couple equations for different $\boldsymbol{k}$ states. This coupling leads to significant complications in the evaluation of (3.17–19) in comparison to the corresponding free-carrier equations, (2.23, 24). As discussed in the introduction to this chapter, if all Coulomb-potential contributions are dropped, i.e, $V_q$ is set to zero, the semiconductor Bloch equations reduce to the undamped inhomogeneously broadened two-level Bloch equations. Of course, the limit $V_q = 0$ is unacceptable for semiconductors.

The Coulomb terms in (3.17) show a large degree of symmetry. To see this more clearly, we write the Hartree-Fock part of (3.17), i.e., without $\partial p_{\boldsymbol{k}}/\partial t|_{\mathrm{col}}$ as

$$\left.\frac{\mathrm{d}p_{\boldsymbol{k}}}{\mathrm{d}t}\right|_{\mathrm{HF}} = \sum_{\boldsymbol{k}'} \Theta_{\boldsymbol{k}\boldsymbol{k}'}p_{\boldsymbol{k}'} - \frac{\mathrm{i}}{\hbar}\mu_{\boldsymbol{k}}E\left(n_{\mathrm{e}\boldsymbol{k}} + n_{\mathrm{h}\boldsymbol{k}} - 1\right) \quad , \tag{3.22}$$

where for $\boldsymbol{k} = \boldsymbol{k}'$

$$\Theta_{\boldsymbol{k}\boldsymbol{k}} = -\,\mathrm{i}\omega'_{\boldsymbol{k}} + \frac{\mathrm{i}}{\hbar}\sum_{\boldsymbol{k}''\neq\boldsymbol{k}} V_{|\boldsymbol{k}-\boldsymbol{k}''|}\left(n_{\mathrm{e}\boldsymbol{k}''} + n_{\mathrm{h}\boldsymbol{k}''}\right) \quad , \tag{3.23}$$

and for $\boldsymbol{k} \neq \boldsymbol{k}'$

$$\Theta_{\boldsymbol{k}\boldsymbol{k}'} = -\frac{\mathrm{i}}{\hbar}V_{|\boldsymbol{k}-\boldsymbol{k}'|}\left(n_{\mathrm{e}\boldsymbol{k}} + n_{\mathrm{h}\boldsymbol{k}} - 1\right) \quad . \tag{3.24}$$

We see that the Coulomb terms appear with opposite signs in the diagonal and nondiagonal elements of the matrix $\underline{\Theta}$. This leads to compensation effects in the influence of these many-body terms, e.g., on aspects of the optical spectra. For example, the high degree of excitation independence of the excitonic resonance frequency, see, e.g., Fig. 4.13, results from the cancellation of the density dependent bandgap renormalization (diagonal part) and the weakening of the exciton binding energy (nondiagonal part). We will illustrate these effects in more detail in the following chapters of this book.

## 3.2 Interband Coulomb Effects

In this section, we examine the low density limit of the semiconductor Bloch equations. In this limit, $n_{e\boldsymbol{k}} = n_{h\boldsymbol{k}} \simeq 0$, and the collision terms vanish because no scattering partners are available. Equation (3.17) reduces to

$$\frac{\mathrm{d}p_{\boldsymbol{k}}}{\mathrm{d}t} = -\mathrm{i}\omega_{\boldsymbol{k}} p_{\boldsymbol{k}} + \mathrm{i}\Omega_{\boldsymbol{k}} \quad , \tag{3.25}$$

which effectively isolates the influence of the renormalized electric-dipole interaction frequency $\Omega_{\boldsymbol{k}}$ (Rabi frequency). Choosing the plane-wave optical field

$$E(\boldsymbol{R},t) = \tfrac{1}{2} E_0 \, \mathrm{e}^{\mathrm{i}(\boldsymbol{K}\cdot\boldsymbol{R}-\nu t)} + \mathrm{c.c.} \quad , \tag{3.26}$$

where $\boldsymbol{R}$ is a center of mass coordinate, and making the rotating-wave approximation [see Sect. 2.2], we obtain the Fourier-transformed (3.25) as

$$(\omega_{\boldsymbol{k}} - \nu + \mathrm{i}\gamma)\, p_{\boldsymbol{k}} = \Omega_{\boldsymbol{k}} \quad , \tag{3.27}$$

where $\gamma$ is a small phenomenological damping coefficient. Fourier transforming (3.27) to coordinate space, we find

$$\left[-\frac{\hbar^2 \nabla_{\boldsymbol{r}}^2}{2m_\mathrm{r}} - \frac{e^2}{\epsilon_\mathrm{b} r} + \varepsilon_\mathrm{g} - \hbar\,(\nu - \mathrm{i}\gamma)\right] p(\boldsymbol{r}) = \mu E_0 \mathrm{e}^{\mathrm{i}(\boldsymbol{K}\cdot\boldsymbol{R}-\nu t)} \delta^3(\boldsymbol{r}) V \quad , \tag{3.28}$$

where we ignore the $\boldsymbol{k}$-dependence of the interband dipole matrix element, which is often a reasonable approximation as long as we are only interested in small $\boldsymbol{k}$-values and frequencies close to the fundamental absorption edge.

Equation (3.28) is an inhomogeneous differential equation, which may be solved by expanding $p(\boldsymbol{r})$ as a linear superposition of the solutions of the homogeneous equation

$$\left(-\frac{\hbar^2 \nabla_{\boldsymbol{r}}^2}{2m_\mathrm{r}} - \frac{e^2}{\epsilon_\mathrm{b} r}\right) \psi_n(\boldsymbol{r}) = \varepsilon_n \psi_n(\boldsymbol{r}) \quad , \tag{3.29}$$

which is the Schrödinger equation for the relative motion of an electron and a hole interacting with the attractive Coulomb potential. In semiconductor physics this equation is known as the *Wannier equation.* As already mentioned in Chap. 1, there is a one-to-one correspondence between the electron-hole problem and the hydrogen atom if one replaces the proton by the valence-band hole. The solutions of the Wannier equation are therefore completely analogous to those of the hydrogen problem, which are discussed in most quantum mechanics textbooks. The bound states in the Wannier equation are called excitons, or more specifically *Wannier excitons*, and there are continuum states.

The bound and continuum eigenfunctions of the Wannier equation form a complete orthonormal basis set, so that we can write

$$p(\boldsymbol{r}) = \sum_n p_n \psi_n(\boldsymbol{r}) \quad . \tag{3.30}$$

Substituting (3.30) into (3.28), multiplying by $\psi_m^*(\boldsymbol{r})$ and integrating over $\boldsymbol{r}$ yields

$$p_m = -\frac{\mu V \psi_m^*(\boldsymbol{r}=0)}{\hbar(\nu - \mathrm{i}\gamma) - \varepsilon_\mathrm{g} - \varepsilon_m} E_0 \mathrm{e}^{\mathrm{i}(\boldsymbol{K}\cdot\boldsymbol{R}-\nu t)} , \tag{3.31}$$

where we used the orthonormality condition

$$\int \mathrm{d}^3 r\, \psi_m^*(\boldsymbol{r})\psi_n(\boldsymbol{r}) = \delta_{m,n} . \tag{3.32}$$

Inserting (3.31) into (3.30) gives

$$p(\boldsymbol{r}) = -\sum_n E_0 \mathrm{e}^{\mathrm{i}(\boldsymbol{K}\cdot\boldsymbol{R}-\nu t)} \frac{\mu V \psi_n^*(\boldsymbol{r}=0)}{\hbar(\nu - \mathrm{i}\gamma) - \varepsilon_\mathrm{g} - \varepsilon_n} \psi_n(\boldsymbol{r}) , \tag{3.33}$$

which has the Fourier transform

$$p_{\boldsymbol{k}} = -\sum_n E_0 \mathrm{e}^{\mathrm{i}(\boldsymbol{K}\cdot\boldsymbol{R}-\nu t)} \frac{\mu \psi_n^*(\boldsymbol{r}=0)}{\hbar(\nu - \mathrm{i}\gamma) - \varepsilon_\mathrm{g} - \varepsilon_n} \int \mathrm{d}^3 r\, \psi_n(\boldsymbol{r})\, \mathrm{e}^{\mathrm{i}\boldsymbol{k}\cdot\boldsymbol{r}} . \tag{3.34}$$

Using an equation of the form of (2.7) for the polarization amplitude, we obtain

$$P(\boldsymbol{R}) = -2\,|\mu|^2 E_0 \sum_n \frac{|\psi_n(\boldsymbol{r}=0)|^2}{\hbar(\nu - \mathrm{i}\gamma) - \varepsilon_\mathrm{g} - \varepsilon_n} , \tag{3.35}$$

where $|\psi_n(\boldsymbol{r}=0)|^2$ is the probability of finding the electron and hole within the same atomic unit cell (zero spatial separation on our coarse grained length scale). The optical susceptibility $\chi(\nu)$ of (1.73) is then given by

$$\chi(\nu) = -\frac{2\,|\mu|^2}{V\epsilon_\mathrm{b}} \sum_n \frac{|\psi_n(\boldsymbol{r}=0)|^2}{\hbar(\nu - \mathrm{i}\gamma) - \varepsilon_\mathrm{g} - \varepsilon_n} . \tag{3.36}$$

In cgs units, the corresponding absorption coefficient $\alpha(\nu)$ is

$$\begin{aligned} \alpha(\nu) &= \frac{4\pi\nu}{n_\mathrm{b}c} \operatorname{Im}[\chi(\nu)] \\ &= \alpha_0 \left[ \sum_{n=1}^{\infty} \frac{4\pi}{n^3} \delta\left(\Delta + \frac{1}{n^2}\right) + \Theta(\Delta)\, \pi \frac{\mathrm{e}^{\pi/\sqrt{\Delta}}}{\sinh\left(\pi/\sqrt{\Delta}\right)} \right] , \end{aligned} \tag{3.37}$$

where

$$\Delta = \frac{\hbar\nu - \varepsilon_\mathrm{g}}{\varepsilon_\mathrm{R}} ,$$

$$\alpha_0 = \frac{2\,|\mu|^2}{\hbar n_\mathrm{b} c a_0^3} ,$$

and we have used the explicit form of the electron-hole pair eigenfunctions [*Haug* and *Koch* (1994)]. Equation (3.37) is known as the *Elliot formula* and describes the bandgap absorption spectrum in an unexcited bulk semiconductor.

Equation (3.37) predicts that the absorption spectrum consists of a series of $\delta$-functions at discrete energies. These resonances are the exciton peaks. The prefactor in front of the $\delta$-functions in (3.37) shows that the exciton resonances have a rapidly decreasing oscillator strength $\propto n^{-3}$. The appearance of the exciton resonances in the absorption spectrum is a unique consequence of the electron-hole Coulomb attraction.

The second term in (3.37), $\alpha_{\text{cont}}$, describes the continuum absorption due to the ionized states. It can be written in terms of the free-carrier absorption,

$$\alpha_{\text{free}}(\omega) = \alpha_0 \sqrt{\Delta}\,\Theta(\Delta) \ , \tag{3.38}$$

as

$$\alpha_{\text{cont}}(\omega) = \alpha_{\text{free}}(\omega) \frac{\pi}{\sqrt{\Delta}} \frac{e^{\pi/\sqrt{\Delta}}}{\sinh\left(\pi/\sqrt{\Delta}\right)} \ , \tag{3.39}$$

where the factor multiplying $\alpha_{\text{free}}$ is called the *Sommerfeld* or *Coulomb enhancement* factor. It is a simple exercise to verify that this factor approaches the value $2\pi/\sqrt{\Delta}$ for $\Delta \to 0$, which cancels the $\sqrt{\Delta}$ factor in the free-carrier absorption of (3.38) and yields a constant value at the bandgap. This is strikingly different from the square-root law of the free-carrier absorption.

If one takes into account the broadening of the exciton resonances caused by, for example, the scattering of electron-hole pairs with phonons, then only a few bound states can be spectrally resolved. An example of an absorption spectrum predicted by the Elliot formula is depicted in Fig. 3.3. In order to plot the spectrum in Fig. 3.3 we introduced a small amount of broadening in (3.37). As a consequence of the prefactor $|\psi_n(\mathbf{r} = 0)|^2$ in (3.36) only the s-like states contribute to the optical response in semiconductors with

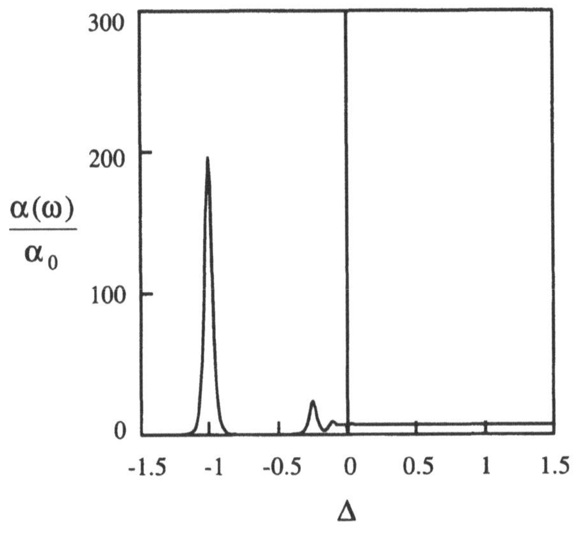


**Fig. 3.3.** Optical spectrum at the absorption edge predicted by the Elliot formula. The absorption coefficient $\alpha(\Delta)$ (in units of $\alpha_0$) is plotted versus $\Delta = (\hbar\nu - \varepsilon_g)/\varepsilon_R$. In order to show the absorption spectrum a small broadening of the exciton resonances has been introduced. An absorption spectrum of this kind can be observed e.g. in high quality GaAs at low temperatures

dipole allowed transitions. Figure 3.3 shows that the dominant feature is the 1*s*-exciton absorption peak. The 2*s*-exciton can also be resolved, but its height is only 1/8-th of that of the 1*s*-resonance. The other exciton states in GaAs materials usually appear only as a collection of unresolvable peaks just below the bandgap. Note that the continuum absorption is almost constant in the spectral region shown. These features have all been observed in spectra measured at low temperatures using sufficiently high-quality semiconductors.

Generally, the existence of resonances and the enhancement of the continuum optical spectrum can be traced back to the renormalization of the electric-dipole interaction energy (3.21). This renormalization is caused by the attractive Coulomb interaction between electrons and holes and is responsible for the pronounced increase in the optical absorption around the absorption edge when compared to the free-carrier predictions. The increased absorption, which is an example of the more general phenomenon of interband Coulomb enhancement, may be explained as follows: Due to Coulomb attraction, an electron and a hole have a greater tendency to be in the vicinity of each other for a longer duration than would be the case if they were noninteracting particles. This increases the interaction time, which in turn leads to a higher probability of an optical transition.

## 3.3 Screened Hartree-Fock Approximation

As soon as we consider semiconductors with elevated carrier densities we need to include the effects of the collision contributions, $\partial/\partial t|_{\mathrm{col}}$. The equations, as written in (3.17–19), are formally exact as long as we do not specify the collision contributions, $\partial/\partial t|_{\mathrm{col}}$. However, approximations are unavoidable in the derivation of explicit expressions for these terms, which give rise to carrier and polarization relaxation, as well as plasma screening. Different approaches will be discussed in this, and in the following chapter.

An approach, whose advantage is that it is rather simple to implement, involves using the phenomenological relaxation time approximation given (2.25, 31) to describe the polarization and carrier relaxation. Furthermore, the effects of plasma screening are included phenomenologically by replacing the bare Coulomb potential $V_q$ in (3.3) by the screened Coulomb potential $V_{sq}$. This treatment of screening effects leads to the *semiconductor Bloch equations* in the *screened Hartree-Fock approximation*, which can be systemically derived using many-body Green's function techniques [*Binder* and *Koch* (1995)]. Note that the Coulomb interaction Hamiltonian with the bare Coulomb potential, i.e., $V_q$, already contains the mechanism for plasma screening. Therefore, one should be concerned that an *ad hoc* replacement of $V_q$ with $V_{sq}$ in (3.3) might count some screening effects twice. Such problems can only be avoided within a systematic many-body approach. An example of that is discussed in the following Chap. 4. For more details, we refer the interested reader to the original literature listed at the end of this chapter.

To implement the screened Hartree-Fock approximation, we need a screening model. One approach is to use a self-consistent quantum theory of plasma screening involving arguments from classical electrodynamics and quantum mechanics. Given an electron at the origin of our coordinate system, we wish to know what effect this electron has on its surroundings. To find out, we introduce a test charge, i.e., a charge sufficiently small as to cause negligible perturbation. In vacuum, the electrostatic potential due to the electron is $\phi(r) = e/r$. However, in a semiconductor there is a background dielectric constant $\epsilon_b$ which is due to everything in the semiconductor in the absence of the carriers themselves. Furthermore, there is the carrier distribution that is changed by the presence of the test electron at the origin (see Fig. 3.4). The new carrier distribution, $\langle n_s(\boldsymbol{r})\rangle$, in turn changes the electrostatic potential.

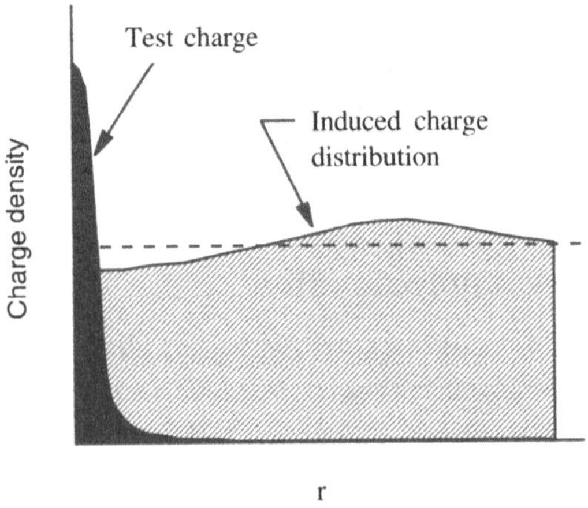


**Fig. 3.4.** An electron at the origin induces a change in the carrier distribution. The electron density is $n_e(\boldsymbol{r}) = \delta^3(\boldsymbol{r})$ and the new carrier distribution is $\langle n_s(\boldsymbol{r})\rangle$, where the subscript s stands for screening, which is the net effect of the induced carrier distribution change

We denote the carrier density distribution as an expectation value since we plan to calculate it quantum mechanically. To derive the induced carrier distribution, we first simplify the problem by assuming that the screening effects of an electron-hole plasma equal the sum of the effects resulting from the separate electron and hole plasmas. As such we neglect excitonic screening, which is not a bad approximation for the elevated carrier densities present in conventional semiconductor lasers. Starting with the electron plasma, we note from (1.96, 97) that the corresponding quantum-mechanical operator for the screened electron charge distribution is $en_s(\boldsymbol{r})$ with

$$\begin{aligned} n_s(\boldsymbol{r}) &= \frac{1}{V}\sum_{\boldsymbol{k},\boldsymbol{k}'} e^{i(\boldsymbol{k}-\boldsymbol{k}')\cdot\boldsymbol{r}} a^\dagger_{\boldsymbol{k}'} a_{\boldsymbol{k}} \\ &= \sum_{\boldsymbol{q}} n_{s\boldsymbol{q}}\, e^{i\boldsymbol{q}\cdot\boldsymbol{r}} \quad . \end{aligned} \tag{3.40}$$

Here, the Fourier transform of the density operator is given by

$$n_{\mathrm{s}q} = \frac{1}{V} \sum_k a_{k-q}^\dagger a_k \ , \tag{3.41}$$

and $V$ (with no argument or subscript) is the volume of the semiconductor medium. In a rigorous treatment we would use the electronic part of the many-body Hamiltonian to obtain an equation of motion for $n_{\mathrm{s}q}$. At the level of a self-consistent Hartree-Fock approach, we can treat screening effects on the basis of an effective single-particle Hamiltonian

$$H_{\mathrm{eff}} = \sum_k \varepsilon_{ek} a_k^\dagger a_k + V \sum_q V_{\mathrm{s}q} n_{\mathrm{s},-q} \ , \tag{3.42}$$

where

$$V_{\mathrm{s}q} = \frac{1}{V} \int \mathrm{d}^3 r \, V_{\mathrm{s}}(r) \, \mathrm{e}^{-\mathrm{i}q\cdot r} \ , \tag{3.43}$$

with

$$V_{\mathrm{s}}(r) = e\phi_{\mathrm{s}}(r) \ , \tag{3.44}$$

and $\phi_{\mathrm{s}}(r)$ is the screened electrostatic potential.

With the effective Hamiltonian (3.42), we get the equation of motion

$$\begin{aligned} \mathrm{i}\hbar \frac{\mathrm{d}}{\mathrm{d}t} a_{k-q}^\dagger a_k &= \left[a_{k-q}^\dagger a_k, H_{\mathrm{eff}}\right] \\ &= \left(\varepsilon_k - \varepsilon_{k-q}\right) a_{k-q}^\dagger a_k \\ &\quad + \sum_p V_{sp} \left(a_{k-q}^\dagger a_{k+p} - a_{k-q-p}^\dagger a_k\right) \ . \end{aligned} \tag{3.45}$$

Taking the expectation value and keeping only slowly varying terms, namely those with $\boldsymbol{p} = -\boldsymbol{q}$, we get

$$\mathrm{i}\hbar \frac{\mathrm{d}}{\mathrm{d}t} \langle a_{k-q}^\dagger a_k \rangle = \left(\varepsilon_k - \varepsilon_{k-q}\right) \langle a_{k-q}^\dagger a_k \rangle + V_{\mathrm{s}q} \left(n_{k-q} - n_k\right) \ . \tag{3.46}$$

We suppose that $\langle a_{k-q}^\dagger a_k \rangle$ has a solution of the form $\mathrm{e}^{(\delta - \mathrm{i}\omega)t}$, where the infinitesimal $\delta$ indicates that the perturbation has been switched on adiabatically, i.e., that we had a homogeneous plasma at $t = -\infty$. We further suppose that the induced charge distribution follows this response. This transforms (3.46) to

$$\langle a_{k-q}^\dagger a_k \rangle = V_{\mathrm{s}q} \frac{n_{k-q} - n_k}{\hbar(\omega + \mathrm{i}\delta) + \varepsilon_{k-q} - \varepsilon_k} \tag{3.47}$$

so that

$$\langle n_{\mathrm{s}q} \rangle = \frac{V_{\mathrm{s}q}}{V} \sum_k \frac{n_{k-q} - n_k}{\hbar(\omega + \mathrm{i}\delta) + \varepsilon_{k-q} - \varepsilon_k} \ . \tag{3.48}$$

The induced charge distribution is a source in Poisson's equation

$$\nabla^2 \phi_{\mathrm{s}}(r) = -\frac{4\pi e}{\epsilon_{\mathrm{b}}} \left[n_{\mathrm{e}}(\boldsymbol{r}) + \langle n_{\mathrm{s}}(\boldsymbol{r}) \rangle\right] \ . \tag{3.49}$$

The Fourier transform of this equation is

$$\phi_{sq} = \frac{4\pi e}{\epsilon_b q^2}\left(\frac{1}{V} + \langle n_{sq}\rangle\right) \quad , \tag{3.50}$$

where for a point charge at the origin

$$n_{eq} = \frac{1}{V}\int d^3 r\, \delta^3(\boldsymbol{r})\, e^{-i\boldsymbol{q}\cdot\boldsymbol{r}} = \frac{1}{V} \quad . \tag{3.51}$$

Using $V_{sq} \equiv e\phi_{sq}$, we substitute (3.48) into (3.50) and solve for $V_{sq}$ to find

$$V_{sq} = V_q\left(1 - V_q \sum_{\boldsymbol{k}} \frac{n_{\boldsymbol{k}-\boldsymbol{q}} - n_{\boldsymbol{k}}}{\hbar(\omega + i\delta) + \varepsilon_{\boldsymbol{k}-\boldsymbol{q}} - \varepsilon_{\boldsymbol{k}}}\right)^{-1} \quad , \tag{3.52}$$

where $V_q$ is the unscreened Coulomb potential, (3.4). Repeating the derivation for the hole plasma, and adding the electron and hole contributions, we find the screened Coulomb potential energy between carriers

$$V_{sq} = \frac{V_q}{\epsilon_q(\omega)} \quad , \tag{3.53}$$

where the longitudinal dielectric function $\epsilon_q(\omega)$ is given by

$$\epsilon_q(\omega) = 1 - V_q \sum_{\boldsymbol{k}} \sum_{\alpha=e,h} \frac{n_{\alpha,\boldsymbol{k}-\boldsymbol{q}} - n_{\alpha,\boldsymbol{k}}}{\hbar(\omega + i\delta) + \varepsilon_{\alpha,\boldsymbol{k}-\boldsymbol{q}} - \varepsilon_{\alpha,\boldsymbol{k}}} \quad . \tag{3.54}$$

This equation is the *Lindhard formula.* It describes a complex retarded dielectric function, i.e., the poles are in the lower complex frequency plane, and it includes spatial dispersion ($q$ dependence) and spectral dispersion ($\omega$ dependence).

In many practical situations, the Lindhard formula (3.54) is numerically too complicated to use because of its continuum of poles. Then one has to make additional approximations leading to a simplified, but also less general version. In general it is dangerous to make uncontrolled approximations to the Lindhard formula because there are important sum rules, which are valid for the Lindhard formula and which a simplified dielectric function also has to obey [*Mahan* (1981)]. To see how to obtain a proper simplification, we look at the long wavelength ($q \to 0$) limit of the Lindhard formula. We assume a quasi-equilibrium system, where $n_{\alpha\boldsymbol{k}} = f_{\alpha\boldsymbol{k}}$ is the Fermi-Dirac distribution function. We expand $\varepsilon_{\alpha,\boldsymbol{k}-\boldsymbol{q}}$ and $f_{\alpha,\boldsymbol{k}-\boldsymbol{q}}$ around $q = 0$ to find

$$\begin{aligned} \varepsilon_{\alpha,\boldsymbol{k}-\boldsymbol{q}} - \varepsilon_{\alpha\boldsymbol{k}} &= \frac{\hbar^2}{2m_\alpha}\left(k^2 - 2\boldsymbol{k}\cdot\boldsymbol{q} + q^2\right) - \frac{\hbar^2 k^2}{2m_\alpha} \\ &\simeq -\sum_i \frac{\hbar^2 k_i q_i}{m_\alpha} \quad , \end{aligned} \tag{3.55}$$

$$\begin{aligned} f_{\alpha,\boldsymbol{k}-\boldsymbol{q}} - f_{\alpha\boldsymbol{k}} &= f_{\alpha\boldsymbol{k}} - \sum_i q_i \frac{\partial f_{\alpha\boldsymbol{k}}}{\partial k_i} + \ldots - f_{\alpha\boldsymbol{k}} \\ &\simeq -\sum_i q_i \frac{\partial f_{\alpha\boldsymbol{k}}}{\partial k_i} \quad . \end{aligned} \tag{3.56}$$

Inserting these expansions into (3.54), expanding the resulting denominator, noticing that $\sum_{\boldsymbol{k}} \partial f_{\alpha \boldsymbol{k}}/\partial k = 0$, and integrating the remaining term by parts, we get the classical (or Drude) dielectric function

$$\epsilon_{q=0}(\omega) = 1 - \frac{\omega_{\mathrm{pl}}^2}{\omega^2} \quad . \tag{3.57}$$

Here the square of the electron-hole plasma frequency is given by

$$\omega_{\mathrm{pl}}^2 = \frac{4\pi N e^2}{\epsilon_{\mathrm{b}} m_{\mathrm{r}}} = 16\pi N a_0^3 \left(\frac{\varepsilon_{\mathrm{R}}}{\hbar}\right)^2 \quad , \tag{3.58}$$

$N = V^{-1} \sum_{\boldsymbol{k}} f_{\boldsymbol{k}}$ is the electron or hole density, and $m_{\mathrm{r}}$ is the reduced electron-hole mass.

The classical result (3.57) is in many aspects too simplistic for a realistic description of many-body plasma screening effects in both active and passive semiconductors. Instead, in the so-called *plasmon-pole approximation* one replaces the continuum of electron-pair excitations, represented by the continuum of poles in the Lindhard formula, by a single effective plasmon pole. In this approximation, the inverse Lindhard dielectric function is replaced by

$$\frac{1}{\epsilon_q(\omega)} = 1 + \frac{\omega_{\mathrm{pl}}^2}{(\omega + \mathrm{i}\delta)^2 - \omega_q^2} \quad , \tag{3.59}$$

which has the same structure as the long-wavelength plasma result (3.57), but instead of $\omega_{\mathrm{pl}}$ in the denominator, it has the effective plasmon frequency $\omega_q$ defined by

$$\omega_q^2 = \omega_{\mathrm{pl}}^2 \left(1 + \frac{q^2}{\kappa^2}\right) + C \left(\frac{\hbar q^2}{4 m_{\mathrm{r}}}\right)^2 \quad . \tag{3.60}$$

Here, $C$ is a numerical constant usually taken between 1 and 4, and $\kappa$ is the inverse static screening length

$$\kappa = \left(\frac{4\pi e^2}{\epsilon_{\mathrm{b}}} \sum_{\alpha=e,h} \frac{\partial N}{\partial \mu}\right)^{1/2} \quad . \tag{3.61}$$

Without discussing details of the derivation of (3.60), we just mention here that $\omega_q$ has been determined such that the dielectric function (3.54) fulfills certain sum rules [*Mahan* (1981), *Lundquist* (1967), *Haug* and *Koch* (1994)].

For many practical applications, one ignores the damped response of the screening represented by $\omega+\mathrm{i}\delta$ in the dielectric function (3.59). In this "static" plasmon-pole approximation, the screened Coulomb potential (3.53) is simply given by

$$V_{\mathrm{s}q} = V_q \left(1 - \frac{\omega_{\mathrm{pl}}^2}{\omega_q^2}\right) \quad , \tag{3.62}$$

where $\omega_q^2$ is defined in (3.60). The approximation (3.62) often allows analytic results, or at least much simpler numerical results for the effects of plasma screening.

## 3.4 Bandgap Renormalization in the Screened Hartree-Fock Approximation

In this section, we apply the results of the previous section to study the bandgap renormalization effect. First we note that by replacing $V_q$ by $V_{sq}$ in (3.3), (3.8) becomes

$$\varepsilon_{\mathrm{h}\boldsymbol{k}}^s = -\varepsilon_{\mathrm{v}\boldsymbol{k}} + \sum_{q \neq 0} V_{\mathrm{s}q} \quad , \tag{3.63}$$

i.e., the hole energy in the presence of the screened Coulomb potential. However, to express the hole-energy in terms of an effective mass, we have to remember that the hole effective mass is taken from low-excitation experiments, that is, for an unscreened Coulomb potential. Hence we rewrite (3.63) as

$$\begin{aligned} \varepsilon_{\mathrm{h}\boldsymbol{k}}^s &= -\varepsilon_{\mathrm{v}\boldsymbol{k}} + \sum_{q \neq 0} V_q + \sum_{q \neq 0} (V_{\mathrm{s}q} - V_q) \\ &= \varepsilon_{\mathrm{h}\boldsymbol{k}} + \sum_{q \neq 0} (V_{\mathrm{s}q} - V_q) \\ &= \varepsilon_{\mathrm{h}\boldsymbol{k}} + \Delta\varepsilon_{\mathrm{CH}} \quad , \end{aligned} \tag{3.64}$$

where we define the *Debye shift*, or *Coulomb-hole (CH) self energy*

$$\Delta\varepsilon_{\mathrm{CH}} \equiv \sum_{q \neq 0} [V_{\mathrm{s}q} - V_q] \quad . \tag{3.65}$$

This term is independent of wave vector and is usually considered as an additional contribution to the bandgap shift. It is due to the increasingly effective plasma screening that occurs with increasing carrier density. To see this, we note that at very low carrier densities, the lack of vacant valence band states together with the exclusion principle limit the ability of the valence electron distribution to effectively screen the Coulomb repulsion between a conduction electron and any one of the valence electrons. The corresponding abundance of vacant conduction band states are energetically inaccessible to a valence electron via the Coulomb interaction. At higher carrier densities, more vacant valence band states are available to allow the redistribution of charges for more effective screening. Since the screening of a repulsive interaction leads to a lowering of the conduction electron energy, the transition energy decreases with increasing density.

Recalling the static plasmon-pole approximation used in simplifying the Lindhard formula, we rewrite the Coulomb-hole self energy as

$$\begin{aligned}\Delta\varepsilon_{\mathrm{CH}} &= \sum_{q\neq 0}(V_{\mathrm{s}q}-V_q)\\ &= \sum_{q\neq 0} V_q\left(\frac{1}{\epsilon_q}-1\right)\\ &= -\sum_{q\neq 0} V_q\omega_{\mathrm{pl}}^2\left[\omega_{\mathrm{pl}}^2\left(1+\frac{q^2}{\kappa^2}\right)+\frac{C}{4}\left(\frac{\hbar q^2}{2m_{\mathrm{r}}}\right)^2\right]^{-1} .\end{aligned} \tag{3.66}$$

Converting the sum to an integral, using formula (2.161) from *Gradshteyn* and *Rhyzhik* (1980), and expressing all parameters in terms of $a_0$ and $\varepsilon_{\mathrm{R}}$ [(1.14, 15)], we write $\Delta\varepsilon_{\mathrm{CH}}$ as

$$\Delta\varepsilon_{\mathrm{CH}} = -2\varepsilon_{\mathrm{R}}a_0\kappa\left(1+\frac{\sqrt{C}a_0^2\kappa^2\varepsilon_{\mathrm{R}}}{\hbar\omega_{\mathrm{pl}}}\right)^{-\frac{1}{2}} . \tag{3.67}$$

In addition to the Coulomb-hole self energy, there is a second contribution to the carrier-density dependence of $\omega_{\boldsymbol{k}}$ that comes from the Hartree-Fock energy correction as described by (3.20). In the screened Hartree-Fock approximation, this contribution, which we now refer to as the screened-exchange (SX) shift, is

$$\Delta\varepsilon_{\mathrm{SX},\boldsymbol{k}} = -\sum_{\boldsymbol{k}'\neq\boldsymbol{k}} V_{s,|\boldsymbol{k}-\boldsymbol{k}'|}\left(n_{\mathrm{e}\boldsymbol{k}'}+n_{\mathrm{h}\boldsymbol{k}'}\right) , \tag{3.68}$$

which is similar to (3.20) except that the bare Coulomb potential is replaced by the screened Coulomb potential. It is often a reasonable approximation to neglect the weak $\boldsymbol{k}$-dependence in $\Delta\varepsilon_{\mathrm{SX},\boldsymbol{k}}$ and just use a $\boldsymbol{k}$-independent $\Delta\varepsilon_{\mathrm{SX}}$. In this case, both contributions to the $\omega_{\boldsymbol{k}}$ renormalization are independent of $\boldsymbol{k}$ and the renormalized bandgap is given simply by

$$\varepsilon_{\mathrm{g}} = \varepsilon_{\mathrm{g0}} + \Delta\varepsilon_{\mathrm{SX}} + \Delta\varepsilon_{\mathrm{CH}} . \tag{3.69}$$

In Fig. 3.5 we plot the different screened Hartree-Fock contributions to the renormalized bandgap as function of scaled interparticle distance $r_{\mathrm{s}}$ defined by

$$\frac{4}{3}\pi r_{\mathrm{s}}^3 = \frac{1}{Na_0^3} . \tag{3.70}$$

The screened Hartree-Fock approximation often overestimates the bandgap reduction. Improvements to this problem as well as other many-body effects will be discussed in Chap. 4.

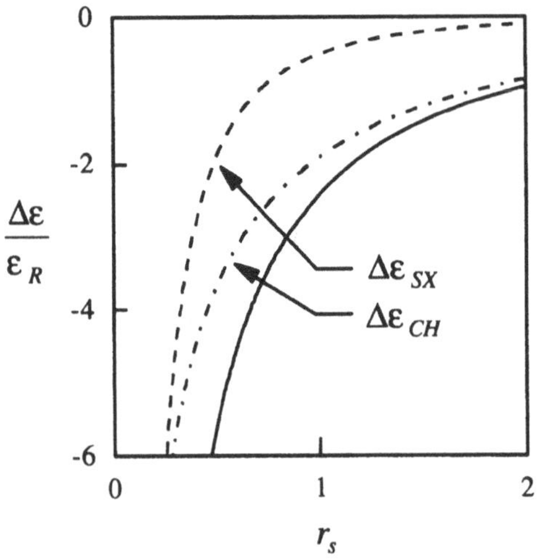


**Fig. 3.5.** Bandgap reduction as function of normalized interparticle distance $r_s$, (3.70), for the bulk-material parameters of Chap. 2. The reduced gap is plotted as *solid curve*, the screened exchange contribution as *dashed curve* and the Coulomb-hole contribution as *dash-dotted curve*

## 3.5 Padé Approximation

Using the relaxation rate approximation and the screened Hartree-Fock approximation, we write for the induced dipole $p_{\boldsymbol{k}}$,

$$\frac{\mathrm{d}p_{\boldsymbol{k}}}{\mathrm{d}t} = -(\mathrm{i}\omega_{\boldsymbol{k}} + \gamma)\, p_{\boldsymbol{k}} - \mathrm{i}\Omega_{\boldsymbol{k}}(z,t)\,(n_{\mathrm{e}\boldsymbol{k}} + n_{\mathrm{h}\boldsymbol{k}} - 1) \quad . \tag{3.71}$$

Formally integrating from $-\infty$ to $t$, gives

$$p_{\boldsymbol{k}}(t) = -\mathrm{i}\int_{-\infty}^{t} \mathrm{d}t'\, \mathrm{e}^{(\mathrm{i}\omega_{\boldsymbol{k}}+\gamma t)(t'-t)}\,\Omega_{\boldsymbol{k}}(z,t')\,[n_{\mathrm{e}\boldsymbol{k}}(t') + n_{\mathrm{h}\boldsymbol{k}}(t') - 1] \quad . \tag{3.72}$$

As in Sect. 2.2 for the free-carrier theory, we make the rate equation approximation which assumes that the carrier probabilities and the electric field envelope vary little in the time $T_2 \equiv 1/\gamma$. Using the field (2.41), we have

$$\begin{aligned} p_{\boldsymbol{k}}(t) &= -\frac{\mathrm{i}}{\hbar}\,[n_{\mathrm{e}\boldsymbol{k}}(t) + n_{\mathrm{h}\boldsymbol{k}}(t) - 1]\left[\frac{1}{2}\mu_{\boldsymbol{k}}E(z)\frac{\mathrm{e}^{\mathrm{i}[Kz-\nu t-\phi(z)]}}{\mathrm{i}\,(\omega_{\boldsymbol{k}}-\nu)+\gamma}\right. \\ &\quad \left. + \sum_{\boldsymbol{k}'\neq\boldsymbol{k}} V_{s,|\boldsymbol{k}-\boldsymbol{k}'|}\int_{-\infty}^{t}\mathrm{d}t'\,\mathrm{e}^{(\mathrm{i}\omega_{\boldsymbol{k}}+\gamma)(t'-t)}p_{\boldsymbol{k}}(t')\right] \quad . \end{aligned} \tag{3.73}$$

We solve (3.73) by iteration in powers of the screened Coulomb interaction energy $V_{s,|\boldsymbol{k}-\boldsymbol{k}'|}$. To lowest order, i.e., setting $V_{s,|\boldsymbol{k}-\boldsymbol{k}'|} = 0$ in (3.73), we find the free-carrier result (2.43)

$$\begin{aligned} p^{(0)}_{\boldsymbol{k}}(t) &= -\frac{\mathrm{i}}{2\hbar}\mu_{\boldsymbol{k}}E(z)\,\mathrm{e}^{\mathrm{i}[Kz-\nu t-\phi(z)]}\frac{n_{\mathrm{e}\boldsymbol{k}} + n_{\mathrm{h}\boldsymbol{k}} - 1}{\mathrm{i}\,(\omega_{\boldsymbol{k}}-\nu)+\gamma} \\ &= \frac{1}{2}E(z)\,\mathrm{e}^{\mathrm{i}[Kz-\nu t-\phi(z)]}\chi^{(0)}_{\boldsymbol{k}} \quad , \end{aligned} \tag{3.74}$$

where for later convenience we introduce the $\boldsymbol{k}$-dependent susceptibility function

$$\chi_{\boldsymbol{k}}^{(0)} = -\frac{\mathrm{i}\mu_{\boldsymbol{k}}}{\hbar}\frac{n_{\mathrm{e}\boldsymbol{k}}+n_{\mathrm{h}\boldsymbol{k}}-1}{\mathrm{i}\,(\omega_{\boldsymbol{k}}-\nu)+\gamma} \quad . \tag{3.75}$$

Substituting the lowest-order result into (3.73) and noting that $\chi_{\boldsymbol{k}}^{(0)}(t)$ varies little in the time $T_2$, we find the first-order contribution

$$\begin{aligned} p_{\boldsymbol{k}}^{(1)}(t) &= -\frac{\mathrm{i}}{2\hbar}E(z)\,\mathrm{e}^{\mathrm{i}[Kz-\nu t-\phi(z)]}\frac{n_{\mathrm{e}\boldsymbol{k}}+n_{\mathrm{h}\boldsymbol{k}}-1}{\mathrm{i}\,(\omega_{\boldsymbol{k}}-\nu)+\gamma}\sum_{\boldsymbol{k}'\neq\boldsymbol{k}} V_{s,|\boldsymbol{k}-\boldsymbol{k}'|}\chi_{\boldsymbol{k}'}^{(0)} \\ &= \frac{1}{2}E(z)\,\mathrm{e}^{\mathrm{i}[Kz-\nu t-\phi(z)]}\chi_{\boldsymbol{k}}^{(0)}q(\boldsymbol{k}) \quad , \end{aligned} \tag{3.76}$$

where the complex dimensionless factor

$$q(\boldsymbol{k}) = \frac{1}{\mu_{\boldsymbol{k}}}\sum_{\boldsymbol{k}'\neq\boldsymbol{k}} V_{s,|\boldsymbol{k}-\boldsymbol{k}'|}\chi_{\boldsymbol{k}'}^{(0)} \quad . \tag{3.77}$$

In principle, we could continue to iterate (3.73) in this way until reaching any desired accuracy. However, this process does not converge rapidly and offers no substantial CPU-time improvement over the direct numerical solution of (3.73) obtained by discretizing the integral and using a matrix inversion [*Haug* and *Koch* (1994)]. On the other hand, we obtain a remarkably accurate result for gain media by treating $p_{\boldsymbol{k}}^{(0)}$ and $p_{\boldsymbol{k}}^{(1)}$ as the first two terms of a geometrical series, which we then "resum" [*Haug* and *Koch* (1989)]. This approximation is the simplest kind of Padé approximation [*Gaves-Morris* (1973)]. Accordingly, adding (3.74, 76) and resumming, we find

$$\begin{aligned} p_{\boldsymbol{k}}(t) &\simeq p_{\boldsymbol{k}}^{(0)}(t)+p_{\boldsymbol{k}}^{(1)}(t) \\ &= \frac{1}{2}E(z)\,\mathrm{e}^{\mathrm{i}[Kz-\nu t-\phi(z)]}\chi_{\boldsymbol{k}}^{(0)}\,(1+q(\boldsymbol{k})) \\ &\simeq \frac{1}{2}E(z)\,\mathrm{e}^{\mathrm{i}[Kz-\nu t-\phi(z)]}\frac{\chi_{\boldsymbol{k}}^{(0)}}{1-q(\boldsymbol{k})} \quad . \end{aligned} \tag{3.78}$$

Substituting this equation into (2.7), and using the result in the self-consistency equations (1.76, 77), we find the gain and carrier-induced phase shift

$$\begin{aligned} g-\mathrm{i}\frac{\mathrm{d}\phi}{\mathrm{d}z} &= \frac{\mathrm{i}K}{2}\chi \\ &= \frac{\nu}{2\epsilon_0 nc\hbar V}\sum_{\boldsymbol{k}}|\mu_{\boldsymbol{k}}|^2\frac{n_{\mathrm{e}\boldsymbol{k}}+n_{\mathrm{h}\boldsymbol{k}}-1}{\mathrm{i}\,(\omega_{\boldsymbol{k}}-\nu)+\gamma}\,\frac{1}{1-q(\boldsymbol{k})} \quad . \end{aligned} \tag{3.79}$$

The carrier-induced phase shift is related to the carrier-induced refractive index by (1.81). Under quasi-equilibrium conditions $n_{\alpha\boldsymbol{k}} = f_{\alpha\boldsymbol{k}}$, where the Fermi-Dirac distributions $f_{\alpha\boldsymbol{k}}$ are given by (1.19). Equation (3.79) is quite convenient for numerical evaluation since it expresses the many-body susceptibility in a form that is very similar to the free-carrier result (2.52, 53).

The many-body effects can be identified explicitly as i) a carrier density dependence of the transition energy, $\omega_{\boldsymbol{k}}(N)$, and ii) the factor $[1-q(\boldsymbol{k},N)]^{-1}$, which represents the Coulomb enhancement in our Padé approximation.

## 3.6 Bulk Semiconductors

Using (3.62) for $V_{sq}$ with $\omega_q^2$ given by (3.60), we can write the screened Coulomb potential for a bulk semiconductor medium as

$$V_{sq} = \frac{8\pi\varepsilon_R a_0}{Vq^2}\left(\frac{\varepsilon_q}{\varepsilon_\kappa}+\frac{C}{4}\frac{\varepsilon_q^2}{\varepsilon_{pl}^2}\right)\left(1+\frac{\varepsilon_q}{\varepsilon_\kappa}+\frac{C}{4}\frac{\varepsilon_q^2}{\varepsilon_{pl}^2}\right)^{-1} , \tag{3.80}$$

where $a_0$ and $\varepsilon_R$ are the exciton Bohr radius and the Rydberg energy. We define

$$\varepsilon_q = \frac{\hbar^2 q^2}{2m_r} ,$$
$$\varepsilon_\kappa = \frac{\hbar^2 \kappa^2}{2m_r} ,$$
$$\varepsilon_{pl} = \hbar\omega_{pl} , \tag{3.81}$$

where $\kappa$, (3.61), is the inverse screening length and $\omega_{pl}$, (3.58), is the electron-hole plasma frequency. The vector

$$\boldsymbol{q} = \boldsymbol{k} - \boldsymbol{k}'$$

so that

$$q^2 = k^2 + k'^2 - 2kk'\cos\theta , \tag{3.82}$$

where $\theta$ is the angle between $\boldsymbol{k}$ and $\boldsymbol{k}'$. Substituting (3.80) into (3.77) and evaluating the summation as an integral, we get under quasi-equilibrium conditions,

$$q(k) = -\frac{i\varepsilon_R^2 a_0^3}{\pi\hbar}\int_0^\infty dk' \frac{k'^2}{\sqrt{\varepsilon_k \varepsilon_{k'}}}\frac{1+\frac{\varepsilon_k}{\varepsilon_g}}{1+\frac{\varepsilon_{k'}}{\varepsilon_g}}\frac{f_{ek'}+f_{hk'}-1}{i(\omega_{k'}-\nu)+\gamma}\Theta(k,k') . \tag{3.83}$$

The angular integration function

$$\Theta(k,k') = \int_0^\pi d\theta\,\sin\theta\frac{1+\frac{C}{4}\frac{\varepsilon_\kappa\varepsilon_q}{\varepsilon_{pl}^2}}{1+\frac{\varepsilon_q}{\varepsilon_\kappa}+\frac{C}{4}\frac{\varepsilon_q^2}{\varepsilon_{pl}^2}}$$

$$= \Theta_1(k,k') + \Theta_2(k,k') \tag{3.84}$$

can be evaluated analytically. For the first term, we obtain

$$\Theta_1(k,k') = \frac{1}{2}\ln\left(\frac{1+\frac{\varepsilon_{k_+}}{\varepsilon_\kappa}+\frac{C}{4}\frac{\varepsilon_{k_+}^2}{\varepsilon_{pl}^2}}{1+\frac{\varepsilon_{k_-}}{\varepsilon_\kappa}+\frac{C}{4}\frac{\varepsilon_{k_-}^2}{\varepsilon_{pl}^2}}\right) , \tag{3.85}$$

where $k_\pm = |\boldsymbol{k} \pm \boldsymbol{k}'|$. The second term is given by

$$\begin{aligned}\Theta_2(k,k') &= \frac{1}{\sqrt{D}}\left\{\arctan\left[\frac{1}{\sqrt{D}}\left(\frac{C\varepsilon_{k_+}\varepsilon_\kappa}{2\varepsilon_{\rm pl}^2}+1\right)\right]\right. \\ &\quad \left. -\arctan\left[\frac{1}{\sqrt{D}}\left(\frac{C\varepsilon_{k_-}\varepsilon_\kappa}{2\varepsilon_{\rm pl}^2}+1\right)\right]\right\} ,\end{aligned} \tag{3.86}$$

for $D \equiv C\varepsilon_\kappa^2/\varepsilon_{\rm pl}^2 - 1 > 0$ and

$$\begin{aligned}\Theta_2(k,k') &= \frac{1}{2\sqrt{-D}}\left[\ln\left(\frac{C\varepsilon_{k_+}\varepsilon_\kappa}{2\varepsilon_{\rm pl}^2}+1+\sqrt{-D}\right)\right. \\ &\quad -\ln\left(\frac{C\varepsilon_{k_-}\varepsilon_\kappa}{2\varepsilon_{\rm pl}^2}+1+\sqrt{-D}\right)+\ln\left(\frac{C\varepsilon_{k_-}\varepsilon_\kappa}{2\varepsilon_{\rm pl}^2}+1-\sqrt{-D}\right) \\ &\quad \left. -\ln\left(\frac{C\varepsilon_{k_+}\varepsilon_\kappa}{2\varepsilon_{\rm pl}^2}+1-\sqrt{-D}\right)\right] ,\end{aligned} \tag{3.87}$$

for $D \le 0$. The $k'$-integration in (3.83) has to be performed numerically.

The Coulomb-hole self energy contribution to the bandgap renormalization is given by (3.66) or (3.67) and the screened-exchange contribution (3.68) may be approximated by the integral

$$\Delta\varepsilon_{\rm SX} = -\frac{4\varepsilon_{\rm R}a_0}{\pi}\int_0^\infty {\rm d}k'\,\frac{\frac{\varepsilon_{k'}}{\varepsilon_\kappa}+\frac{C}{4}\frac{\varepsilon_{k'}^2}{\varepsilon_{\rm pl}^2}}{1+\frac{\varepsilon_{k'}}{\varepsilon_\kappa}+\frac{C}{4}\frac{\varepsilon_{k'}^2}{\varepsilon_{\rm pl}^2}}\,(f_{{\rm e}k'}+f_{{\rm h}k'}) \ . \tag{3.88}$$

Using (3.85–87) and (3.68) in (3.83), and the result in (3.79), we calculate the gain and carrier-induced phase shift. For bulk semiconductors, we convert the summation over states in (3.79) to a three dimensional integral and find

$$g - {\rm i}\frac{{\rm d}\phi}{{\rm d}z} = \frac{\nu|\mu|^2}{2\epsilon_0 nc\hbar\pi^2}\int_0^\infty \frac{{\rm d}k\,k^2}{\left(1+\frac{\varepsilon_k}{\varepsilon_{\rm g}}\right)^2}\,\frac{f_{{\rm e}k}+f_{{\rm h}k}-1}{{\rm i}(\omega_k-\nu)+\gamma}\,\frac{1}{1-q(k)} , \tag{3.89}$$

where we again used the quasi-equilibrium approximation.

Figure 3.6 plots gain and carrier-induced phase shift spectra calculated using (3.89) for an undoped bulk GaAs medium and different carrier densities. To make a better connection to the experiment we plotted the intensity gain, $G = 2g$. In the free-carrier results (dashed curves), band filling is the only cause for the density dependence of gain and phase shift, while in the many-body results bandgap renormalization and Coulomb enhancement also contribute. An obvious difference between the two theories is the overall frequency shift due to the bandgap renormalization. The many-body spectra for the different carrier densities are also frequency shifted relative to one another since the bandgap renormalization is a carrier density dependent function. Note that Fig. 3.6 (top) shows the existence of gain at frequencies

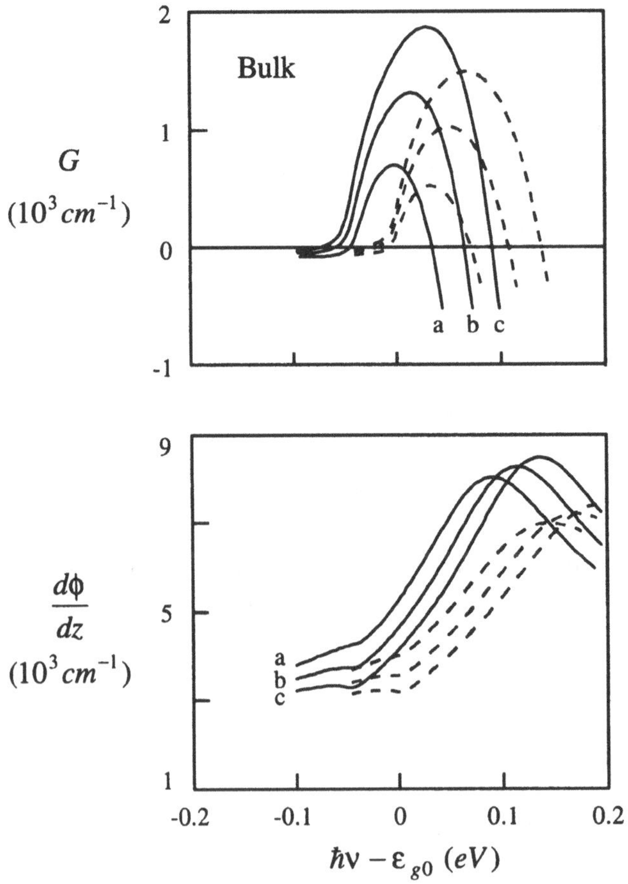


**Fig. 3.6.** Intensity gain $G$ (**top**) and carrier-induced phase shift (**bottom**) versus detuning $\hbar\nu - \varepsilon_{g0}$ according to the Padé approximation (*solid lines*) and free-carrier theory (*dashed lines*). The carrier densities are $N$ = ($a$) $3 \times 10^{18}\mathrm{cm}^{-3}$, ($b$) $4 \times 10^{18}\,\mathrm{cm}^{-3}$, and ($c$) $5 \times 10^{18}\,\mathrm{cm}^{-3}$. We use the bulk medium parameters of Chap. 2, and $\gamma = 10^{13}\,\mathrm{s}^{-1}$

below the unexcited semiconductor bandgap, which for GaAs is 1.42 eV at room temperature. This feature is a consequence of the bandgap renormalization, which decreases the bandgap when the carrier density is increased in GaAs-type materials. The spectral region of optical gain is basically bounded by the renormalized bandgap from below and by the total chemical potential from above

$$\varepsilon_g \le \hbar\nu \le \varepsilon_g + \mu_e + \mu_h \quad , \tag{3.90}$$

which replaces $\varepsilon_{g0}$ in (2.74) by $\varepsilon_g$.

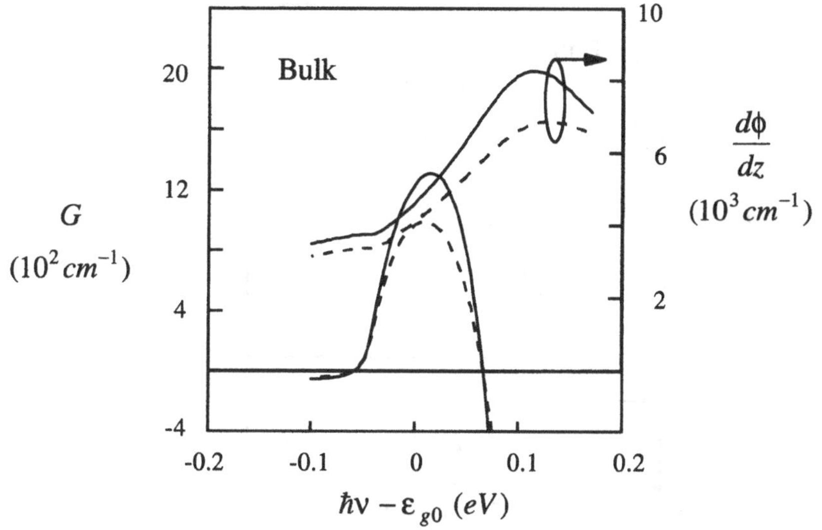


**Fig. 3.7.** Intensity gain $G$ and carrier-induced phase shift (*top curves*) versus detuning $\hbar\nu - \varepsilon_{g0}$ for bulk-material as predicted by the Padé approximation (*solid curve*) and by the free-carrier theory with ad hoc bandgap renormalization (*dashed line*). The carrier density is $N = 4 \times 10^{18}$ cm$^{-3}$

Coulomb enhancement has the effect of reshaping and increasing the magnitude of the gain and absorption spectra. It is most noticeable at low carrier densities, especially when it is possible to resolve the exciton absorption peaks. At the elevated densities needed for gain, Coulomb enhancement effects are not as drastic because plasma screening mitigates the electron-hole Coulomb attraction. Figure 3.7 shows the effects of Coulomb enhancement at a carrier density of $N = 4 \times 10^{18}$ cm$^{-3}$. The dashed curves are obtained using the renormalized bandgap in the free carrier gain formula. Even though this ad hoc inclusion of bandgap renormalization into the free-carrier theory is sometimes used in the literature, it is actually inconsistent because it neglects the Coulomb attraction between an electron and a hole, while it takes into account the exchange interaction and Coulomb repulsion between two electrons or two holes. To be consistent, Coulomb attraction and repulsion have to be treated at the same level of approximation. When using (3.79), we need to keep in mind that the factor $1/[1 - q(k)]$ is only valid at high carrier densities. It ignores the existence of the bound electron-hole states (excitons), and is therefore not correct for densities below the Mott density.

Figure 3.8 (top) shows the peak gains ($G_{\rm pk}$) as predicted by the Padé approximation to the many-body theory and the free-carrier theory for bulk GaAs. The difference between the two curves is due to Coulomb enhancement, which causes an approximately 20 % increase in the peak gain. This difference can increase to a factor of 2 in the wide-bandgap laser compounds,

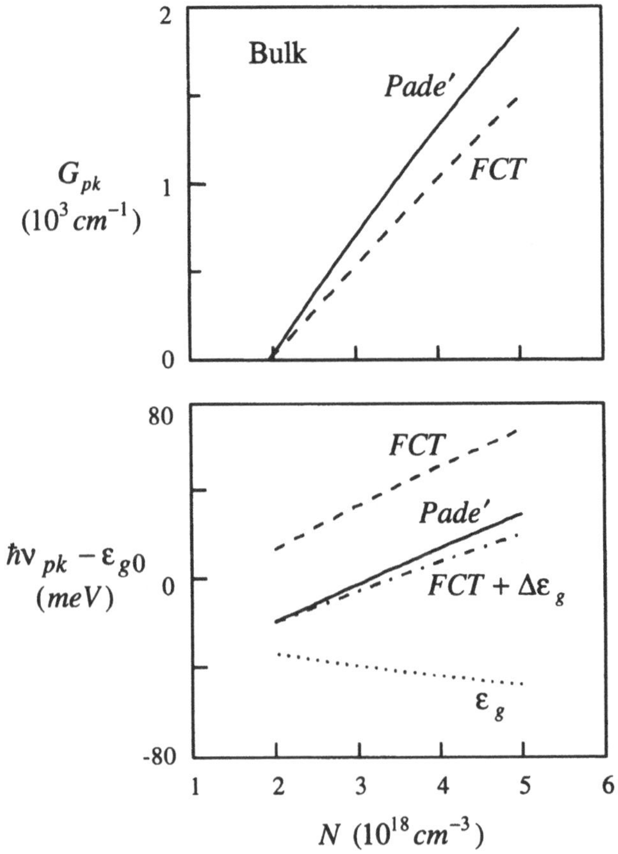


**Fig. 3.8.** (**top**) Bulk-material peak gain $G_{\mathrm{pk}}$ and (**bottom**) peak-gain energy $\hbar\nu_{\mathrm{pk}} - \varepsilon_{g0}$ versus carrier density $N$ according to the Padé approximation (*solid curves*), free-carrier theory FCT (*dashed curves*), and free-carrier theory with a renormalized bandgap (*dot-dashed lines*). For the peak gain, the two free-carrier models coincide. The renormalized bandgap is indicated by the *dotted curve*

where Coulomb interactions are known to be significantly stronger. Figure 3.8 (bottom) shows the dependence of the peak gain energy $\hbar\nu_{\mathrm{pk}}$ on carrier density. We reference the gain peak to the unexcited bandgap energy, $\varepsilon_{g\phi}$. Both band filling and Coulomb enhancement lead to a blue shift of the peak-gain frequency, whereas bandgap renormalization leads to a red shift. In a bulk semiconductor with GaAs-type effective masses, band-filling effects dominate the density-dependent peak-gain shift. The difference between the solid and dot-dashed $\hbar\nu_{\mathrm{pk}}$ curves in Fig. 3.8 is due to the reshaping of the gain spectra by Coulomb enhancement. Note that according to the many-body theory, the peak gain for a carrier density between $2 \times 10^{18}\,\mathrm{cm}^{-3}$ and slightly above

$3 \times 10^{18}\,\mathrm{cm}^{-3}$ occurs below the unrenormalized bandgap, around the frequency of the exciton resonance of the unexcited material.

As discussed in Sect. 2.7, the linewidth enhancement or antiguiding factor $\alpha$ is a useful quantity in semiconductor laser theory. Here we present some numerical results for the $\alpha$ factor, mainly to emphasize the importance of the carrier-induced refractive-index (phase shift) changes. Examples of the computed $\alpha$ spectra for bulk GaAs are shown in Fig. 3.9. Note the appearance of a saddle in the many-body results at high carrier densities. This saddle is observed in experiments in bulk and quantum-well gain media. Our theory associates the existence of this saddle with the many-body interactions.

Figure 3.9 (bottom) shows the different many-body contributions to the $\alpha$ spectra. Both bandgap renormalization and Coulomb enhancement have noticeable effects. A quantitative comparison between the experimental and theoretical spectra requires a more accurate description of the bandstructure than the two band model. Therefore, we postpone all experiment/theory comparisons to Chap. 7, where we include realistic band structures.

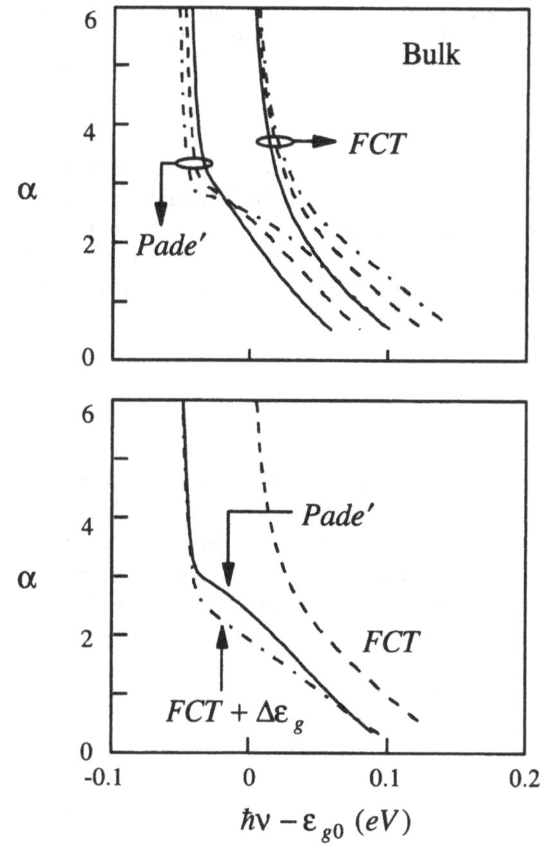


**Fig. 3.9.** The **top figure** shows the linewidth enhancement factor versus detuning $\hbar\nu - \varepsilon_{g0}$ according to the Padé approximation and the free-carrier theory. The carrier densities are $N = 3\times$ (*solid curves*), $4\times$ (*dashed curves*), and $5 \times 10^{18}\,\mathrm{cm}^{-3}$ (*dot-dashed curves*). The **bottom figure** shows the spectrum of the linewidth enhancement factor according to the Padé approximation (*solid line*), free-carrier theory (*dashed line*), and free-carrier theory with ad hoc bandgap renormalization (no Coulomb enhancement – *dash-dotted line*) at a carrier density $4 \times 10^{18}\,\mathrm{cm}^{-3}$

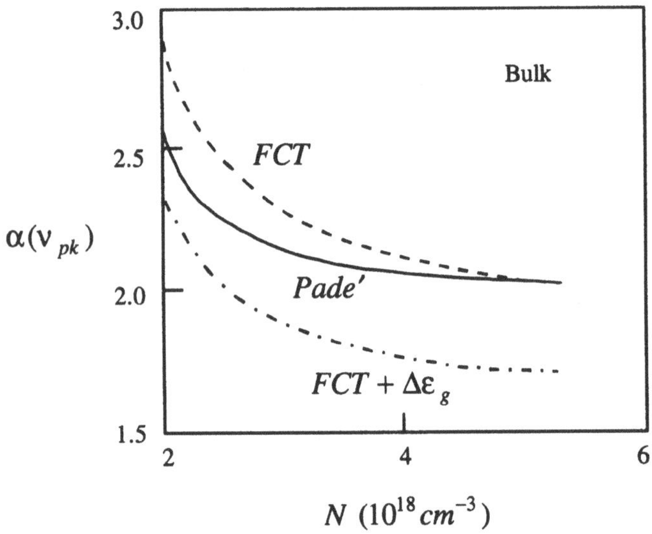


**Fig. 3.10.** Bulk-material linewidth enhancement factor $\alpha$ at the gain peak versus carrier density. The *solid curve* is the Padé approximation result, the *dash curve* is the free-carrier result, and the *dot-dashed curve* is obtained using the renormalized bandgap in the free-carrier theory

In Fig. 3.10, we show the value of $\alpha$ at the peak gain frequency as a function of carrier density. We see that the inclusion of the bandgap renormalization in the free-carrier theory actually yields worse agreement with the many-body results than the unmodified free-carrier theory. By ignoring the Coulomb enhancement, we significantly underestimate the values of $\alpha$ at the peak gain.

## 3.7 Quantum-Wells

To treat the many-body Coulomb interactions in a quantum well, we follow a procedure that is similar to our analysis of the bulk medium. Typically quantum-well widths for lasers are less than 15 nm, which is sufficiently narrow to approximate the carriers as a two-dimensional plasma. A derivation similar to Sect. 3.3 gives the screened Coulomb potential (3.62) with the two dimensional plasma frequency $\omega_{\mathrm{pl}}$,

$$\omega_{\mathrm{pl}}^2 = \frac{2\pi e^2 N_{2\mathrm{d}} w q}{\epsilon_{\mathrm{b}} m_{\mathrm{r}}} = \frac{8\pi N_{2\mathrm{d}} w a_0^3 \varepsilon_{\mathrm{R}}^2 q}{\hbar^2} \quad , \tag{3.91}$$

the effective plasmon frequency $\omega_q$,

$$\omega_q^2 = \omega_{\mathrm{pl}}^2 \left(1 + \frac{q}{\kappa}\right) + \frac{C}{4}\left(\frac{\hbar q^2}{2m_{\mathrm{r}}}\right) \quad , \tag{3.92}$$

and the inverse screening length,

$$\kappa = \frac{2\pi e^2}{\epsilon_b}\left(\frac{\partial N_{2d,e}}{\partial \mu_e} + \frac{\partial N_{2d,h}}{\partial \mu_h}\right) . \tag{3.93}$$

Here $N_{2d}$ is the two-dimensional carrier density introduced in (1.61). In the two-band and quasi-equilibrium approximations near the band edge, (3.93) yields

$$\kappa = \frac{2}{a_0}\left(\frac{m_e}{m_r} f_{e0} + \frac{m_h}{m_r} f_{h0}\right) , \tag{3.94}$$

where $f_{\alpha 0}$ denotes the Fermi-Dirac distribution at $k = 0$. This analytic result can be verified easily by using the explicit 2D expression (1.36) for the chemical potential and the Fermi-Dirac distribution for $k = 0$.

From (3.91–94) and (3.62), we find the screened Coulomb potential

$$V_{sq} = \frac{4\pi\varepsilon_R a_0}{Aq}\left(\frac{q}{\kappa} + \frac{C a_0 q^3}{32\pi N_{2d}}\right)\left(1 + \frac{q}{\kappa} + \frac{C a_0 q^3}{32\pi N_{2d}}\right)^{-1} , \tag{3.95}$$

where $q = |\boldsymbol{k} - \boldsymbol{k}'|$. Substituting (3.95) into (3.77) and evaluating the summation as a two-dimensional integral, we obtain

$$q(k) = -\frac{\mathrm{i}\varepsilon_R a_0}{\pi\hbar\kappa\mu_k}\int_0^\infty \mathrm{d}k'\, k' \mu_{k'} \frac{f_{ek'} + f_{hk'} - 1}{\mathrm{i}(\omega_{k'} - \nu) + \gamma}\Theta(k, k') , \tag{3.96}$$

where, unlike the bulk case (3.84), the angular integration function

$$\Theta(k, k') = \int_0^{2\pi} \mathrm{d}\theta \left(1 + \frac{C\kappa a_0 q^2}{32\pi N_{2d}}\right)\left(1 + \frac{q}{\kappa} + \frac{C a_0 q^3}{32\pi N_{2d}}\right)^{-1} \tag{3.97}$$

cannot be evaluated analytically.

Equations (3.66) and (3.68) for the quantum well give

$$\begin{aligned} \Delta\varepsilon_{CH} &= -2\varepsilon_R a_0 \int_0^\infty \frac{\mathrm{d}q}{1 + \frac{q}{\kappa} + \frac{C a_0 q^3}{32\pi N_{2d}}} \\ &\simeq -2\varepsilon_R a_0 \kappa \ln\left(1 + \sqrt{\frac{32\pi N_{2d}}{C\kappa^3 a_0}}\right) \end{aligned} \tag{3.98}$$

and

$$\Delta\varepsilon_{SX} = -\frac{2\varepsilon_R a_0}{\kappa}\int_0^\infty \mathrm{d}k\, k \frac{1 + \frac{C\kappa a_0 k^2}{32\pi N_{2d}}}{1 + \frac{k}{\kappa} + \frac{C a_0 k^3}{32\pi N_{2d}}}(f_{ek} + f_{hk}) , \tag{3.99}$$

respectively. Equations (3.96), (3.98) and (3.99) are then used in (3.79) to calculate the gain and carrier-induced phase shift. Converting the summation over states in (3.79) to a two-dimensional integral, we find

$$g - \mathrm{i}\frac{\mathrm{d}\phi}{\mathrm{d}z} = \frac{\nu}{2\pi\epsilon_0 n c\hbar w}\int_0^\infty \mathrm{d}k\, k\, |\mu_k|^2 \frac{f_{ek} + f_{hk} - 1}{\mathrm{i}(\omega_k - \nu) + \gamma}\frac{1}{1 - q(k)} . \tag{3.100}$$

Here, we used the quasi-equilibrium approximation, so that $f_{ek}$ and $f_{hk}$ are Fermi-Dirac distributions.

To study the many-body effects in a quantum-well gain medium, we first look at the simple two-band case. To isolate the quantum confinement effects, we choose the electron and hole effective masses to be the same as those used in the bulk case. Figure 3.11 is a plot of the gain and carrier-induced phase shift spectra for different carrier densities. The solid (dashed) curves are from the many-body (free-carrier) theory. As discussed in Chap. 2, band-edge effects are more pronounced in the quantum-well medium because of the step function instead of square root energy dependence of the density of states. Figure 3.11 (top) shows that the sharpness in the leading edge of the

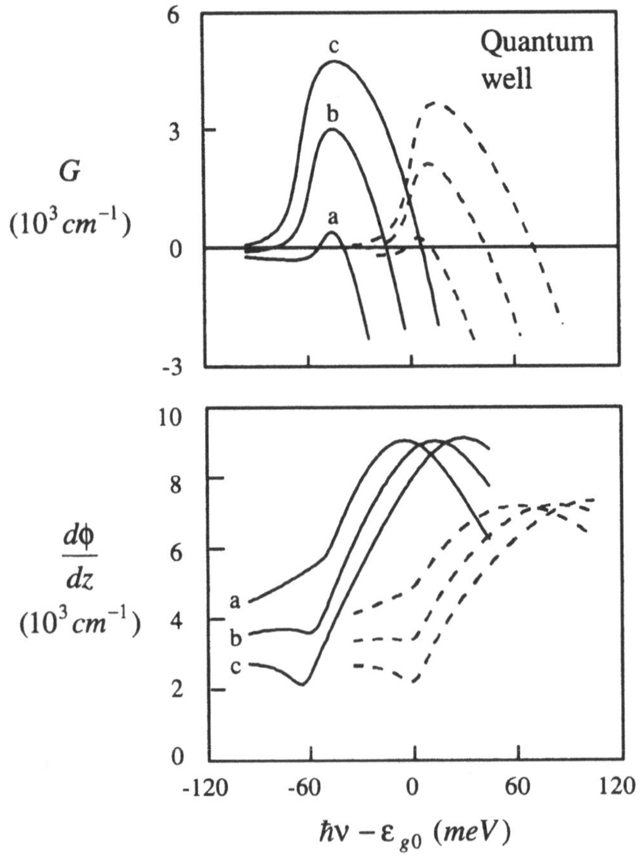


**Fig. 3.11.** Quantum-well gain (**top**) and carrier-induced phase shift (**bottom**) spectra for carrier densities $N_{2d}$ = (*a*) $1.5 \times 10^{12}\,\text{cm}^{-2}$, (*b*) $2 \times 10^{12}\,\text{cm}^{-2}$, and (*c*) $2.5 \times 10^{12}\,\text{cm}^{-2}$ the Padé approximation (*solid curves*), and free-carrier theory (*dashed curves*). We use the parameters for the quantum-well structure of Chap. 2

gain spectrum is smoothed out by the dephasing and Coulomb enhancement effects.

The effect of Coulomb enhancement is responsible for the differences between the solid (many-body) and dashed (free-carrier with bandgap renormalization) curves in Fig. 3.12. As expected, the many-body results (solid curve) yield a higher peak gain than the free carrier results (dashed curve). Both theories show gain rollover, which is not surprising since this rollover effect is caused by band filling with a two-dimensional density of states. Similar to the bulk situation, using the renormalized bandgap in the free-carrier theory only leads to a frequency shift in the free-carrier spectrum. In general, the presence of other subbands, which have different renormalized bandgaps, can lead to further reshaping of the spectrum and consequently to a different peak gain, (Chap. 7).

Figure 3.13 (top) plots the peak gain as a function of carrier density. Coulomb enhancement is responsible for the difference between the solid and dashed curves. Figure 3.13 (bottom) shows the dependence of the peak gain energy on carrier density. The results differ somewhat from those for the bulk medium (Fig. 3.8) because the step function energy density of states in the quantum-well medium reduces the band-filling contribution to the peak-gain frequency shift, so that the many-body interactions play a greater role. Consequently, the many-body (solid curve) and free-carrier (dashed curve)

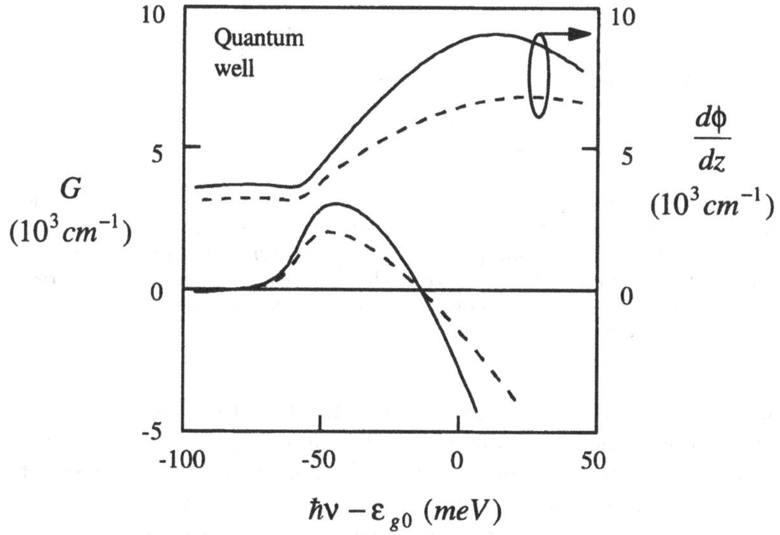


**Fig. 3.12.** Coulomb enhancement effects on gain (*lower curves*) and carrier-induced phase shift (*top curves*) spectra in a quantum-well medium according to the Padé approximation (*solid lines*) and the free-carrier theory with an ad hoc inclusion of bandgap renormalization (*dashed lines*). The curves are for a carrier density of $N_{2d} = 2 \times 10^{12}\,\mathrm{cm}^{-2}$

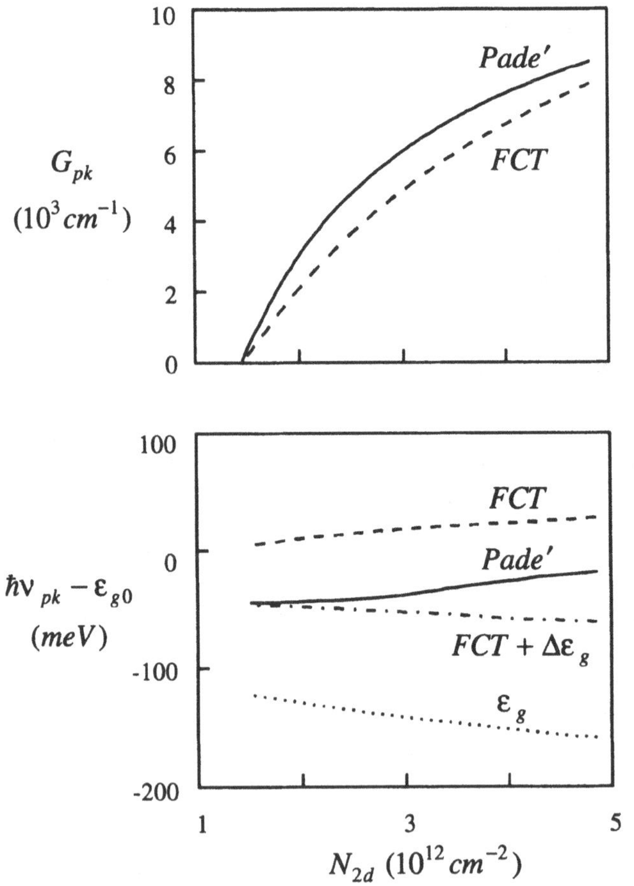


**Fig. 3.13.** The **top figure** shows the Padé approximation (*solid curve*) and free-carrier (*dashed curve*) predictions of quantum-well peak gain versus carrier density. The **bottom figure** shows the gain peak energy versus carrier density for the Padé approximation (*solid curve*), the free-carrier theory (*dashed curve*), and the free-carrier theory with renormalized bandgap (*dot-dashed curve*). The dotted curve is the renormalized bandgap

predictions of $\mathrm{d}\hbar\nu_{\mathrm{pk}}/\mathrm{d}N$ differ more than in the bulk case. Comparison of the solid and dot-dashed curves illustrates the opposing effects of Coulomb enhancement and bandgap renormalization. Without Coulomb enhancement the peak-gain frequency shifts red with increasing carrier density.

As alluded to earlier, the more pronounced features in the quantum-well gain and phase-shift spectra lead to noticeable differences in the $\alpha$ spectra, (2.110), for bulk and quantum-well materials. The Padé curves in Fig. 3.14 (top) show substantial structure in the quantum-well $\alpha$ spectra, especially at high carrier densities. Comparison with the free-carrier spectra (FCT)

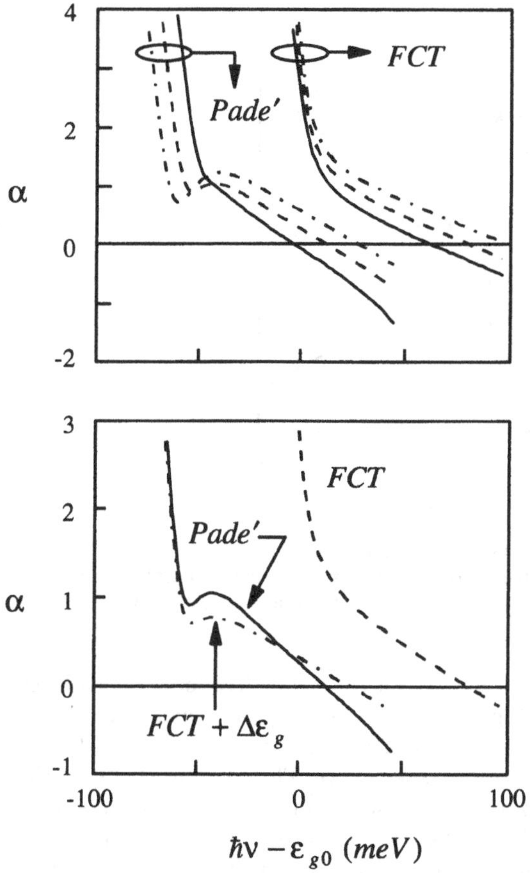


**Fig. 3.14.** The **top figure** shows the quantum-well linewidth-enhancement factor versus detuning, $\hbar\nu - \varepsilon_{g0}$ for carrier densities $N_{2d} = 1.5\times$ (*solid curve*), $2.0\times$ (*dashed curve*), and $2.5 \times 10^{12}\,\mathrm{cm}^{-2}$ (*dot-dashed curve*). The **bottom figure** shows the different screened Hartree-Fock contributions for a carrier density of $2.0 \times 10^{12}\,\mathrm{cm}^{-2}$. The solid curve is the Padé approximation result, the dashed curve is the free carrier result, and the dot-dashed curve is obtained using the renormalized bandgap in the free-carrier theory

reveals that the structures are completely due to the many-body interactions. Figure 3.14 (bottom) shows that both bandgap renormalization and Coulomb enhancement contribute to the $\alpha$ structure.

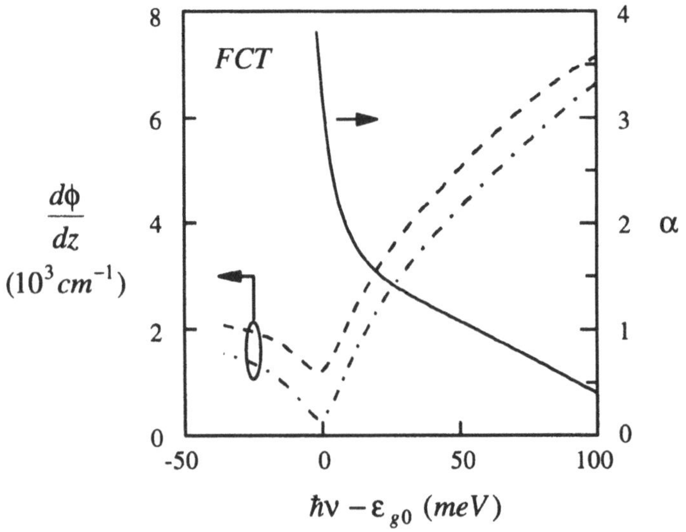


**Fig. 3.15.** Phase shift spectra for the carrier densities $N_{2d} = 3 \times 10^{12}\,cm^{-2}$ (*dashed curve*) and $3.5 \times 10^{12}\,cm^{-2}$ (*dot-dashed curve*), and the corresponding linewidth enhancement $\alpha$ factor spectrum (*solid curve*) for the free-carrier theory

The physical mechanisms responsible for the strong frequency dependence of $\alpha$ is analyzed in Fig. 3.15. Figure 3.15 shows the phase change $d\phi/dz$ for two carrier densities as predicted by the free-carrier theory. The dot-dashed curve is for a higher density than the dashed curve. The difference between the two curves gives the numerator of $\alpha$. The $\alpha$ spectrum is shown as the solid line. Bandgap renormalization shifts the two $d\phi/dz$ curves by different amounts. The higher density curve is shifted more and Fig. 3.16 (top) shows that this leads to variations in the separation between the two curves. Combined with the sharpness of features in $d\phi/dz$, we get the pronounced structure in the $\alpha$ spectrum. Figure 3.16 (bottom) shows that Coulomb enhancement tends to separate the curves, giving higher values for $\alpha$ than in Fig. 3.16 (top), but the structure in the spectrum remains [Fig. 3.16 (bottom)].

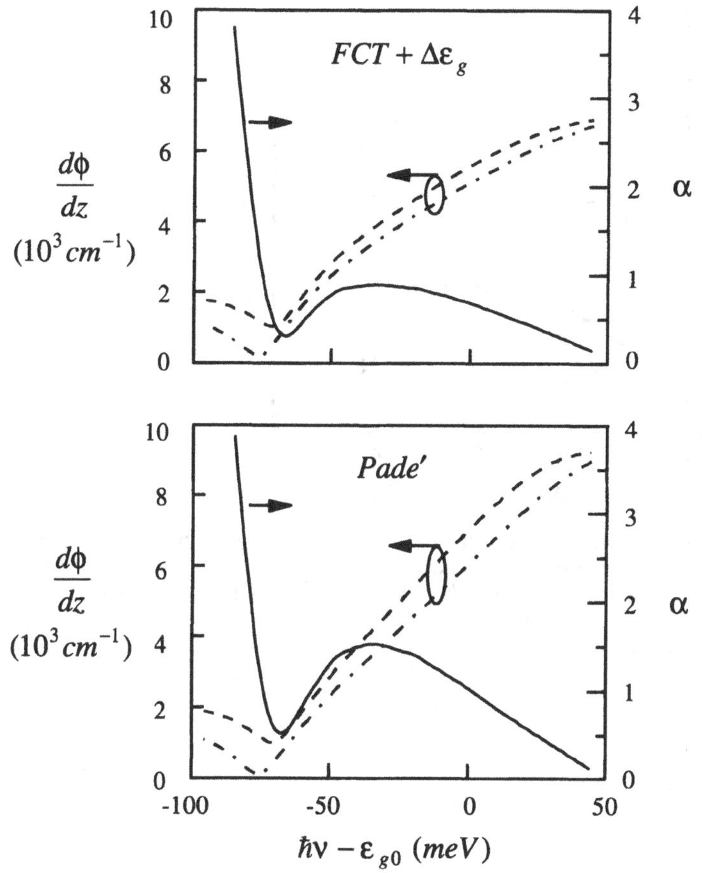


**Fig. 3.16.** Phase shift spectra for the carrier densities $N_{2d} = 3 \times 10^{12}\,\mathrm{cm}^{-2}$ (*dashed curve*) and $3.5 \times 10^{12}\,\mathrm{cm}^{-2}$ (*dot-dashed curve*), and the corresponding linewidth enhancement factor spectrum (*solid curve*). In the **top figure** we use the free-carrier theory with *ad hoc* inclusion of bandgap renormalization, and in the **bottom figure** we use the Padé approximation

Finally, Fig. 3.17 depicts the dependence of $\alpha$ at the peak gain frequency on carrier density. As in the bulk case, the ad hoc inclusion of bandgap renormalization to the free-carrier theory results in worse agreement with the many-body results.

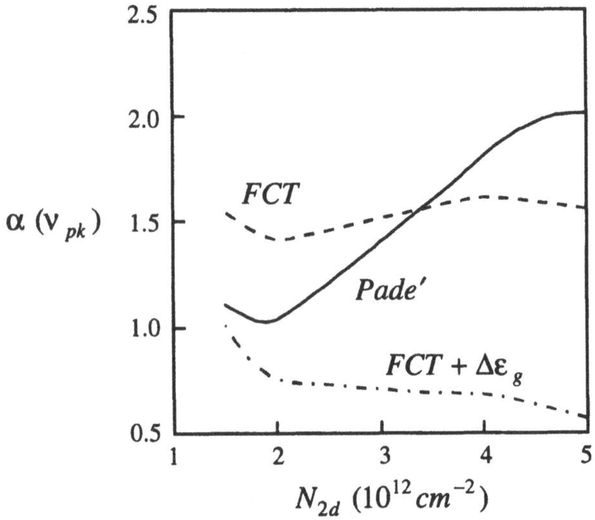


**Fig. 3.17.** Linewidth enhancement factor $\alpha$ at the peak gain frequency versus carrier density. The *solid curve* is the Padé approximation result, the *dashed curve* is the free-carrier result, and the *dot-dashed curve* is obtained using the renormalized bandgap in the free-carrier theory

# 4. Correlation Effects

The previous chapter introduced a systematic procedure to obtain the semiconductor Bloch equations, (3.17–19), with an unscreened Coulomb interaction and without explicit expressions for the collision terms. Systematic approximations to collision contributions, which lead to carrier and polarization relaxation, as well as plasma screening, can be computed at the next higher level of approximation. In this chapter, we outline an analysis of these contributions which, in the context of many-body theory, are customarily referred to as Coulomb correlations.

The detailed derivation of the correlation contributions requires aspects of many-body theory that are beyond the scope of this book [*Binder* and *Koch* (1995), and *Jahnke* et al. (1997)]. What we will present instead in Sect. 4.1 is a schematic outline. The intent is to give the reader a feeling for the origin of the different terms. Section 4.2 examines the results for the carrier density equations, which give the simplest version of the famous quantum Boltzmann equation for carrier-carrier scattering. Section 4.3 describes the corresponding result for the polarization equation.

Section 4.4 shows how the correlation contributions may be incorporated into the gain and refractive index calculations. We describe in some detail the conversion of the general polarization and associated equations to the ones describing specifically quantum-well or bulk structures. The subtleties involving the evaluation of some of the Coulomb contributions, and ways to reduce computation time are discussed. Also in this section, we use the two-band approximation to illustrate the contributions from the Coulomb correlations. We demonstrate the interplay between diagonal and nondiagonal contributions, and document the origin of the problems resulting from using an effective decay rate approximation. Computations of gain and refractive index spectra for actual experimental structures will be presented in Chap. 7, after we discuss the calculation of more realistic band structures in Chaps. 5, 6.

Section 4.5 extends the discussion on collisions to carrier-phonon scattering. The chapter ends with a simplified general discussion of characteristic relaxation times in Sect. 4.6.

## 4.1 Coulomb Correlation Effects

To study the correlation contributions in the semiconductor Bloch equations (3.17–19), we start by deriving equations of motion for quantities describing the deviations of the full correlation terms from their corresponding Hartree-Fock factorized parts. For example,

$$\delta\langle a_k^\dagger a_{k'-q}^\dagger a_{k-q} a_{k'}\rangle = \langle a_k^\dagger a_{k'-q}^\dagger a_{k-q} a_{k'}\rangle - \langle a_k^\dagger a_k\rangle\langle a_{k-q}^\dagger a_{k-q}\rangle\delta_{k,k'} \ , \quad (4.1)$$

whose time derivative is

$$\frac{\mathrm{d}}{\mathrm{d}t}\delta\langle a_k^\dagger a_{k'-q}^\dagger a_{k-q} a_{k'}\rangle = \left\langle \frac{\mathrm{d}}{\mathrm{d}t}\left(a_k^\dagger a_{k'-q}^\dagger a_{k-q}\, a_{k'}\right)\right\rangle$$
$$-\left[\left\langle \frac{\mathrm{d}}{\mathrm{d}t}\left(a_k^\dagger a_k\right)\right\rangle \langle a_{k-q}^\dagger a_{k-q}\rangle + \langle a_k^\dagger a_k\rangle \left\langle \frac{\mathrm{d}}{\mathrm{d}t}\left(a_{k-q}^\dagger a_{k-q}\right)\right\rangle\right]\delta_{k,k'} \ . \quad (4.2)$$

From the Heisenberg equation of motion, we find

$$\frac{\mathrm{d}}{\mathrm{d}t}\delta\langle a_k^\dagger a_{k'-q}^\dagger a_{k-q}\, a_{k'}\rangle$$
$$= \frac{\mathrm{i}}{\hbar}\delta\langle a_k^\dagger a_{k'-q}^\dagger a_{k-q}\, a_{k'}\rangle\Delta\varepsilon_{ekk'q} + \frac{\mathrm{d}}{\mathrm{d}t}\delta\langle a_k^\dagger a_{k'-q}^\dagger a_{k-q} a_{k'}\rangle\bigg|_{\mathrm{Coul}} \ , \quad (4.3)$$

where we used the full electron-hole Hamiltonian (3.6) except for the interaction term for the carriers with the laser field, since it does not play a role in the collisions. Furthermore, we introduced the abbreviations

$$\Delta\varepsilon_{ekk'q} = \varepsilon_{ek} + \varepsilon_{e,k'-q} - \varepsilon_{e,k-q} - \varepsilon_{ek'} \quad (4.4)$$

and

$$\mathrm{i}\hbar\frac{\mathrm{d}}{\mathrm{d}t}\delta\langle a_k^\dagger a_{k'-q}^\dagger a_{k-q} a_{k'}\rangle\bigg|_{\mathrm{Coul}} = \langle [H_C, a_k^\dagger a_{k'-q}^\dagger a_{k-q} a_{k'}]\rangle \ . \quad (4.5)$$

Evaluation of the commutator in (4.5) leads to expressions containing products of up to six operators that are too lengthy to show here.

Formally integrating (4.3), we get

$$\delta\langle a_k^\dagger a_{k'-q}^\dagger a_{k-q} a_{k'}\rangle(t)$$
$$= \int_{-\infty}^{t} \mathrm{d}t' \exp\left[\left(\frac{\mathrm{i}}{\hbar}\Delta\varepsilon_{ekk'q} - \gamma\right)(t-t')\right]\frac{\mathrm{d}}{\mathrm{d}t'}\delta\langle a_k^\dagger a_{k'-q}^\dagger a_{k-q} a_{k'}\rangle\bigg|_{\mathrm{Coul}} \ , \quad (4.6)$$

where $\gamma$ is a phenomenological decay constant that has been added so that the integral vanishes at the lower boundary. In general, the correlation at time $t$ depends on the evolution of the system from $-\infty$ to $t$. This is the Coulombic *memory effect.* At the high carrier densities needed to obtain plasma gain, it is reasonable to assume that the memory time is not very long. Therefore, as the simplest approximation, we neglect memory effects altogether and make the *Markov approximation.* Technically, this amounts to assuming that the Coulomb contribution is slowly varying compared to the exponential, so that we can move it outside the integral in (4.6). The resulting integral can be readily evaluated to give

$$\delta\langle a_k^\dagger a_{k'-q}^\dagger a_{k-q}\, a_{k'}\rangle \simeq \frac{\mathrm{d}}{\mathrm{d}t}\delta\langle a_k^\dagger a_{k'-q}^\dagger a_{k-q} a_{k'}\rangle\Big|_{\mathrm{Coul}} \frac{1}{\mathrm{i}\Delta\varepsilon_{ekk'q}/\hbar - \gamma} \quad . \tag{4.7}$$

A similar result is obtained for the other four-operator terms which appear in the equations of motion (3.9–11).

At this stage we still do not have a closed set of equations because of the six-operator expectation values occuring in $\mathrm{d}\langle a_k^\dagger a_{k'-q}^\dagger a_{k-q} a_{k'}\rangle/\mathrm{d}t|_{\mathrm{Coul}}$. The equation of motion for these six-operator terms introduces eight-operator terms, i.e., the many-body hierarchy problem mentioned earlier. In order to close the equations we now make again a factorization approximation. We factorize all the six- and four-operator terms which occur in the Coulomb parts to obtain the simplest possible expression for the scattering terms. The detailed calculation yields the following results for the electron population equation:

$$\frac{\partial n_{ek}}{\partial t}\Big|_{\mathrm{col}} = -n_{ek}\Sigma_{ek}^{\mathrm{out}}\{n\} + (1-n_{ek})\,\Sigma_{ek}^{\mathrm{in}}\{n\} \quad , \tag{4.8}$$

where we ignored terms containing scattering contributions involving interband polarizations since in this book we mainly deal with situations where higher order coherent polarization contributions are of minor importance. The rates $\Sigma_{ek}^{\mathrm{out}}\{n\}$ and $\Sigma_{ek}^{\mathrm{in}}\{n\}$, describing the effective scattering out of and into the state $\boldsymbol{k}$, are given by

$$\begin{aligned}\Sigma_{ek}^{\mathrm{out}}\{n\} &= \frac{\pi}{\hbar}\sum_{b=e,h}\sum_{q\neq 0}\sum_{k'}\left(2V_q^2 - \delta_{e,b}V_q V_{|k-k'+q|}\right)\\ &\quad\times\delta(\varepsilon_{e,k}+\varepsilon_{b,k'}-\varepsilon_{e,k+q}-\varepsilon_{b,k'-q})\\ &\quad\times(1-n_{e,k+q})\,n_{b,k'}(1-n_{b,k'-q}) \quad ,\end{aligned} \tag{4.9}$$

and

$$\begin{aligned}\Sigma_{ek}^{\mathrm{in}}\{n\} &= \frac{\pi}{\hbar}\sum_{b=e,h}\sum_{q\neq 0}\sum_{k'}\left(2V_q^2 - \delta_{e,b}V_q V_{|k-k'+q|}\right)\\ &\quad\times\delta(\varepsilon_{e,k}+\varepsilon_{b,k'}-\varepsilon_{e,k+q}-\varepsilon_{b,k'-q})\\ &\quad\times n_{e,k+q}(1-n_{b,k'})\,n_{b,k'-q} \quad .\end{aligned} \tag{4.10}$$

The notation $\Sigma\{n\}$ symbolizes the functional dependence of these rates on the electron and hole distribution functions. The corresponding equations for the hole population $n_{hk}$ is obtained by the interchange $e \rightleftharpoons h$ in (4.8–10).

It is sometimes convenient to write the total relaxation rate of $n_{\alpha k}$ as a single decay rate

$$\gamma_{\alpha k}\{n\} \equiv \Sigma_{\alpha k}^{\mathrm{out}}\{n\} + \Sigma_{\alpha k}^{\mathrm{in}}\{n\} \quad , \tag{4.11}$$

in terms of which the scattering integral can be written as

$$\left.\frac{\partial n_{ek}}{\partial t}\right|_{\rm col} = -\gamma_{\alpha k}\{n\} n_{\alpha k} + \Sigma_{\alpha k}^{\rm in}\{n\} \ . \tag{4.12}$$

Here we see explicitly that scattering both into and out of the state $\boldsymbol{k}$ are relaxation processes, although one increases the probability $n_{\alpha k}$, while the other one decreases it.

For the collision terms in the equation for the interband polarization, we obtain

$$\left.\frac{\partial p_{k}}{\partial t}\right|_{\rm col} = -\sum_{k'} \Lambda_{kk'} p_{k'} \ , \tag{4.13}$$

where we again kept only the terms that are linear in the polarization. For $\boldsymbol{k} = \boldsymbol{k}'$,

$$\begin{aligned} \Lambda_{kk} = & \frac{1}{\hbar} \sum_{a,b=e,h} \sum_{k''} \sum_{q\neq 0} \left(2V_q^2 - \delta_{a,b} V_q V_{|k-k''+q|}\right) g(\delta\varepsilon) \\ & \times \left[ n_{{\rm e},k+q} \left(1 - n_{{\rm b}k''}\right) n_{{\rm b},k''-q} + \left(1 - n_{{\rm e},k+q}\right) n_{{\rm b}k''} \left(1 - n_{{\rm b},k''-q}\right) \right] \ . \end{aligned} \tag{4.14}$$

Here, we used the abbreviation

$$\delta\varepsilon = \varepsilon_{ak} + \varepsilon_{{\rm b}k''} - \varepsilon_{a,k+q} - \varepsilon_{{\rm b},k''-q} \ , \tag{4.15}$$

and the generalized $\delta$-function (Heitler Zeta function) is

$$g(x) = \lim_{\gamma=0} \frac{\rm i}{x + {\rm i}\gamma} = \pi\delta\left(x\right) + {\rm i}P\left(\frac{1}{x}\right) \ . \tag{4.16}$$

The expression $P\left(1/x\right)$ stands for the principal value (PV) integral and we ignore a pure exchange term in $\underline{\Lambda}$ which gives only minor corrections under laser conditions. The full scattering terms, including also the higher order polarization contributions, can be found in *Jahnke* et al. (1997). For $\boldsymbol{k} \neq \boldsymbol{k}'$,

$$\begin{aligned} \Lambda_{kk'} = & \frac{1}{\hbar} \sum_{a,b=e,h} \sum_{k''} \left(2V_q^2 - \delta_{a,b} V_q V_{|k-k''+q|}\right) g(-\delta\varepsilon) \\ & \times \left[ \left(1 - n_{ak}\right)\left(1 - n_{{\rm b}k''}\right) n_{{\rm b},k''-q} + n_{ak} n_{{\rm b}k''} \left(1 - n_{{\rm b},k''-q}\right) \right] \ , \end{aligned} \tag{4.17}$$

where $\boldsymbol{q} = \boldsymbol{k}' - \boldsymbol{k}$.

Equation (4.13) has been written already in a form showing that the scattering matrix $\underline{\Lambda}$ adds to the Hartree-Fock matrix $\underline{\Theta}$ of (3.22). In fact, the diagonal and nondiagonal terms in $\underline{\Lambda}$ are the second order (in the Coulomb potential) contributions to the energy and field renormalization, respectively. As mentioned before, the general matrix $\underline{\Lambda}$ contains also terms where one or more of the population factors are replaced by polarizations [*Jahnke* et al. (1997)]. Those terms are important under coherent nonlinear excitation conditions, where a large induced interband polarization is present. However, for most semiconductor laser applications where an incoherent electron-hole plasma exists, these contributions can be ignored.

## 4.2 Carrier Quantum Boltzmann Equation

The collision contributions in the carrier equation (4.8) are actually the simplest version of the famous *quantum Boltzmann scattering integral* for carrier-carrier collisions. The different terms are illustrated schematically in Fig. 4.1. One remarkable feature of (4.8) is that for $n_{\alpha k} = f_{\alpha k}$, i.e., if the carriers are in Fermi-Dirac distributions, the Boltzmann scattering integral is identically zero. This implies that

$$f_{\alpha k} \Sigma_{\alpha k}^{\text{out}}\{f\} = (1 - f_{\alpha k})\, \Sigma_{\alpha k}^{\text{in}}\{f\} \quad , \tag{4.18}$$

for nonvanishing in- and out-scattering rates. Equation (4.18) describes a condition called *detailed balance*, for which the scattering into each state is exactly balanced by the scattering out of that state. This is the quasi-equilibrium situation mentioned in the earlier chapters. It is important to realize that even though the distribution functions are time independent, i.e., Fermi-Dirac, this does not imply the absence of scattering events. The individual terms in (4.18) are nonzero and rather large. However, they exactly balance each other.

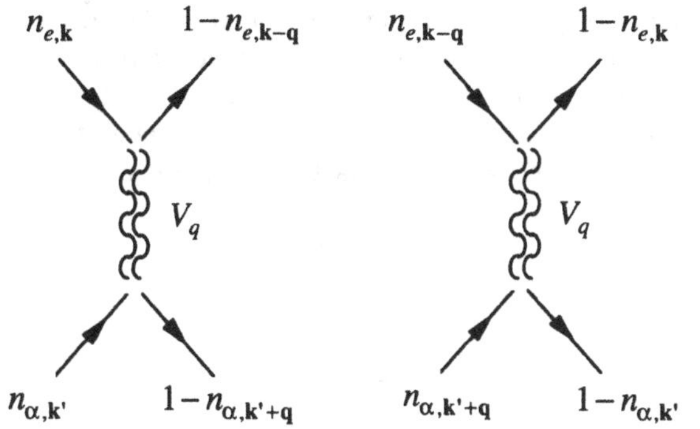


**Fig. 4.1.** Scattering into (**right diagram**) and out of (**left diagram**) the $\boldsymbol{k}^{th}$ electronic state. A scattering partner is needed and the final states have to be available

Generally, there are a number of physical quantities that are conserved in the carrier-carrier scattering processes. These conservation rules can be written as

$$\left.\frac{\partial}{\partial t}\right|_{\text{col}} \left[\sum_{k} F_i\,(k)\, n_{\alpha k}\right] = 0 \quad , \qquad i = 1, 2, \ldots 5 \quad ; \quad \alpha = e, h \tag{4.19}$$

with

$$F_1 = 1 \tag{4.20}$$

$$F_2 = k_x \quad , \qquad F_3 = k_y \quad , \qquad F_4 = k_z \tag{4.21}$$

$$F_5 = k^2 \quad . \tag{4.22}$$

Equations (4.19, 20) correspond to total particle number conservation, (4.19, 21) to total momentum conservation, and (4.19, 22) to total kinetic energy conservation.

Since one typically does not encounter a drifting plasma in a semiconductor gain medium, the total momentum is originally zero, and because of (4.19, 21) it will remain zero. To see the implications of the other conservation rules, let us consider the example of nonequilibrium carrier relaxation experiments performed using femtosecond ($10^{-15}$ s) pulse excitation of semiconductor interband transitions. In these investigations, the initially prepared nonequilibrium carrier distribution is rapidly modified by carrier-carrier collisions so that it approaches the Fermi-Dirac distribution. Under most conditions, the carrier-carrier equilibration processes occur very rapidly, at a sub-picosecond timescale. This situation is depicted in Fig. 4.2, where the electron distributions are plotted at different times after the excitation. The curves are numerical solutions of the full carrier-carrier Boltzmann equation (4.8), including electron-electron, electron-hole, and hole-hole scattering. The carrier distributions immediately after the excitation pulse is shown by the solid curve in Fig. 4.2. The other curves are snapshots during the evolution of the carrier populations towards a Fermi-Dirac distribution. Because of kinetic energy conservation, the plasma temperatures of the relaxed distribution is determined by the kinetic energy of the earlier nonequilibrium distribution. As a result, depending on the excess energy of the excitiation pulse (i.e., how high above the band minimum the carriers are generated), the effective plasma temperature is well above the lattice temperature.

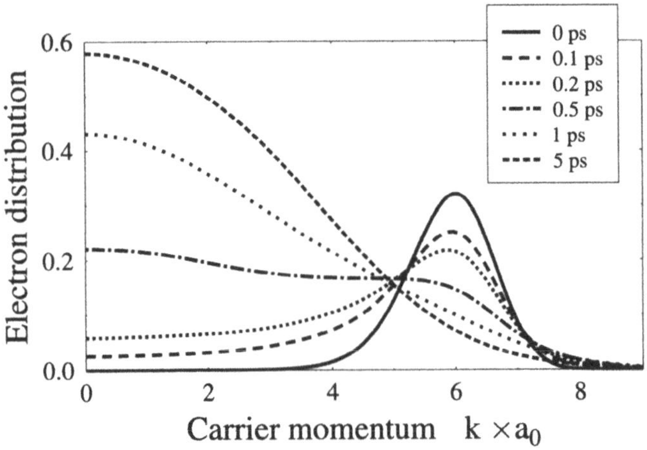


**Fig. 4.2.** Relaxation of femtosecond pulse excited electron distribution. The times measured relative to the center of the excitation pulse are shown in the inset. Only carrier-carrier scattering is included in the calculations. From [*Jahnke* and *Koch* (1995)]

Relaxation of the electron and hole kinetic energies (plasma cooling) happens only by collisons with other quasi-particles, most importantly with phonons. The corresponding carrier-phonon Boltzmann equation will be discussed in Sect. 5 of this chapter. However, to illustrate the effect we plot already here the results obtained by solving the carrier-phonon Boltzmann equation (4.93) alone (Fig. 4.3) and together with the carrier-carrier Boltzmann equation (4.8) (Fig. 4.4). We include only scattering of carriers with longitudinal optical (LO) phonons since this is the most efficient energy relaxation mechanism for carrier distributions with large excess energy. Since the LO phonon energy $\hbar\omega_{\mathrm{LO}}$ has a discrete value, the carrier relaxation of an initial nonequilibrium distribution (solid line in Fig. 4.3), which occurs via successive emission of LO phonons, leads to the occurence of sidebands in the carrier distribution at the energies $\varepsilon - n\hbar\omega_{\mathrm{LO}}$, $n = 1, 2, \ldots$. If one simultaneously includes carrier-carrier scattering, as it occurs under realistic conditions, the discrete phonon sidebands disappear, see Fig. 4.4. However, now the carrier distributions relax to a Fermi-Dirac distribution at the lattice (LO-phonon) temperature.

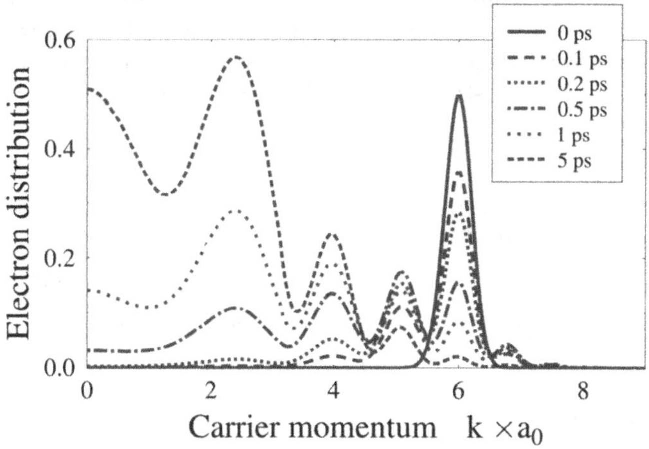


**Fig. 4.3.** Same as Fig. 4.2 but with only carrier-LO-phonon scattering included. From [*Jahnke* and *Koch* (1995)]

In Fig. 4.5 we study a situation where the initial carrier distribution is basically a Fermi-Dirac distribution which however, is locally disturbed in $\boldsymbol{k}$-space. Such a situation is relevant for lasers if we consider, e.g., the situation of a single mode burning a kinetic hole into the distribution function. Figure 4.5a shows the rapid relaxation of the disturbed distribution function back to quasi-equilibrium once the perturbation is switched off. Figure 4.5b shows the corresponding carrier-carrier scattering rates $\gamma_{\alpha k}$ defined in (4.11). We see that typical scattering times are of the order of 50–100 fs.

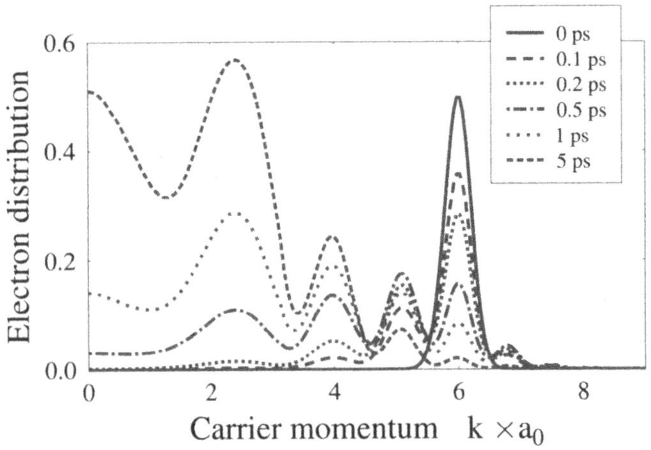


**Fig. 4.4.** Same as Fig. 4.2 but with carrier-carrier and carrier-LO-phonon scattering included. From [*Jahnke* and *Koch* (1995)]

To get an approximate expression for the intraband relaxation rate of (2.31), we take a closer look at a general situation where, as in Fig. 4.5, we have a nonequilibrium carrier distribution that is sufficiently close to the quasi-equilibrium Fermi-Dirac distributions. Under this condition we can set

$$\Sigma^{\text{in}}_{\alpha\boldsymbol{k}}\{n\} \simeq \Sigma^{\text{in}}_{\alpha\boldsymbol{k}}\{f\} \quad ,$$
$$\Sigma^{\text{out}}_{\alpha\boldsymbol{k}}\{n\} \simeq \Sigma^{\text{out}}_{\alpha\boldsymbol{k}}\{f\} \quad , \tag{4.23}$$

in (4.8). Furthermore, substituting

$$\Sigma^{\text{in}}_{\alpha\boldsymbol{k}}\{f\} = \gamma_{\alpha\boldsymbol{k}}\{f\} f_{\alpha\boldsymbol{k}} \quad , \tag{4.24}$$

which is a simple rearrangement of (4.12) under detailed balance conditions, we find

$$\left.\frac{\partial n_{\alpha\boldsymbol{k}}}{\partial t}\right|_{\text{col}} \simeq -\gamma_{\alpha\boldsymbol{k}}\{f\}\,(n_{\alpha\boldsymbol{k}} - f_{\alpha\boldsymbol{k}}) \quad . \tag{4.25}$$

This approximation fails (barely) to preserve the total carrier density $N$. To remedy this defect, we study interactions in the neighborhood of $\boldsymbol{k}_0$, and choose $\gamma_{\alpha\boldsymbol{k}_0}\{f\}$ instead of $\gamma_{\alpha\boldsymbol{k}}\{f\}$, that is

$$\left.\frac{\partial n_{\alpha\boldsymbol{k}}}{\partial t}\right|_{\text{col}} \simeq -\gamma_{\alpha\boldsymbol{k}_0}\{f\}\,(n_{\alpha\boldsymbol{k}} - f_{\alpha\boldsymbol{k}}) \quad . \tag{4.26}$$

which is basically the result given in (2.31). This expression conserves the total carrier density since $\sum_{\boldsymbol{k}} n_{\alpha\boldsymbol{k}} = \sum_{\boldsymbol{k}} f_{\alpha\boldsymbol{k}} = VN$, and is often used to approximate collision terms in the carrier distribution equations of motion.

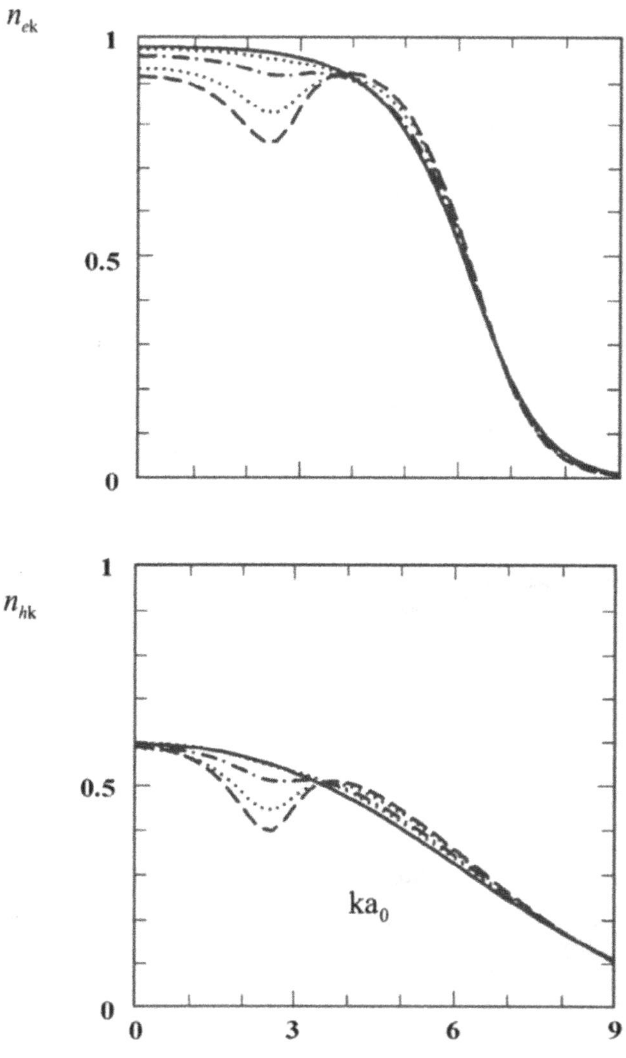


**Fig. 4.5a.** Relaxation of disturbed Fermi distribution functions for electrons (**top**) and holes (**bottom**) at a density $N = 3 \times 10^{18}\,\mathrm{cm}^{-3}$ and temperature $T \simeq 300\,\mathrm{K}$ obtained by numerically solving the Boltzmann equation using the dynamically screened Coulomb potential in RPA approximation. The times are: $t = 0$ (*long dashed*), 21 fs (*dotted*), 75 fs (*dash-dotted*), 147 fs (*dotted*), 796 fs (*solid*). From [*Binder* et al. (1992)]

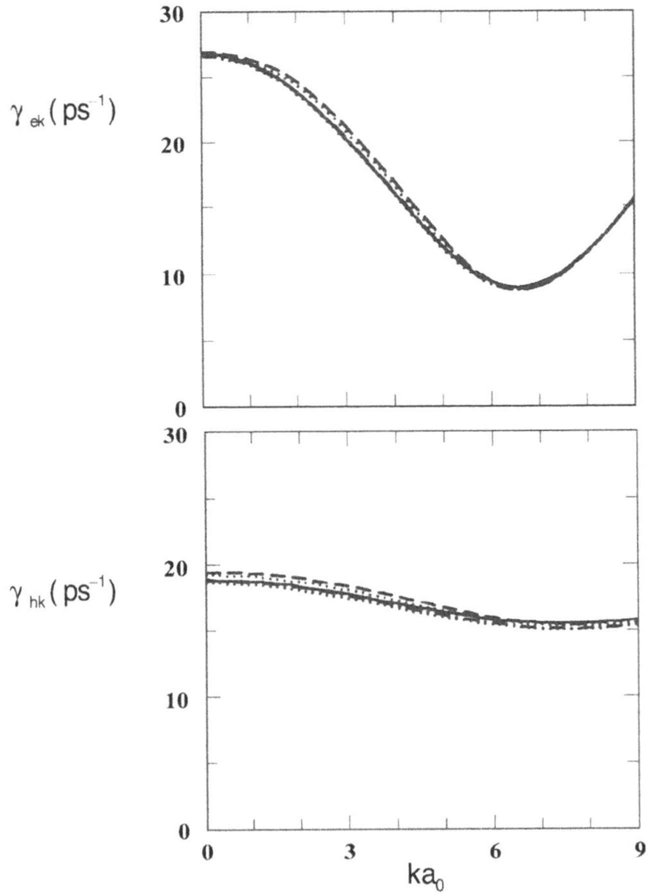


**Fig. 4.5b.** Carrier-carrier scattering rates extracted from Fig. 4.5a. (**top**) electrons, (**bottom**) holes. From [*Binder* et al. (1992)]

## 4.3 Dephasing and Screening

The collision terms in the dynamic equation for the interband polarization include such effects as screening of the Hartree-Fock terms and decay of the total polarization, i.e., optical dephasing. Using (4.16) in (4.14), we can write the diagonal part of $\underline{\Lambda}$ as

$$\Lambda_{\boldsymbol{k}\boldsymbol{k}} = -\mathrm{i}\Delta_{\boldsymbol{k}} + \Gamma_{\boldsymbol{k}} \quad , \tag{4.27}$$

and the nondiagonal part becomes

$$\Lambda_{\boldsymbol{k}\boldsymbol{k}'} = \mathrm{i}\Delta_{\boldsymbol{k}\boldsymbol{k}'} + \Gamma_{\boldsymbol{k}\boldsymbol{k}'} \quad . \tag{4.28}$$

Since $\Lambda_{kk}$ has to be added to $\Theta_{kk}$, (3.23), in the full dynamic (3.17) for the interband polarization, we see that formally, $\Gamma_k$ describes a momentum dependent diagonal dephasing rate and $\Delta_k$ yields the corresponding corrections ($\propto V_q^2$) to the Hartree-Fock renormalizations of the free-particle energies.

In the same way, $\Delta_{kk'}$ and $\Gamma_{kk'}$ yield momentum dependent nondiagonal damping and shift contributions. It is interesting to note that

$$\sum_k \left.\frac{\partial p_k}{\partial t}\right|_{\text{col}} = -\sum_{kk'} \Lambda_{kk'} p_{k'} = 0 \quad , \tag{4.29}$$

which demonstrates that Coulomb dephasing of the interband polarization is a pure interference phenomenon.

The detailed analysis shows, see e.g. *Jahnke* et al. (1997), that the nondiagonal dephasing contributions have the effect of partially compensating the influence of the diagonal part. Because this compensation is significant, it is crucial to treat both diagonal and nondiagonal terms symmetrically. The relaxation rate approximation used in Chaps. 2, 3 cannot be justified at the microscopic level, so that a gain expression based on a purely diagonal description of dephasing can only be regarded as a *fit* of the microscopic results, with the dephasing rate treated as a phenomenological input parameter. As such, it is important that one *should not* use (4.27) to compute a dephasing rate. Rather, the dephasing rate should be chosen to account for the effects of both the $\Gamma_k$ and the $\Gamma_{k,k'}$ contributions.

## 4.4 Formulation of Numerical Problem

Under small signal gain conditions,

$$\frac{\mathrm{d}n_{ek}}{\mathrm{d}t} = \frac{\mathrm{d}n_{hk}}{\mathrm{d}t} = 0 \quad . \tag{4.30}$$

These conditions simplify the calculations considerably, since we only have to deal with the polarization equation of motion. Combining (3.22) and (4.13) gives

$$\frac{\mathrm{d}p_k}{\mathrm{d}t} = -\mathrm{i}\omega_k' p_k - \frac{\mathrm{i}}{\hbar}\mu_k E(\boldsymbol{R},t)\left(f_{ek} + f_{hk} - 1\right) - \mathrm{i}\sum_{k'}\left(\Theta_{kk'} - \mathrm{i}\Lambda_{kk'}\right)p_{k'} \quad , \tag{4.31}$$

where $\omega_k'$ is the unrenormalized transition energy of (3.12), and we have assumed quasi-equilibrium conditions. The Fermi-Dirac distributions $f_{ek}$ and $f_{hk}$ are given by (1.19), using (1.20) for a fixed value of the total carrier density $N$, which is an input parameter. For the numerical analysis it is convenient to remove the rapidly varying phase factor from $p_k$ and to work with a slowly varying polarization amplitude

$$s_k = p_k \, \mathrm{e}^{\mathrm{i}\nu t} \quad . \tag{4.32}$$

Substituting (4.32, 3.26) with $\boldsymbol{R}=0$ for $E(\boldsymbol{R},t)$, into (4.31) gives

$$\frac{\mathrm{d}s_{\boldsymbol{k}}}{\mathrm{d}t} = \mathrm{i}\left(\nu - \omega'_{\boldsymbol{k}}\right) s_{\boldsymbol{k}} - \frac{\mathrm{i}}{\hbar}\mu_{\boldsymbol{k}} E_0 \left(f_{\mathrm{e}k} + f_{\mathrm{h}k} - 1\right) - \mathrm{i}\sum_{\boldsymbol{k}'} \left(\Theta_{\boldsymbol{k}\boldsymbol{k}'} - \mathrm{i}\Lambda_{\boldsymbol{k}\boldsymbol{k}'}\right) s_{\boldsymbol{k}'} \ , \tag{4.33}$$

where the laser field amplitude $E_0$ is an input to the computations.

Figure 4.6 shows a flow diagram for computing the gain and refractive index at some given carrier density and laser frequency. The approach is based on the numerical solution of (4.33). First, we read the input parameters, from which the first two terms on the right-hand side of (4.33) are computed. In the next steps, we evaluate $\Theta_{\boldsymbol{k}\boldsymbol{k}'}$ and $\Lambda_{\boldsymbol{k}\boldsymbol{k}'}$. In the numerical analysis, the functions $f_{\mathrm{e}k}$, $f_{\mathrm{h}k}$ and $s_{\boldsymbol{k}}$ are stored for a finite number of discrete $\boldsymbol{k}$ points. To discretize these functions, one needs to choose the appropriate step sizes and cut-offs. After doing so, $\Theta_{\boldsymbol{k}\boldsymbol{k}'}$ and $\Lambda_{\boldsymbol{k}\boldsymbol{k}'}$ are evaluated, which yields the last

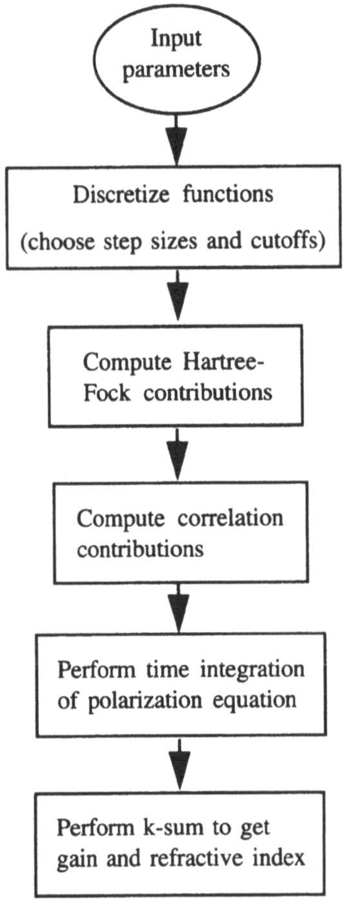


**Fig. 4.6.** Flow diagram for the numerical calculation of gain and refractive index

two terms in (4.33). Equation (4.33) is solved for the steady steady solution which is then used in (2.7, 1.76, 79) to determine the gain and refractive index change.

To begin the numerical analysis, we first convert the carrier momentum summations contained in the various terms in (4.33) into integrals. As in the previous chapters, this conversion separates the bulk and quantum-well analysis. We first consider the quantum-well case.

### 4.4.1 Quantum-Wells

In a quantum-well structure, the assumption of a continuous carrier momentum distribution leads to the replacement

$$\sum_{k'} \to \frac{4\pi^2}{A} \int_0^\infty \mathrm{d}k'\, k' \int_0^{2\pi} \mathrm{d}\phi' \ , \tag{4.34}$$

which is a generalization of (1.59), to allow for angular dependences in the integrand. To evaluate the integrals one can use, for example, the trapezoidal rule. This simple method involves discretizing the integration variable and then approximating the integral as a sum of trapezoidal areas. An integral $S$ of the function $y(x)$ then becomes

$$\begin{aligned} S &= \int_{x_{\min}}^{x_{\max}} \mathrm{d}x\, y(x) \\ &\simeq \sum_{k=2}^{N} \frac{1}{2}(y_k + y_{k-1})(x_k - x_{k-1}) \ , \end{aligned} \tag{4.35}$$

where $x_N = x_{\max}$ and $x_1 = x_{\min}$. We can streamline the computations by rewriting (4.35) as

$$\begin{aligned} S &= \sum_{k=2}^{N} \frac{1}{2} y_k (x_k - x_{k-1}) + \sum_{k=1}^{N-1} \frac{1}{2} y_k (x_{k+1} - x_k) \\ &= \sum_{k=2}^{N-1} y_k \frac{x_{k+1} - x_{k-1}}{2} + y_N \frac{x_N - x_{N-1}}{2} + y_1 \frac{x_2 - x_1}{2} \ , \end{aligned} \tag{4.36}$$

which is computationally faster because each summation step requires only $y_k$ instead of $y_k$ and $y_{k-1}$. For equal step sizes, i.e.,

$$\Delta x = \frac{x_k - x_{k-1}}{2} \qquad \text{for } 2 \le k \le N \ , \tag{4.37}$$

(4.36) becomes

$$S = \Delta x \left( \sum_{k=2}^{N-1} y_k + \frac{y_N + y_1}{2} \right) \ . \tag{4.38}$$

**Step Sizes.** To apply the trapezoidal rule, one begins by choosing an appropriate step size $\Delta x$, and in the case of an indefinite integral, the cut-offs $x_1$ and $x_N$ as well. The $\phi'$ integration in (4.34) is often straightforward, since it involves a definite integral, where the integrand usually does not have a strong angular dependence. On the other hand, the $k'$ integration requires some care. Here, the desire to minimize computation time by maximizing the step size $\mathrm{d}k$ and minimizing the cut-off $k_{\max}$, should be balanced by having a sufficiently small $\mathrm{d}k$ and a sufficiently large $k_{\max}$ in order to accurately describe $f_{\mathrm{e}k}$, $f_{\mathrm{h}k}$ and $s_{\boldsymbol{k}}$.

Since $s_{\boldsymbol{k}}$ is not known at the start of a calculation, one has to base the initial guesses for $dk$ and $k_{\max}$ on $f_{\mathrm{e}k}$ or $f_{\mathrm{h}k}$. For each distribution $f_{\alpha k}$, we calculate a momentum

$$k_{\alpha f} = \frac{\sqrt{2m_\alpha \mu_\alpha}}{\hbar} \ , \tag{4.39}$$

where $m_\alpha$ and $\mu_\alpha$ are the effective mass and chemical potential, respectively. Since $k_{\mathrm{e}f}$ and $k_{\mathrm{h}f}$ are usually different, we define a representative $k_f$, which may, for example, be the larger one of $k_{\mathrm{e}f}$ and $k_{\mathrm{h}f}$. The step size $\mathrm{d}k$ is then determined by specifying the number of intervals $n_f$ within $k_f$, i.e.,

$$\mathrm{d}k = \frac{k_f}{n_f} \ . \tag{4.40}$$

Obviously, the above procedure does not work well for small carrier densities when both chemical potentials are negative. In these situations, we further specify a minimum limit for $k_f$ based on experience from calculations at higher densities. In calculating the spectra shown in this book, we impose the condition $k_f w/\pi \geq 0.2$, where $w$ is the quantum-well width.

To determine the cut-off $k_{\max}$, we specify $n_{\max}$, the total number of $k$-points used in the numerical analysis. The requirement on $n_{\max}$ is that the range $0 \leq k \leq k_{\max}$, where

$$k_{\max} = n_{\max}\,\mathrm{d}k \ , \tag{4.41}$$

is sufficiently large so that $f_{\mathrm{e}k}$, $f_{\mathrm{h}k}$ and $s_{\boldsymbol{k}}$ vanish as one approaches the upper limit.

Figure 4.7 shows an example of the relationships between $k_{\mathrm{e}f}$, $k_{\mathrm{h}f}$, $dk$ and $k_{\max}$. The $x$-axis is the normalized carrier momentum $kw/\pi$, where the quantum-well width is $w = 4\,\mathrm{nm}$. We plotted the carrier distributions $f_{\mathrm{e}k}$ (solid curve) and $f_{\mathrm{h}k}$ (dashed curve) for a carrier density $N_{\mathrm{2d}} = 1.6 \times 10^{12}\,\mathrm{cm}^{-2}$, and a two-band system with effective masses, $m_{\mathrm{e}} = 0.071 m_0$ and $m_{\mathrm{h}} = 0.167 m_0$. The chemical potentials are $\mu_{\mathrm{e}} = 0.110\,\mathrm{eV}$ and $\mu_{\mathrm{h}} = 0.027\,\mathrm{eV}$, which according to (4.39) gives $k_{\mathrm{e}f} w/\pi = 0.394$ and $k_{\mathrm{h}f} w/\pi = 0.260$. Choosing $k_f = k_{\mathrm{e}f}$ together with $n_f = 32$ and $n_{\max} = 81$, gives $\mathrm{d}k = 0.012\pi/w$ and $k_{\max} = 0.998\pi/w$, respectively. Other band-stucture properties are: unexcited bandgap energy $\varepsilon_{\mathrm{g}0} = 1.376\,\mathrm{eV}$, dipole moment $\mu = e \times 3.41\,\text{Å}$ and, with the exception of Fig. 4.17, we use a 3-d exciton binding energy $\varepsilon_{\mathrm{r}}^{\mathrm{3d}} = 3.287\,\mathrm{meV}$. The same band structure is used in subsequent calculations in this chapter.

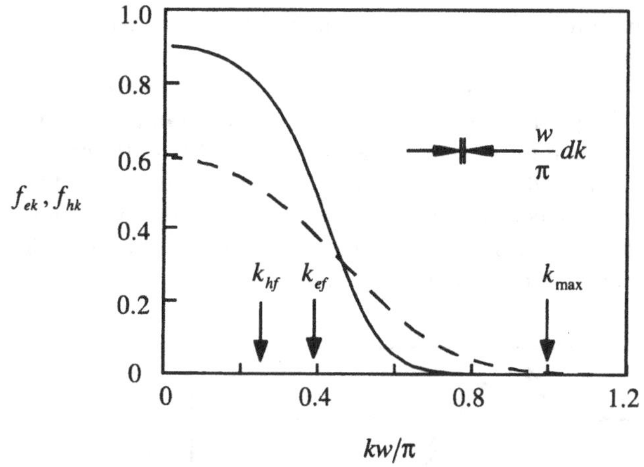


**Fig. 4.7.** Example of $k_{ef}$, $k_{hf}$, $dk$ and $k_{max}$ for a gain calculation involving a two-band system with effective masses, $m_e = 0.071 m_0$ and $m_h = 0.167 m_0$. The *solid* and *dashed curves* show the quasi-equilibrium carrier distributions $f_{ek}$ and $f_{hk}$, respectively, for the carrier density $N_{2d} = 1.6 \times 10^{12}\,\mathrm{cm}^{-2}$

When choosing the parameters $n_f$ and $n_{max}$, one should be aware that the polarization amplitude often has more structure and vanishes at higher $k$ than the carrier distributions $f_{ek}$ and $f_{hk}$. This is clearly seen in Fig. 4.8, where we plotted the real and imaginary parts of the polarization amplitude from a typical gain calculation. When performing the calculations, we assumed that there is no angular dependence in the polarization amplitude, i.e., $s_{\mathbf{k}} = s_k$. With this assumption, the memory requirement is significantly

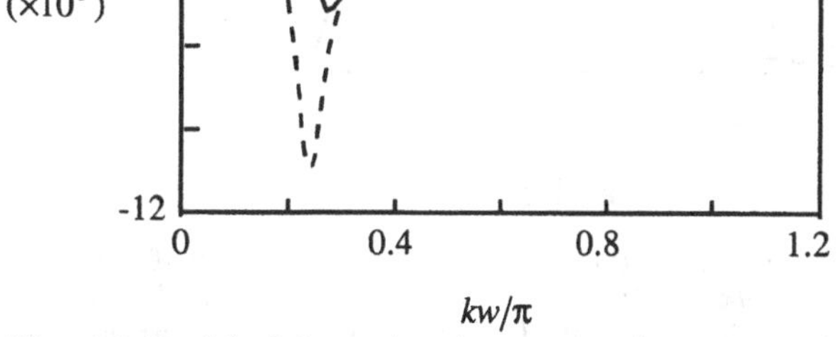

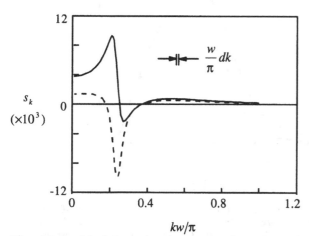

**Fig. 4.8.** Real (*solid curve*) and imaginary (*dashed curve*) parts of the polarization amplitude from a gain calculation. The parameters are the same as in Fig. 4.7

reduced, and the evaluation of the summations in the last term of (4.33) is computationally less demanding.

A lower resolution or a smaller cut-off can result in the incorrect prediction of structure in the gain spectrum. Figure 4.9 shows examples of the error incurred in the spectrum for the carrier density $N_{2d} = 1.2 \times 10^{12}\,\mathrm{cm}^{-2}$. Comparison of the solid and dotted curves show that the choice of a too large d$k$ gives rise to the erroneous prediction of oscillations in the spectrum, especially at high photon energies. The error caused by a too small cut-off is illustrated by the dashed curve. Here, the reduction of $k_{max}$ by a factor

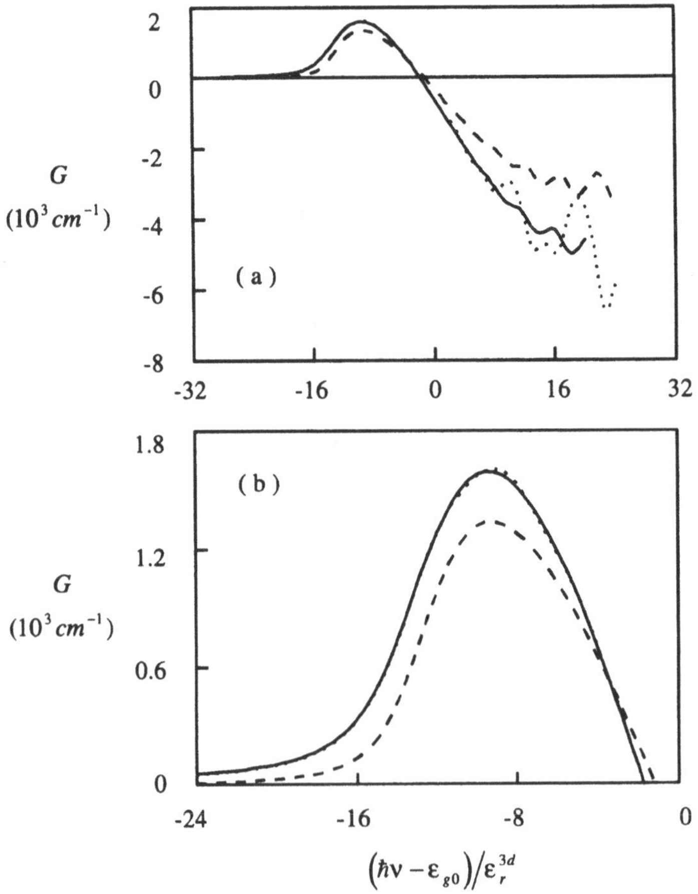


**Fig. 4.9a, b.** (**a**) Spectra computed for the carrier density $N_{2d} = 1.2 \times 10^{12}\,\mathrm{cm}^{-2}$. The curves are calculated using $dk = 0.016\pi/w$ and $k_{max} = 1.004\pi/w$ (*solid curve*), d$k$ = $0.033\pi/w$ and $k_{max} = 1.004\pi/w$ (*dotted curve*), and d$k$ = $0.016\pi/w$ and $k_{max} = 0.575\pi/w$ (*dashed curve*). (**b**) Magnified view of the gain portion of the spectra

of two causes inaccuracies in the calculation of the correlation contributions, leading to an underestimation of the gain (Fig. 4.9b).

**Hartree-Fock Contributions.** Once the step sizes and integration limits are determined, we can proceed with the computation of the Hartree-Fock terms. Using (4.34, 3.24) for the nondiagonal Hartree-Fock contribution in (4.33) becomes

$$\sum_{k'} \Theta_{k'k} s_{k'} = \frac{w_k}{\hbar} \frac{A}{(2\pi)^2} \int_0^\infty \mathrm{d}k'\, k' \int_0^{2\pi} \mathrm{d}\phi' \frac{e^2}{2\epsilon_\mathrm{b} A q\,(k, k', \phi')} s_{k'} \quad . \tag{4.42}$$

Here, we use MKS units and define

$$w_k \equiv f_{\mathrm{e}k} + f_{\mathrm{h}k} - 1 \quad . \tag{4.43}$$

Furthermore, we assume ideal two-dimensional confinement, so that the Fourier transform of the Coulomb potential is given by (3.5). In (4.42),

$$\boldsymbol{q} = \boldsymbol{k}' - \boldsymbol{k} \quad , \tag{4.44}$$

which gives

$$q(k, k', \phi') = \sqrt{k^2 + k'^2 - 2kk' \cos\phi'} \quad . \tag{4.45}$$

Again, assuming that there is no angular $\phi'$ dependence in the polarization, (4.42) becomes,

$$\begin{aligned} \sum_{k'} \Theta_{k'k} s_{k'} &= w_k \int_0^\infty \mathrm{d}k'\, k' \int_0^{2\pi} \mathrm{d}\phi' \frac{e^2}{8\pi^2 \epsilon_\mathrm{b} \hbar q(k, k', \phi')} s_{k'} \\ &= w_k \int_0^\infty \mathrm{d}k'\, \vartheta(k', k) s_{k'} \quad , \end{aligned} \tag{4.46}$$

where we define the function

$$\vartheta(k', k) \equiv \frac{e^2}{8\pi^2 \epsilon_\mathrm{b} \hbar} k' \int_0^{2\pi} \mathrm{d}\phi' \frac{1}{q(k, k', \phi')} \quad . \tag{4.47}$$

In (4.46, 47) $k' \neq k$ since we are dealing with the nondiagonal contributions.

Moving to the diagonal Hartree-Fock contribution, we write (3.23) in terms of a continuous momentum distribution as

$$\begin{aligned} \Theta_{kk} &= -\frac{1}{\hbar} \frac{A}{(2\pi)^2} \int_0^\infty \mathrm{d}k''\, k'' \int_0^{2\pi} \mathrm{d}\phi'' \frac{e^2}{2\epsilon_\mathrm{b} A q(k, k'', \phi'')} (w_{k''} + 1) \\ &= -\int_0^\infty \mathrm{d}k''\, k'' \int_0^{2\pi} \mathrm{d}\phi'' \frac{e^2}{8\pi^2 \epsilon_\mathrm{b} \hbar q(k, k'', \phi'')} (w_{k''} + 1) \\ &= -\int_0^\infty \mathrm{d}k''\, \vartheta(k'', k)(w_{k''} + 1) \quad , \end{aligned} \tag{4.48}$$

where we recall that implicit in the integration is the condition $\boldsymbol{k}'' \neq \boldsymbol{k}$. Nevertheless, considerable care must be taken when integrating around the singularity at $\boldsymbol{k}'' = \boldsymbol{k}$. We can do so without decreasing the grid size d$k$, and hence increasing the dimensions of $w_k$ and $s_k$, by realizing that the variations in these quantities are sufficiently smooth. Therefore, (4.48) may be approximated by

$$\begin{aligned} \Theta_{kk} \simeq & -\int_{-\infty}^{k-\delta} \mathrm{d}k''\, \vartheta(k'',k)(w_{k''}+1) - \int_{k+\delta}^{\infty} \mathrm{d}k''\, \vartheta(k'',k)(w_{k''}+1) \\ & -(w_k+1)\int_{k-\delta}^{k+\delta} \mathrm{d}k''\, \vartheta(k'',k) \ . \end{aligned} \tag{4.49}$$

Here, we choose $\delta$ to be sufficiently small so that $w_{k''} \approx w_k$ for $k-\delta \leq k'' \leq k+\delta$. In our actual calculations we choose $\delta = \Delta$, where $\Delta$ corresponds to the step size used in the $k'$ integration in (4.46). Defining the diagonal element

$$\vartheta(k,k) = \frac{e^2}{8\pi^2 \epsilon_{\mathrm{b}} \hbar} \frac{1}{2\Delta} \int_{k-\Delta}^{k+\Delta} \mathrm{d}k''\, k'' \int_{>0}^{2\pi} \mathrm{d}\phi''\, \frac{1}{q(k,k'',\phi'')} \ , \tag{4.50}$$

(4.49) becomes

$$\begin{aligned} \Theta_{kk} = & -\int_{-\infty}^{k-\Delta} \mathrm{d}k''\, \vartheta(k'',k)(w_{k''}+1) - \int_{k+\Delta}^{\infty} \mathrm{d}k''\, \vartheta(k'',k)(w_{k''}+1) \\ & -(w_k+1)\vartheta(k,k) \ , \end{aligned} \tag{4.51}$$

where the nondiagonal elements $\vartheta(k'',k)$ in the first two terms remain as defined by (4.47).

Figure 4.10 illustrates why (4.49) is a good approximation. Plotted in the figure is $\vartheta(k',k)$ as a function of $k'w/\pi$ at various $k$. Also shown are the carrier distributions $f_{ek}$ and $f_{hk}$ (long- and short-dashed curves, respectively) from Fig. 4.7. The sharply peaked behavior of $\vartheta(k',k)$ around $k' \simeq k$ compared to the smoothly varying $f_{ek}$ and $f_{hk}$ supports the approximation used to obtain (4.49). Furthermore, the figure shows why significantly smaller step sizes d$k''$ and $d\phi''$ are necessary when computing $\vartheta(k',k)$ in the neighborhood of the singularity.

We plot in Fig. 4.11 the diagonal element $\vartheta(k,k)$ given by (4.50) as a function of $k$. The figure shows that the step size for the $k''$ integration should be twenty to hundred times smaller than the step size of $\Delta$ of the $k'$ integration. Furthermore, the step size for the $\phi''$ integration in the diagonal elements of $\Theta$ should be at least two orders of magnitude smaller than that for the nondiagonal elements. While the evaluation of $\vartheta(k',k)$ may be time consuming because of these small step sizes, they need to be performed only rarely because, except for the background dielectric constant $\epsilon_{\mathrm{b}}$, $\vartheta(k',k)$ is independent of most of the other parameter values. One instance when we need to recompute $\vartheta(k',k)$ is when we change the discretization of $f_{ek}$, $f_{hk}$ and $s_k$.

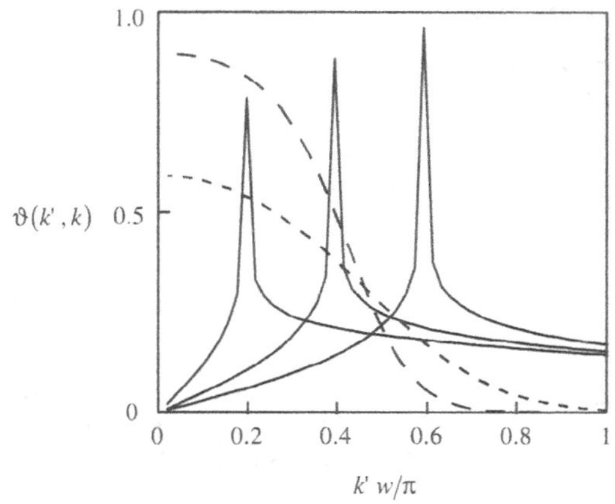


**Fig. 4.10.** Nondiagonal element of $\vartheta(k',k)$ (*solid curves*) versus $k'w/\pi$ for $kw/\pi = 0.2$, 0.4 and 0.6, from left to right. The parameters are the same as in Fig. 4.7. Also shown are the carrier distributions $f_{ek}$ and $f_{hk}$ (*long-* and *short-dashed curves*, respectively)

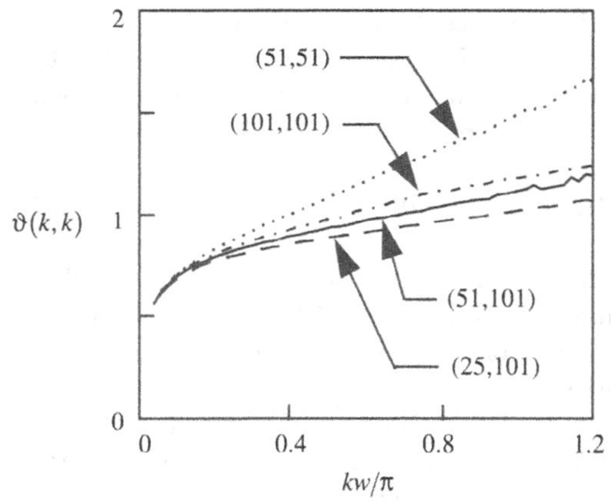


**Fig. 4.11.** $\vartheta(k,k)$ versus $kw/\pi$ for different $dk''$ and $d\phi''$. The other parameters are the same as those in Fig. 4.7. The curves are labeled by $(n_{k''}, n_{\phi''})$ where the step sizes, $dk'' = 2\Delta/(n_{k''}-1)$ and $d\phi'' = \pi/(n_{\phi''}-1)$

With the knowledge of $\vartheta(k',k)$, we can evaluate the Hartree-Fock contributions in (4.33). The exchange shift in (3.20) is

$$\begin{aligned} \Delta\varepsilon_{X,k} &= -\int_0^\infty dk'\, k' \int_0^{2\pi} d\phi'\, V_{|\boldsymbol{k}-\boldsymbol{k}'|}(w_{k'}+1) \\ &= -\int_0^\infty dk'\, \vartheta(k',k)(w_{k'}+1) \ . \end{aligned} \tag{4.52}$$

Figure 4.12 shows the exchange shift calculated for a range of carrier densities using the same parameters as in Fig. 4.7. We note a $k$ dependence in the exchange shift, which may be thought of as a density dependent modification of the effective mass. For numerical simplicity, one sometimes ignores the k-dependence of $\Delta\varepsilon_x$, as in Sect. 3.4. As discussed in Sect. 3.4, the exchange shift is only one component of the bandgap renormalization. To estimate the total effects of bandgap renormalization, one also has to evaluate the diagonal and nondiagonal correlation contributions, as will be discussed below.

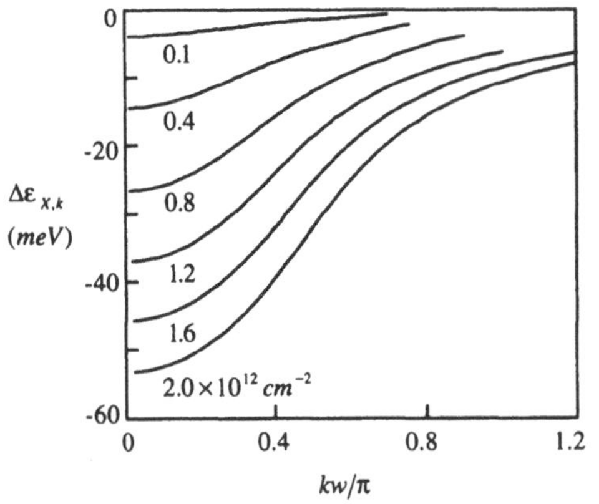


**Fig. 4.12.** Exchange shift versus $k$ for different carrier densities. The other parameters are the same as in Fig. 4.7

The other Hartree-Fock contribution besides the exchange shift is the excitonic or Coulomb enhancement, see (3.21). As discussed in Sect. 3.2 this contribution has a dominating influence on the absorption spectra, particularly at low carrier densities. In the present formulation, (3.17) becomes

$$\frac{\mathrm{d}s_k}{\mathrm{d}t} = \mathrm{i}(\nu - \omega_k' + \Delta\varepsilon_{X,k} + \mathrm{i}\gamma)s_k + \frac{\mathrm{i}}{\hbar}\mu_k E_0 - \mathrm{i}w_k \int_0^\infty \mathrm{d}k'\, \vartheta(k',k) s_{k'} \quad , (4.53)$$

where we use (4.33) without the correlation contributions and introduce a small phenomenological damping coefficient $\gamma = 10^{-13}$ s to facilitate the numerical solution. The steady state solutions of (4.53) for a range of laser frequencies $\nu$ and carrier densities $N$ give the absorption spectra shown in Fig. 4.13. On the $x$-axis, we referenced the photon energy to the unexcited quantum-well bandgap energy $\varepsilon_{g0}$. The quantum-well exciton binding energy is the energy difference between $\varepsilon_{g0}$ and the exciton absorption resonance. The spectra show a well-known consequence of two-dimensional quantum confinement, which is the factor of four higher exciton binding energy than the bulk case, i.e., $\varepsilon_{\mathrm{R}}^{2\mathrm{d}} = 4 \times \varepsilon_{\mathrm{R}}^{3\mathrm{d}}$. Furthermore, they show that the spectral position of the excitonic absorption resonance is pretty much independent of carrier density. At the present level of approximation (i.e., neglecting correlation effects), this result is a consequence of the cancellation of two effects: phase-space filling and bandgap renormalization. The phase-space filling contribution to the weakening of the exciton resonance is included through the factor $w_k$ in the last term in (4.53) which describes the effects of the attractive interband Coulomb interaction. Coulomb attraction is strongest in the absence of carriers when $w_k = -1$, and weakens with increasing carrier density, when $w_k > -1$. To illustrate the phase-space filling contribution, we solve (4.53) with $\Delta\varepsilon_{X,k} = 0$. The result is shown in Fig. 4.14a, which shows a reduction of the exciton binding energy and absorption amplitude with increasing carrier density. In the full equation, this reduction in exciton binding energy is balanced by bandgap renormalization. Figure 4.14b shows the exchange energy shift at zone center $\Delta\varepsilon_{X,0}$ as a function of carrier den-

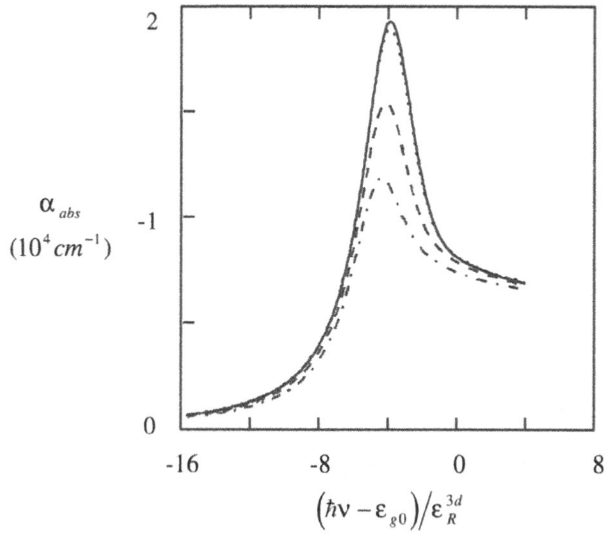


**Fig. 4.13.** Absorption spectra for the carrier densities $N_{2\mathrm{d}} = 0$ (*solid curve*), $10^{10}$ (*dotted curve*), $10^{11}$ (*dashed curve*) and $2 \times 10^{11}\,\mathrm{cm}^{-2}$ (*dot-dashed curve*)

sity. The partial cancellation between the two effects, i.e., the reduction of the bandgap and of the exciton binding energy, leads to the feature that the exciton absorption resonance decreases in amplitude but remains constant in energy with increasing carrier density, as seen in Fig. 4.13.

**Correlation Contributions.** The correlation contributions to the polarization equation are discussed in Sect. 4.3. Rather than using exactly (4.14, 17) in the numerical analysis we account for the screening effects from the correlations of still higher order, that are left out in the derivations of Sect. 4.1, by replacing the unscreened Coulomb potential in $\underline{\Lambda}$ by the screened one, i.e.,

$$V_q \to V_{sq} = \frac{V_q}{\epsilon_q} \ , \tag{4.54}$$

where $\epsilon_q$ is the static Lindhard dielectric function discussed in Sect. 3.3.

First, we consider the diagonal collision terms, (4.14), in the interband polarization dynamics. Again assuming a continuous momentum distribution and using MKS units, (4.14) becomes

$$\begin{aligned} \Lambda_{kk} &= \Gamma_k - \mathrm{i}\Delta_k \\ &= \frac{1}{2\hbar}\left(\frac{e}{2\pi}\right)^4 \int_0^\infty \mathrm{d}q\, q \sum_{a=e,h} \sum_{b=e,h} \int_0^{2\pi} \mathrm{d}\phi \int_0^{2\pi} \mathrm{d}\phi' \int_0^\infty \mathrm{d}k'\, k' \\ &\quad \times \left( \frac{1}{q^2 \epsilon^2(q)} - \frac{\delta_{a,b}}{2} \frac{1}{q\epsilon(q)} \frac{1}{q'\epsilon(q')} \right) \frac{\Delta + \mathrm{i}\delta\varepsilon}{\Delta^2 + \delta\varepsilon^2} \\ &\quad \times \left[ (1 - f_{ak_f}) f_{bk'} (1 - f_{bk'_f}) + f_{ak_f} (1 - f_{bk'}) f_{bk'_f} \right] \ , \end{aligned} \tag{4.55}$$

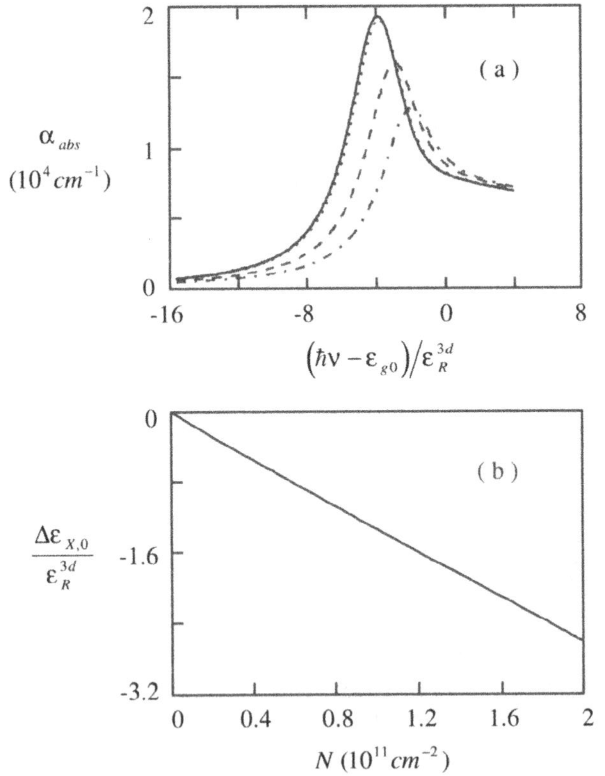


**Fig. 4.14a, b.** (**a**) Field renormalization contribution to the absorption spectrum of Fig. 4.13. The carrier densities are $N_{2d} = 0$ (*solid curve*), $10^{10}\,\mathrm{cm}^{-2}$ (*dotted curve*), $10^{11}\,\mathrm{cm}^{-2}$ (*dashed curve*) and $2 \times 10^{11}\,\mathrm{cm}^{-2}$ (*dot-dashed curve*). (**b**) Exchange shift at $k = 0$ versus carrier density. The combined effects of field renormalization and exchange shift give the spectra in Fig. 4.13

where

$$\boldsymbol{k}_f = \boldsymbol{k} + \boldsymbol{q} \ , \tag{4.56}$$

$$\boldsymbol{k}'_f = \boldsymbol{k}' - \boldsymbol{q} \ , \tag{4.57}$$

$$\boldsymbol{q}' = \boldsymbol{k} - \boldsymbol{k}' + \boldsymbol{q} \ . \tag{4.58}$$

With the help of the sketch in Fig. 4.15 we can convince ourselves that the following relations hold:

$$k_f^2 = k^2 + q^2 + 2kq\cos\phi \ , \tag{4.59}$$

$$k_f'^2 = k'^2 + q^2 - 2k'q\cos(\phi' - \phi) \ , \tag{4.60}$$

$$q'^2 = k_f^2 + k'^2 - 2kk'\cos\phi' - 2qk'\cos(\phi' - \phi) \ . \tag{4.61}$$

For parabolic bands, the energy difference in (4.15) can be written as

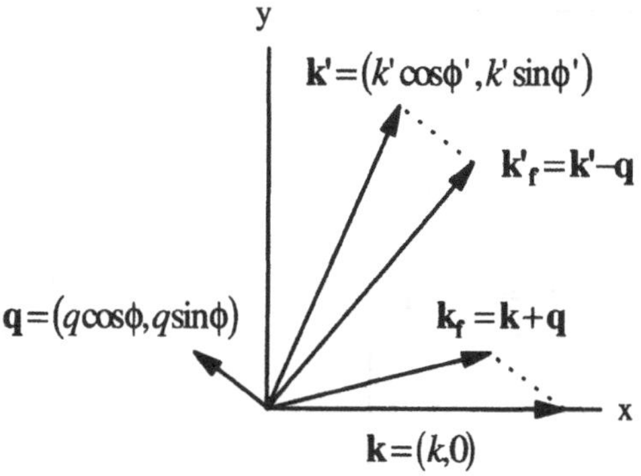


**Fig. 4.15.** Momenta involved in calculation of carrier-carrier collisions

$$\delta\varepsilon = \frac{\hbar^2}{2m_a}(k^2 - k_f^2) + \frac{\hbar^2}{2m_b}(k'^2 - k_f'^2) \ . \qquad (4.62)$$

To compute the dielectric function, we convert the summations in the static ($\omega = 0$) version of the Lindhard expression (3.54) to integrals. This gives

$$\epsilon(q) = 1 - \frac{e^2}{8\pi^2\epsilon_b q} 2 \sum_{a=e,h} \int_0^\infty dk''\, k'' \int_0^{2\pi} d\phi'' \, \frac{f_{a,|\boldsymbol{k}''+\boldsymbol{q}|} - f_{ak''}}{\varepsilon_{a,|\boldsymbol{k}''+\boldsymbol{q}|} - \varepsilon_{ak''}} \ , \qquad (4.63)$$

where the factor of 2 comes from the spin summation, and

$$|\boldsymbol{k}' + \boldsymbol{q}|^2 = k'^2 + q^2 + 2kq\cos\phi' \ . \qquad (4.64)$$

One may also use the numerically less demanding single plasmon pole result, (3.59), instead of the full Lindhard expression. The disadvantage is that it contains phenomenological parameters, such as $C$.

In (4.55), the diagonal contributions are defined for the limit $\Delta \to 0$. Figure 4.16 shows the sensitivity of $\Gamma_k$ and $\Delta_k$ to $\Delta$. Formally, the $dq$ integration in (4.55) could just as well be replaced with a $dk_f$ integration. However, the actual numerical integrations can be very different. This is due to the fact that the Coulomb potential is $\propto q^{-1}$. Therefore the upper limit of $q$ needed to reach a desired accuracy is more straightforward to estimate than the upper limit in $k_f$ which depends on $k$.

Figure 4.17 illustrates the dependences of the diagonal dephasing contribution $\Gamma_k$ and energy shift $\Delta_k$ on experimental conditions. Shown are two sets of curves that are separated because of the difference in the respective exciton binding energy used. Within each set, the different curves correspond to different carrier densities. For one set of curves, we use an exciton binding energy that is representative of conventional III–V laser compounds. For the second set of curves, describing a more rapid polarization dephasing, we use an exciton binding energy that is more typical for the wide bandgap, II–VI and group-III nitride compounds. Fig. 4.17 also illustrates the dependences of $\Gamma_k$ and $\Delta_k$ on carrier density and momentum.

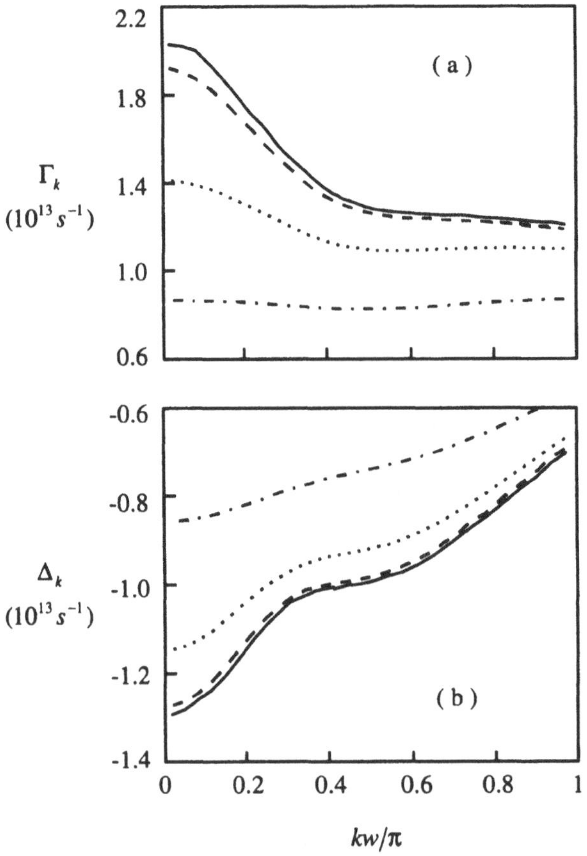


**Fig. 4.16a, b.** (**a**) $\Gamma_k$ and (**b**) $\Delta_k$ versus $kw/\pi$ for $N = 1.2 \times 10^{12}\,\mathrm{cm}^{-2}$. The different curves are for $\Delta = 0.1$ (*solid curve*), 1 (*dashed curve*), 10 (*dotted curve*), and 50 meV (*dot-dashed curve*)

For the nondiagonal correlation contributions we introduce

$$\begin{aligned}\lambda'(q,k) &= \frac{1}{2\hbar}\left(\frac{e}{2\pi}\right)^4 q \sum_{a=e,h}\sum_{b=e,h}\int_0^{2\pi} \mathrm{d}\phi \int_0^{2\pi} \mathrm{d}\phi' \int_0^{\infty} \mathrm{d}k'\, k' \\ &\times \left(\frac{1}{q^2\epsilon^2(q)} - \frac{\delta_{a,b}}{2}\frac{1}{q\epsilon(q)}\frac{1}{q'\epsilon(q')}\right)\frac{\Delta - \mathrm{i}\delta\varepsilon}{\Delta^2 + \delta\varepsilon^2} \\ &\times \left[f_{a,k} f_{bk'}(1 - f_{bk'_f}) + (1 - f_{a,k})(1 - f_{bk'}) f_{bk'_f}\right] \quad . \qquad (4.65)\end{aligned}$$

Since in our numerical evaluations we discretize the $\boldsymbol{k}$ continuum, $s_k$ and $w_k$ are stored only for a discrete set of $k$ values. In order to evaluate the subsequent integrations [see ahead to (4.74, 75)], we have to interpolate between the stored values. In practice, we convert from $\lambda'(q,k)$ to $\lambda(k_f,k)$, with an algorithm that allocates $\lambda'(q,k)$ to the appropriate $\lambda(k_f,k)'s$. One such al-

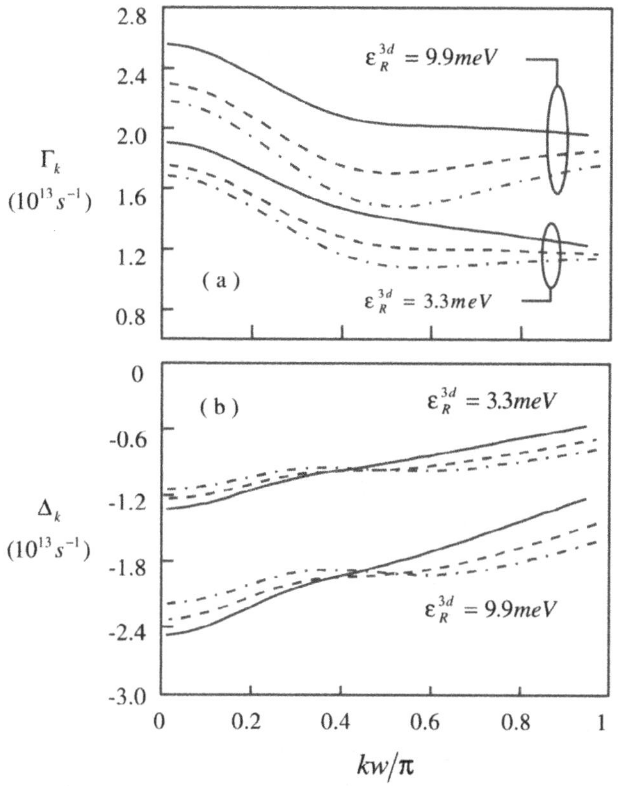


**Fig. 4.17a, b.** **(a)** $\Gamma_k$ and **(b)** $\Delta_k$ versus $kw/\pi$ for carrier densities $N = 8 \times 10^{11}$ (*solid curve*), $1.2 \times 10^{12}$ (*dashed curve*), and $1.6 \times 10^{12}\,\mathrm{cm}^{-2}$ (*dot-dashed curve*). The two sets of curves correspond to exciton binding energies, $\varepsilon_r^{3\mathrm{d}} = 3.3$ and $9.9\,\mathrm{meV}$

gorithm involves first using (4.65) to calculate $\lambda'(q,k)$. For a given value of $q$, we get a value for $k_f = |\boldsymbol{k}+\boldsymbol{q}|$. Suppose $k_f$ falls between two momenta where we have stored $s_k$ and $w_k$, i.e., $k_{n-1} \le k_f \le k_n$, then we make the allocation to $\Lambda(k_{n-1},k_m)$ and $\Lambda(k_n,k_m)$ according to

$$\lambda''(q,k_{n-1},k_m) = \frac{|k_f - k_n|}{k_n - k_{n-1}}\lambda(q,k) \quad ,$$

$$\lambda''(q,k_n,k_m) = \frac{|k_f - k_{n-1}|}{k_n - k_{n-1}}\lambda(q,k) \quad . \tag{4.66}$$

The steps are repeated for a range of $0 < q \le q_{\max}$, where $q_{\max}$ is sufficiently large so that $\lambda'(q_{\max},k) \to 0$. Then, we sum over all the contributions to get

$$\lambda(k_n,k_m) = \sum_q \lambda''(q,k_n,k_m) \quad . \tag{4.67}$$

Figure 4.18 shows examples of $\lambda(k_n,k_m)$.

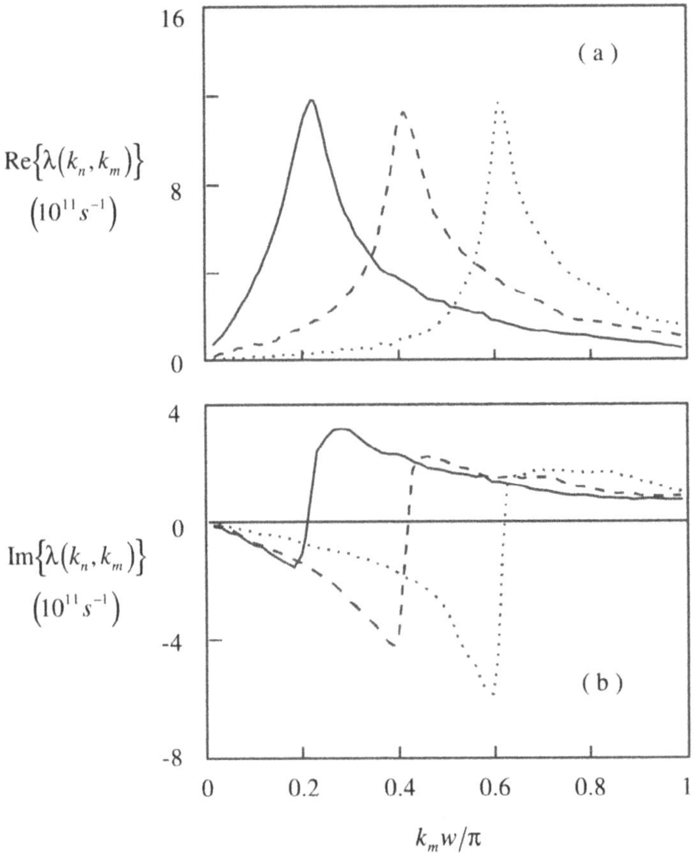


**Fig. 4.18a, b.** (**a**) $\text{Re}\{\lambda(k_n, k_m)\}$ and (**b**) $\text{Im}\{\lambda(k_n, k_m)\}$ versus $k_n w/\pi$ for $N = 1.2 \times 10^{12}\,\text{cm}^{-2}$, and $k_m w/\pi = 0.2$ (*solid line*), 0.4 (*long dashed line*), and 0.6 (*dotted line*)

**Polarization Equations** We are now in the position to solve (4.33). To put this equation into a form that is suited for numerical evaluation, we write explicitly the real and imaginary parts introducing the following quantities:

$$s_k = u_k + \mathrm{i}v_k \quad , \tag{4.68}$$

$$E_0 = E_0^r + \mathrm{i}E_0^i \quad , \tag{4.69}$$

$$w_k = f_{ek} + f_{hk} - 1 \quad . \tag{4.70}$$

With these definitions we convert (4.33) to

$$\frac{\mathrm{d}u_k}{\mathrm{d}t} = -\delta_k v_k + \mu_k E_0^i w_k - \Gamma_k u_k + Q_k \quad , \tag{4.71}$$

$$\frac{\mathrm{d}v_k}{\mathrm{d}t} = \delta_k u_k - \mu_k E_0^r w_k - \Gamma_k u_k + R_k \quad . \tag{4.72}$$

Here, the detuning $\delta_k$ is given by

$$\hbar\delta_k = \hbar(\nu - \omega'_k) - \Delta\varepsilon_{xk} - \Delta_k \quad , \tag{4.73}$$

and

$$Q_k = \int_0^\infty \mathrm{d}k' \left[v_{k'}(\vartheta_{k'k}w_k - \Delta_{k'k}) + u_{k'}\Gamma_{k'k}\right] \quad , \tag{4.74}$$

$$R_k = \int_0^\infty \mathrm{d}k' \left[u_{k'}(-\vartheta_{k'k}w_k + \Delta_{k'k}) + v_{k'}\Gamma_{k'k}\right] \quad . \tag{4.75}$$

The fourth-order Runge-Kutta method provides a stable finite difference method for solving differential equations such as (4.71, 72). To illustrate this method, we look at the generic equation

$$\frac{\mathrm{d}y}{\mathrm{d}t} = F(y) \quad . \tag{4.76}$$

Let us step from time $t_1$ to $t_2 = t_1 + \mathrm{d}t$, starting at $t = t_1$ with $y = y_0$. The first step involves moving to the mid-point $t' = t_1 + \mathrm{d}t/2$ by using a derivative that is evaluated with $y = y_0$. This gives the mid-point value

$$y(t') = y_1 = y_0 + F(y_0)\frac{\mathrm{d}t}{2} \quad . \tag{4.77}$$

We can also estimate a value for $y(t')$ using the derivative evaluated with $y = y_1$,

$$y(t') = y_2 = y_0 + F(y_1)\frac{\mathrm{d}t}{2} \quad . \tag{4.78}$$

This yields another value $F(y_2)$ for the derivative, which we use to propagate from $t_1$ to $t_2$

$$y(t_2) = y_3 = y_0 + F(y_2)\,\mathrm{d}t \quad . \tag{4.79}$$

This last propagation gives us yet another value $F(y_3)$ for the derivative. Now we perform the step from $t_1$ to $t_2$ using a statistical average of all the derivatives obtained [*Press* et al. (1988)]:

$$y(t_2) = y_0 + \frac{\mathrm{d}t}{6}\left\{F(y_0) + 2\left[F(y_1) + F(y_2)\right] + F(y_3)\right\} \quad . \tag{4.80}$$

The adaptation of (4.80) to coupled differential equations, such as (4.71, 72), is relatively straightforward. Figure 4.19 shows an example of the evolution of $u_k$ and $v_k$.

Several opportunities exist for reducing the complexity and time needed for the numerical computations. The most important simplification comes from the small signal condition, (4.30), which allows us to neglect changes in the carrier distributions. As a result, the calculations of $\vartheta(k', k)$ and $\lambda(k', k)$, which are computationally intensive, need to be performed only once in the evaluation of an entire spectrum (of fixed carrier density).

Other possibilities for speeding up the calculations come from approximating certain contributions with equivalent functions. For example, from

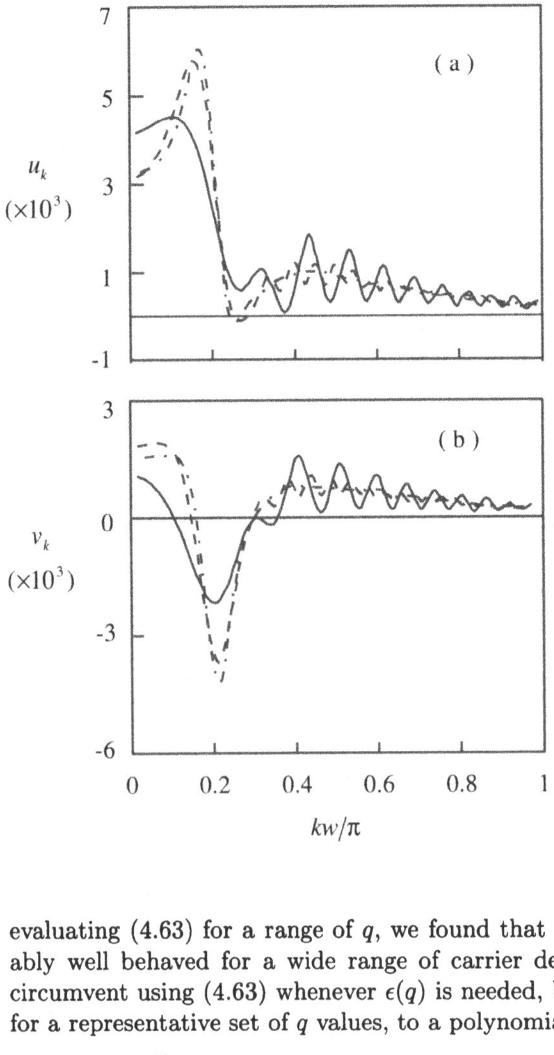


**Fig. 4.19a, b.** Polarization contributions $u_k$ (**a**) and $v_k$ (**b**) at times $t = 8$ (*solid curve*), 16 (*dashed curve*), and 24 ps (*dot-dashed curve*). The carrier density is $N = 1.2 \times 10^{12}$ cm$^{-2}$ and the detuning is $(\hbar\nu - \varepsilon_{g0})/\varepsilon_r^{2d} = -2.0$

evaluating (4.63) for a range of $q$, we found that the product $q\epsilon_q$ is reasonably well behaved for a wide range of carrier densities. This allows us to circumvent using (4.63) whenever $\epsilon(q)$ is needed, by fitting $q\epsilon(q)$, calculated for a representative set of $q$ values, to a polynomial,

$$q\epsilon(q) = \sum_{n=0}^{N} a_n q^n \quad , \tag{4.81}$$

where the coefficients $a_n$ are obtained from a least-squares fit. Similarly, the diagonal correlation contributions $\Gamma_k$ and $\Delta_k$ are sufficiently smooth functions of $k$ that they may also be fitted with polynomials.

The calculation of $Q_k$ and $R_k$ consumes a substantial amount of computer time because it has to be performed for each time step. In many of our numerical runs, we obtained noticeable time savings by computing $Q_k$ and $R_k$ for a sampling of $k$, and interpolate the intermediate values.

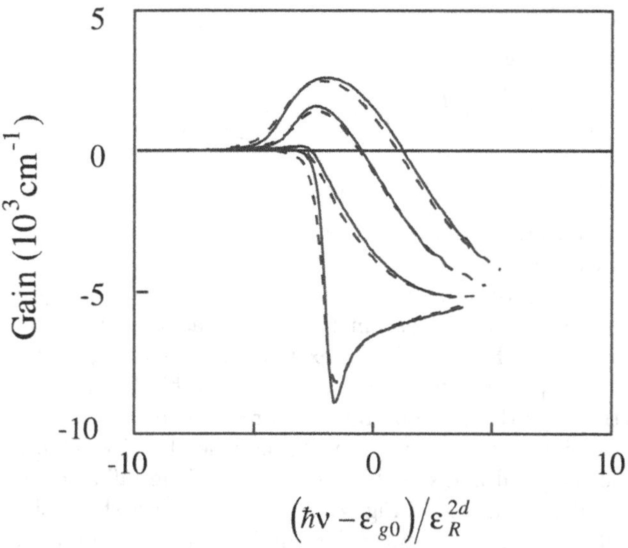


**Fig. 4.20.** Gain/absorption spectra with (*solid curves*) and without (*dashed curves*) exchange contributions. The densities are $N_{2d} = 4 \times 10^{11}$ (*lowest curves*), $8 \times 10^{11}$, $1.2 \times 10^{12}$ and $1.6 \times 10^{12}\,\mathrm{cm}^{-2}$ (*top curves*)

Finally, Fig. 4.20 illustrates the influence of the exchange contributions in $\underline{\Lambda}$ to the gain spectrum. For situations where the differences shown in the figure can be considered as unimportant, one can save computing time by dropping the exchange terms in $\underline{\Lambda}$. To see the reduction in computation time, we refer to (4.59–61). With the exchange contribution, the arguments in the cosines are $\phi$, $\phi'$ and $\phi' - \phi$. The angular integrations should then be performed from $0 \leq \phi \leq 2\pi$ and $0 \leq \phi' \leq 2\pi$. On the other hand, if the exchange contribution is neglected, the angular arguments are $\phi$ and $\phi' - \phi$, and considerable time saving is possible by performing the angular integrations from $0 \leq \phi \leq \pi$ and $0 \leq \phi' - \phi \leq \pi$. Some time saving also results from the Coulomb potential entering later in the nested integrations, as can be seen in the equations below:

$$\begin{aligned}
\Lambda_{kk} &= \Gamma_k - \mathrm{i}\Delta_k \\
&= \frac{1}{2\hbar}\left(\frac{e}{2\pi}\right)^4 \int_0^\infty \mathrm{d}q\, q \frac{1}{q^2\epsilon^2(q)} \\
&\quad \times \sum_{a=e,h}\sum_{b=e,h} \int_0^{2\pi} \mathrm{d}\phi \int_0^{2\pi} \mathrm{d}\phi' \int_0^\infty \mathrm{d}k'\, k' \frac{\Delta + \mathrm{i}\delta\varepsilon}{\Delta^2 + \delta\varepsilon^2} \\
&\quad \times \left[(1 - f_{ak_f}) f_{bk'} (1 - f_{bk'_f}) + f_{ak_f}(1 - f_{bk'}) f_{bk'_f}\right] \qquad (4.82)
\end{aligned}$$

and

$$\lambda'(q,k) = \frac{1}{2\hbar}\left(\frac{e}{2\pi}\right)^4 q \frac{1}{[q\epsilon(q)]^2} \times \sum_{a=e,h}\sum_{b=e,h}\int_0^{2\pi} d\phi \int_0^{2\pi} d\phi' \int_0^{\infty} dk'\, k' \frac{\Delta - i\delta\varepsilon}{\Delta^2+\delta\varepsilon^2} \times \left[f_{a,k} f_{bk'}(1-f_{bk'_f}) + (1-f_{a,k})(1-f_{bk'})f_{bk'_f}\right] \quad . \tag{4.83}$$

Of course, whether we can ignore the exchange contributions has to be determined on a case by case basis.

Figure 4.21 shows gain/absorption spectra obtained using the procedure outlined in this section. The material parameters as well as step sizes and cut-offs are the same as those in Fig. 4.7. The unexcited bandgap energy is denoted as $\varepsilon_{g0}$ and $\varepsilon_R^{2d} = 4 \times \varepsilon_R^{3d}$ is the two-dimensional exciton binding energy. The low density spectra show the exciton absorption resonance, whose location remains constant with changing carrier density because of the balancing of the diagonal and off diagonal many-body contributions. The influence of the correlation effects is demonstrated in Fig. 4.22. The solid curve shows the absorption spectrum at $N_{2d} = 2 \times 10^{11}$ cm$^{-2}$. The dashed curve is computed by neglecting the nondiagonal correlation contribution. Comparison of this and the solid curve shows the importance of the nondiagonal contribution in the presence of the exciton resonance at this density. The dotted curve

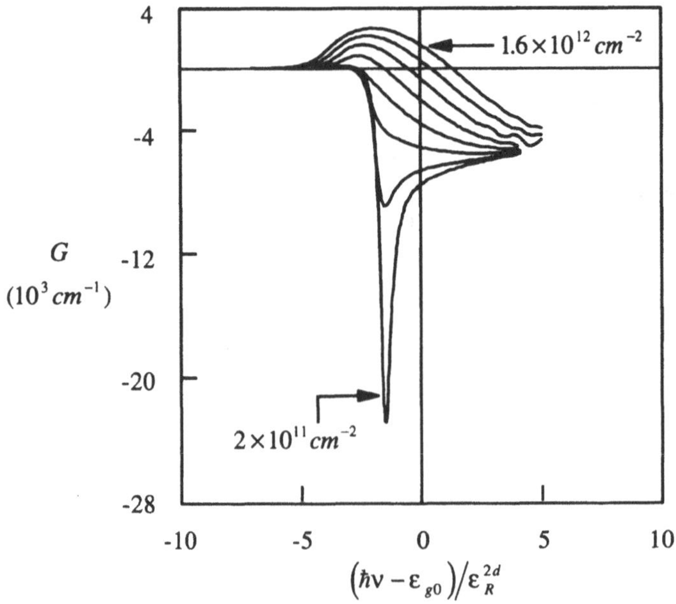


**Fig. 4.21.** Gain/absorption spectra for a range of carrier densities $N_{2d} = 2 \times 10^{11}$ (*bottom curve*) – $1.6 \times 10^{12}$ cm$^{-2}$ (*top curve*). The density is increased in increments of $2 \times 10^{11}$ cm$^{-2}$

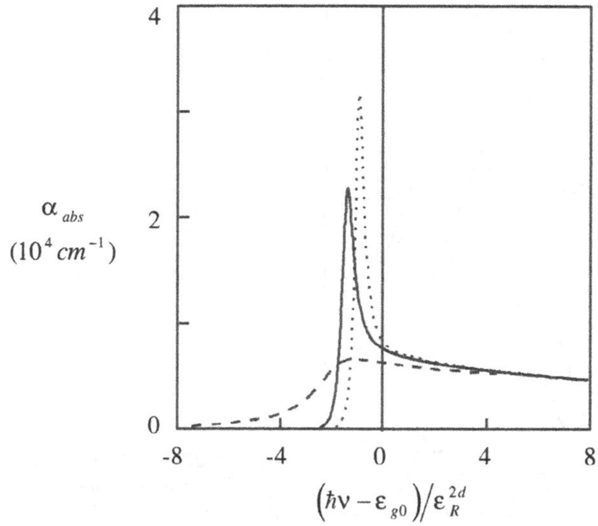


**Fig. 4.22.** Absorption spectrum (*solid curve*) for the carrier density $N_{2d} = 2 \times 10^{11}\,\mathrm{cm}^{-2}$. The difference between the *dashed* and *solid curves* shows the nondiagonal correlation contribution, while the difference between the *dotted* and *solid curves* shows the imaginary part of the correlation contribution

is computed by neglecting only the imaginary part of the correlation (both diagonal and nondiagonal) contributions. Comparison with the solid curve shows that screening effects cause a reduction of the exciton absorption as well as a red shift that, together with the exchange shift, compensates the blue shifts due to the field renormalization and the real part of the correlation contributions.

In Fig. 4.23a, we concentrate on elevated carrier densities where the optical spectra show gain. Comparison of these spectra and those in Figs. 4.23b, c illustrates the differences between the present gain calculation and the ones using the gain formula (3.100) based on the relaxation rate approximation. The results of the relaxation rate approximation for both Lorentzian (b) and sech (c) lineshape functions show a steep rise in gain from the bandgap edge, indicating the strong influence from the step function like two-dimensional density of states. The full calculation, on the other hand, shows a gentler, more bulk-like slope. The figures also demonstrates that correlation effects lead to a different density dependence of the gain peak. The full calculation predicts a blue shift in the gain peak with increasing carrier density, while the relaxation rate results show negligible shift (see dotted lines).

To understand the origin of the differences depicted in Fig. 4.23, we plot in Fig. 4.24 the individual contributions to the gain spectrum. The short-dashed curve shows the gain spectrum for $N_{2d} = 1.2 \times 10^{12}\,\mathrm{cm}^{-2}$ in the absence of Coulomb (Hartree-Fock and scattering) effects. The shape of the spectrum is

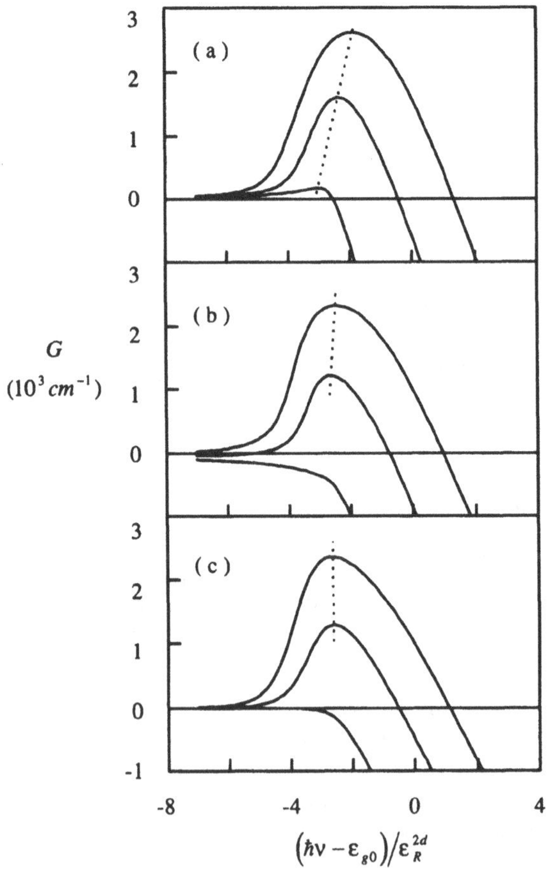


**Fig. 4.23a–c.** Gain spectra for carrier densities $N_{2d} = 8 \times 10^{11}$, $1.2 \times 10^{12}$ and $1.6 \times 10^{12}\ \mathrm{cm}^{-2}$ (from *bottom* to *top*). (**a**) shows the results of the full calculations. (**b**) and (**c**) show the results obtained using (3.100) with Lorentzian and sech lineshape functions, respectively, and a dephasing rate $\gamma = 10^{13}\ \mathrm{s}^{-1}$. The dotted lines show the changes in the gain peaks with carrier density

completely determined by the band filling factor. Since we have no homogeneous broadening of the optical transitions, one sees the sharp gain increase at the band edge because of the step function-like two-dimensional density of states. The long-dashed curve shows the pure Hartree-Fock results. There is a frequency shift and a reshaping of the gain spectrum because of bandgap renormalization and Coulomb enhancement, respectively. The solid curve is the final result, obtained by additionally including the correlation effects. Because of dephasing (real part of correlations), the peak gain decreases and the gain width increases. There is also an energy shift due to plasma screening (imaginary part of correlations). In particular, our analysis shows that the interplay of diagonal and nondiagonal contributions shifts the gain peak towards higher energy and leads to a long wing on the low energy side. These modifications clearly demonstrate the strong influence of carrier correlations on the gain spectrum.

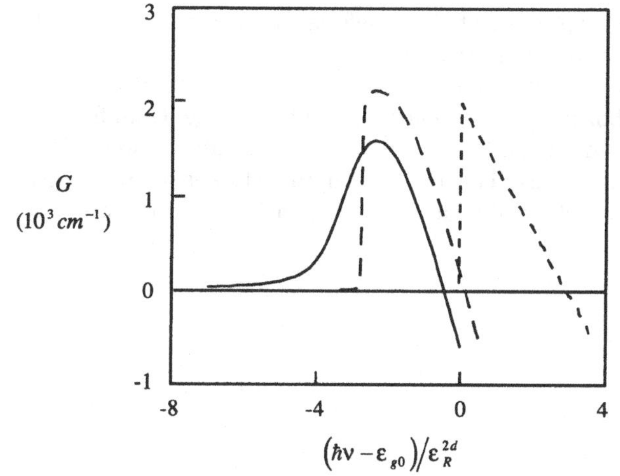


**Fig. 4.24.** Computed gain spectrum (solid curve) for the carrier density $N_{2d} = 1.2 \times 10^{12}\,\mathrm{cm}^{-2}$. The *short-dashed curve* is the spectrum when Coulomb effects are neglected. The *long-dashed curve* shows the effects of including only the Hartree-Fock contributions

Figure 4.25 documents the separate modifications caused by the diagonal and nondiagonal scattering contributions. The dashed curve in Fig. 4.25 shows that the diagonal contribution alone predicts a significantly smaller gain peak, an increased width (full width at half maximum), and a crossover back to absorption energetically below the gain spectrum. The nondiago-

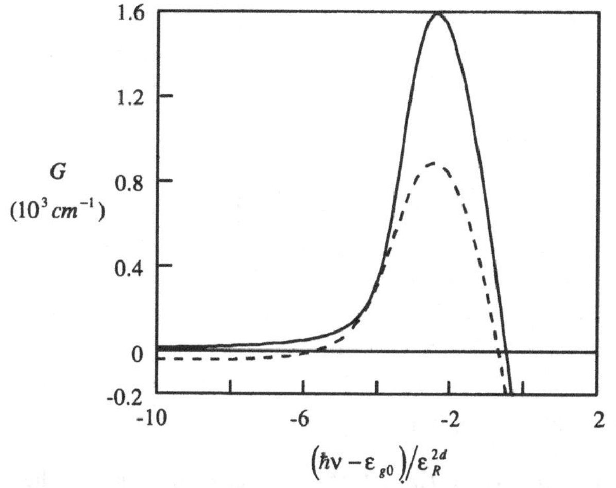


**Fig. 4.25.** Gain spectrum with (*solid curve*) and without (*dashed curve*) nondiagonal Coulomb correlation contributions. All parameters are similar to those in Fig. 4.24

nal contribution compensates most of these defects and leads to important modifications of the spectral gain shape.

A long standing problem with gain calculations is the inaccuracy of gain predictions in the neighborhood of the renormalized band edge. Gain formulas based on Lorentzian lineshape functions tend to yield absorption below the renormalized bandgap energy. This unphysical prediction is often blamed on the oversimplified description of polarization dephasing. The argument is

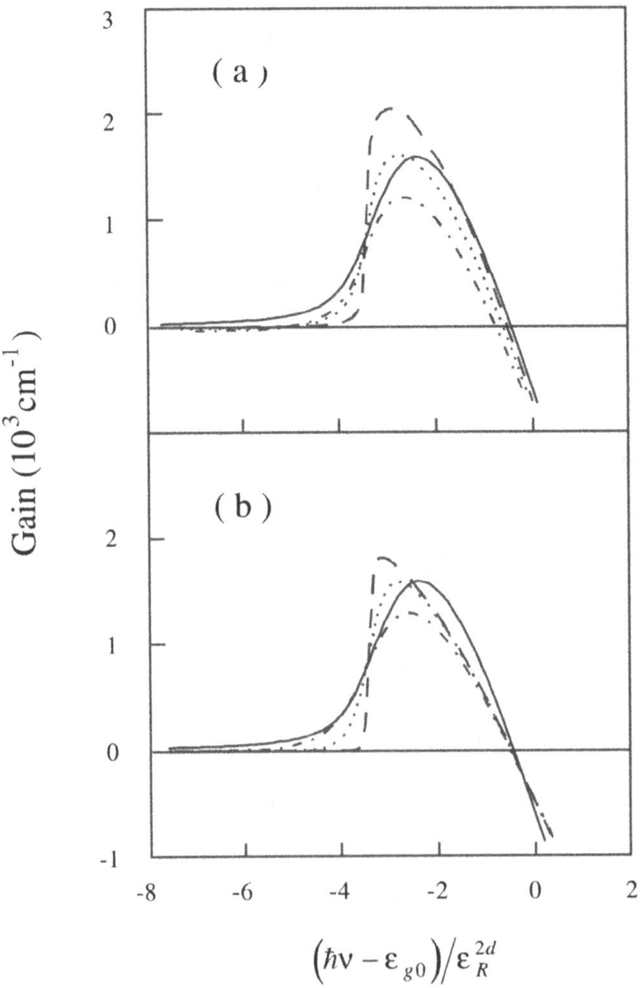


**Fig. 4.26a, b.** Gain spectrum for $N_{2d} = 1.2 \times 10^{12}\,\mathrm{cm}^{-2}$. The *solid curve* is the result of the full calculations. The *other curves* are obtained using (3.100) with (**a**) Lorentzian and (**b**) sech lineshape functions. The dephasing rates are $\gamma = 10^{12}\,\mathrm{s}^{-1}$ (*dashed curves*), $5 \times 10^{12}\,\mathrm{s}^{-1}$ (*dotted curves*), and $10^{13}\,\mathrm{s}^{-1}$ (*dot-dashed curves*)

that the assumption of $\partial p_{\boldsymbol{k}}/\partial t|_{\mathrm{col}} = -\gamma p_{\boldsymbol{k}}$, where $\gamma$ is a constant dephasing rate, overestimates the contributions from high-$\boldsymbol{k}$ (absorbing) states. However, the dashed curve in Fig. 4.25 shows that the problem remains even with a more precise treatment of the diagonal contribution to polarization dephasing, which uses the $\boldsymbol{k}$-dependent dephasing rate given by the real part of (4.55). Figure 4.25 demonstrates clearly that the solution of the lineshape problem is obtained through the consistent treatment of correlation effects, which involves the evaluation of both diagonal and nondiagonal correlation terms in the polarization equation. With the inclusion of the nondiagonal contribution the absorption below the bandgap energy is eliminated.

Figures 4.26, 27 summarize attempts to approximate the full results with gain expressions based on purely diagonal descriptions of dephasing. In Fig. 4.26, the dashed curves are computed using (a) Lorentzian and (b) sech lineshape functions, where we treat the widths of the lineshape functions as a fitting parameter. We see that because of the significant nondiagonal Coulomb correlation effects, it is not possible to accurately reproduce the gain spectrum regardless of the value for the dephasing rate. On the other hand, if one also treats the carrier density as a fitting parameter, then the comparison improves noticeably. Figure 4.27 shows that such a two parameter fit (dephasing time and density) can work sufficiently well for some situations. Under these conditions, we can get reasonable agreement with experiments, using a purely diagonal dephasing approximation. This may explain past successes in describing experimental results with gain models that we presently

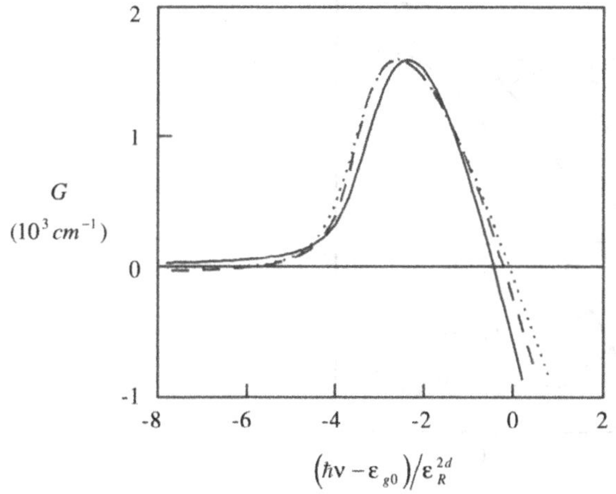


**Fig. 4.27.** Quantum-well gain spectrum obtained from the full calculation (*solid curve*), and from (3.100) with Lorentzian (*dashed curve*) and sech (*dotted curve*) lineshape functions. The carrier density for the solid curve is $N_{2\mathrm{d}} = 1.2 \times 10^{12}\,\mathrm{cm}^{-2}$. The densities and dephasing rates for the other curves are chosen to give the best fit to the solid curve. They are $N_{2\mathrm{d}} = 1.32 \times 10^{12}\,\mathrm{cm}^{-2}$ and $\gamma = 10^{13}\,\mathrm{s}^{-1}$ for the Lorentzian lineshape function and $N_{2\mathrm{d}} = 1.30 \times 10^{12}\,\mathrm{cm}^{-2}$ and $\gamma = 10^{13}\,\mathrm{s}^{-1}$ for the sech lineshape function, respectively

know are not fully consistent. The present and past analyses, however, predict different carrier densities. For the example treated here, the differences, $N_{2d} = 1.20 \times 10^{12}\,\mathrm{cm}^{-2}$ versus $1.30 \times 10^{12}\,\mathrm{cm}^{-2}$ and $1.32 \times 10^{12}\,\mathrm{cm}^{-2}$, for the sech and Lorentzian lineshape functions, respectively, are most likely within the experimental uncertainty.

Figure 4.28 shows the spectra for the carrier-induced refractive index change, $\mathrm{d}\phi/\mathrm{d}z$, for different carrier densities. The dashed curves are computed using the relaxation rate approximation with a Lorentzian lineshape function and a dephasing rate, $\gamma = 10^{13}\,\mathrm{s}^{-1}$. Clearly seen is the smoothing of the curves in the full calculation, a similar effect as the smoothing occurring in the gain spectrum. Figure 4.29 shows the antiguiding ($\alpha$) spectra for different carrier densities. The points indicate the locations of the peak gain. Note that for the low density of $9 \times 10^{11}\,\mathrm{cm}^{-2}$, there is considerable difference between the present calculation and the relaxation rate calculation for $\alpha$ at the gain peak. This results from the fact that interband Coulomb correlation effects are more pronounced at low densities due to a partial cancellation of the diagonal and nondiagional scattering contributions. The presence of structure in the spectra at higher densities is mostly a consequence of the Hartree-Fock many-body contributions.

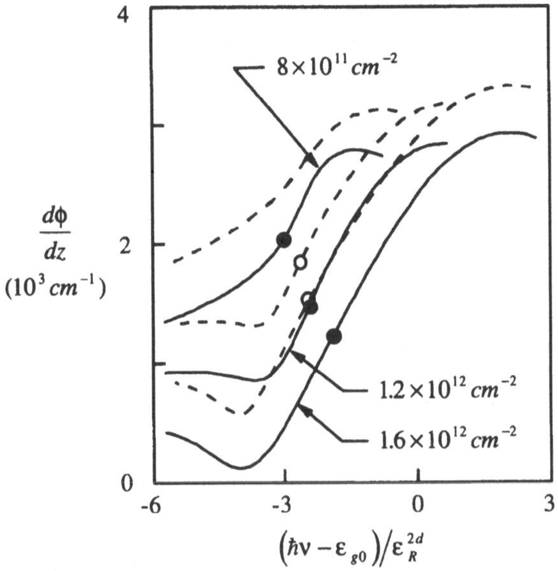


**Fig. 4.28.** Carrier induced phase shift spectra for different carrier densities. The *solid lines* show the results of the full calculation. The *dashed curves* are obtained using (3.100) and a Lorentzian lineshape function with the dephasing rate, $\gamma = 10^{13}\,\mathrm{s}^{-1}$. The *points* indicate the values at the gain peaks. The *lowest solid* and *dashed curve* is for the density $N_{2d} = 1.6 \times 10^{12}\,\mathrm{cm}^{-2}$, the *middle curves* are for $N_{2d} = 1.2 \times 10^{12}\,\mathrm{cm}^{-2}$, and the *top curves* are for $N_{2d} = 8 \times 10^{11}\,\mathrm{cm}^{-2}$, respectively

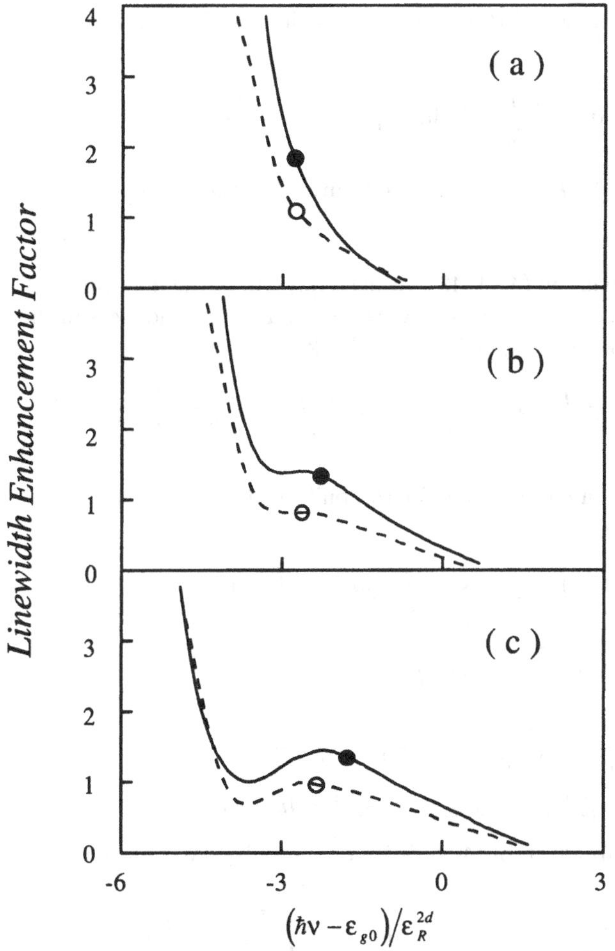


**Fig. 4.29a–c.** Linewidth enhancement (or antiguiding factor) spectra for the carrier densities (**a**) $9 \times 10^{11}$, (**b**) $1.3 \times 10^{12}$, and (**c**) $1.7 \times 10^{12}\,\mathrm{cm}^{-2}$. The *solid curves* are the results of the full calculations. The *dashed curves* are obtained using (3.100) and a Lorentzian lineshape function with the dephasing rate, $\gamma = 10^{13}\,\mathrm{s}^{-1}$. The points indicate the values at the peaks of the respective gain spectra

### 4.4.2 Bulk-Material

The approach used to set up the numerical analysis for bulk semiconductor media is very similar to that for the quantum-well problem. The basic differences are the use of (3.4) instead of (3.5) for the Coulomb potential, and

$$\sum_{\boldsymbol{k}'} \to \int_0^\infty \mathrm{d}k'\, k'^2 \int_0^\pi \mathrm{d}\theta' \sin\theta' \int_0^{2\pi} \mathrm{d}\phi' \tag{4.84}$$

instead of (4.34). The diagonal Hartree-Fock contribution is then

$$\begin{aligned}\Delta\varepsilon_x &= \Theta_{kk}\\ &= -\frac{1}{\hbar}\frac{V}{(2\pi)^3}\int_0^\infty \mathrm{d}k'\,k'^2\int_0^\pi \mathrm{d}\theta'\sin\theta'\int_0^{2\pi}\mathrm{d}\phi'\,\frac{e^2(f_{\mathrm{e}k}+f_{\mathrm{h}k})}{\epsilon_{\mathrm{b}}Vq(k',\theta',\phi')^2}\\ &= -\frac{e^2}{8\pi^3\epsilon_{\mathrm{b}}\hbar}\int_0^\infty \mathrm{d}k'\,k'^2(w_{k'}+1)\int_0^\pi \mathrm{d}\theta'\sin\theta'\int_0^{2\pi}\mathrm{d}\phi'\,\frac{1}{q(k',\theta',\phi')^2}\ ,\end{aligned} \tag{4.85}$$

where $\boldsymbol{q}$ and $w_k$ are given by (4.44, 43), respectively. As in the quantum-well problem, we assume that there are no angular ($\phi'$ and $\theta'$) dependences in the polarization. Then, the nondiagonal Hartree-Fock contribution is

$$\sum_{k'}\Theta_{k'k}p_{k'} = w_k\frac{e^2}{8\pi^3\epsilon_{\mathrm{b}}\hbar}\int_0^\infty \mathrm{d}k'\,k'^2\int_0^\pi \mathrm{d}\theta'\sin\theta'\int_0^{2\pi}\mathrm{d}\phi'\,\frac{1}{q(k',\theta',\phi')^2}\,p_{k'}\ . \tag{4.86}$$

The diagonal correlation contribution for the bulk medium is

$$\begin{aligned}\Lambda_{kk} &= \Gamma_k - \mathrm{i}\Delta_k\\ &= \frac{1}{2\pi^2\hbar}\left(\frac{e}{2\pi}\right)^4\int_0^\infty \mathrm{d}q\sum_{a,b=e,h}\int_0^\pi \mathrm{d}\theta\sin\theta\int_0^{2\pi}\mathrm{d}\phi\\ &\quad\times\int_0^\infty \mathrm{d}k'\,k'^2\int_0^\pi \mathrm{d}\theta'\sin\theta'\int_0^{2\pi}\mathrm{d}\phi'\\ &\quad\times\left(\frac{1}{q^2\epsilon^2(q)}-\frac{\delta_{a,b}}{2}\frac{1}{q'^2\epsilon(q)\epsilon(q')}\right)\frac{\Delta+\mathrm{i}\delta\varepsilon}{\Delta^2+\delta\varepsilon^2}\\ &\quad\times\left[(1-f_{ak_f})f_{bk'}(1-f_{bk'_f})+f_{ak_f}(1-f_{bk'})f_{bk'_f}\right]\ ,\end{aligned} \tag{4.87}$$

where $\boldsymbol{k}_f$, $\boldsymbol{k}'_f$ and $\boldsymbol{q}'$ are given in (4.56–58). For the bulk-material, the static Lindhard formula is

$$\epsilon(q) = 1-\frac{e^2}{8\pi^3\epsilon_{\mathrm{b}}q^2}2\sum_{a=e,h}\int_0^\infty \mathrm{d}k'\,k'^2\int_0^\pi \mathrm{d}\theta'\sin\theta'\int_0^{2\pi}\mathrm{d}\phi'\,\frac{f_{a,|\boldsymbol{k}'+\boldsymbol{q}|}-n_{ak'}}{\varepsilon_{a,|\boldsymbol{k}'+\boldsymbol{q}|}-\varepsilon_{ak'}}\ . \tag{4.88}$$

Finally, the nondiagonal correlation contribution is

$$\begin{aligned}\lambda'(q,k) &= \frac{1}{2\pi^2\hbar}\left(\frac{e}{2\pi}\right)^4\sum_{a,b=e,h}\int_0^\pi \mathrm{d}\theta\sin\theta\int_0^{2\pi}\mathrm{d}\phi\\ &\quad\times\int_0^\infty \mathrm{d}k'\,k'^2\int_0^\pi \mathrm{d}\theta'\sin\theta'\int_0^{2\pi}\mathrm{d}\phi'\\ &\quad\times\left(\frac{1}{q^2\epsilon^2(q)}-\frac{\delta_{a,b}}{2}\frac{1}{q'^2\epsilon(q)\epsilon(q')}\right)\frac{\Delta-i\delta\varepsilon}{\Delta^2+\delta\varepsilon^2}\\ &\quad\times\left[f_{a,k}f_{bk'}(1-f_{bk'_f})+(1-f_{a,k})(1-f_{bk'})f_{bk'_f}\right]\ .\end{aligned} \tag{4.89}$$

At this point, the remaining equations needed for the numerical analysis of the bulk-material are identical to (4.66, 67, 71, 72, 74, 75).

## 4.5 Carrier-Phonon Scattering

Even though the carrier-carrier scattering process may dominate the fast carrier redistribution under typical laser densities, this process does not dissipate energy since the kinetic energy is one of the conserved quantities in carrier-carrier collisions, see Sect. 4.2.

The most important source of energy dissipation, i.e., "carrier cooling" is caused by the coupling of the electronic system to the lattice. The dominant part of this carrier-phonon coupling can be modeled by the interaction Hamiltonian

$$H_{\mathrm{e-}p} = \sum_{\boldsymbol{k},\boldsymbol{q}} \hbar G_q a^{\dagger}_{\boldsymbol{k}+\boldsymbol{q}} a_{\boldsymbol{k}} \left( b_{\boldsymbol{q}} + b^{\dagger}_{-\boldsymbol{q}} \right) \quad , \tag{4.90}$$

where $b_{\boldsymbol{q}}$ and $b^{\dagger}_{\boldsymbol{q}}$ are the annihilation and creation operators of longitudinal optical (LO) phonons, which are the quanta of the longitudinal polarization oscillations due to ionic displacements in a polar semiconductor. Equation (4.90) describes how an electron can be scattered in its band by emitting or absorbing one LO phonon.

Generally, scattering between electron (hole) and other optical as well as acoustical phonons exists in semiconductors. All these effects can be treated reasonably well at the level of quantum Boltzmann equations. Here, we concentrate on the LO phonon coupling as a representative, and often dominant scattering mechanism.

The matrix element $G_q$ for the linear interaction of the electrons with the lattice polarization is

$$G_q^2 = \frac{\omega_{\mathrm{LO}} V_{sq}}{2\hbar} \left( \frac{1}{\epsilon_\infty} - \frac{1}{\epsilon_0} \right) \quad , \tag{4.91}$$

where $\omega_{\mathrm{LO}}$ is the LO-phonon frequency, and $\epsilon_0$ and $\epsilon_\infty$ are the low- and high-frequency background dielectric constants of the medium. The coupling described by (4.90, 91) is usually called Fröhlich electron-LO phonon coupling. This coupling influences both, the carrier intraband relaxation and the electron-hole interband kinetics.

For example, the electron-LO phonon coupling is one of the mechanisms behind the low-frequency lineshape (Urbach tail of the band-edge absorption) because it gives rise to (not spectrally resolved) phonon side bands in which not only a photon but also one or several thermal phonons are absorbed. However, for our present discussion it is more important to study the electron intraband relaxation due to electron-LO phonon coupling. For this purpose we evaluate

$$\left. \frac{\mathrm{d}n_{\mathrm{e}\boldsymbol{k}}}{\mathrm{d}t} \right|_{\mathrm{e-}p} = \frac{\mathrm{i}}{\hbar} \left[ H_{\mathrm{e-}p}, n_{\mathrm{e}\boldsymbol{k}} \right] \quad , \tag{4.92}$$

using second-order perturbation theory, i.e., we solve (4.92) by formal integration and iterate the result twice. In this way we obtain

$$\left.\frac{dn_{ek}}{dt}\right|_{e-p} = -2\pi \sum_{q,\pm} G_q^2 \delta(\Delta^{\pm}_{k,q}) \left[ n_{ek}(1-n_{e,k-q}) \left( n_{ph,\pm q} + \frac{1}{2} \pm \frac{1}{2} \right) - (1-n_{ek})\, n_{e,k-q} \left( n_{ph,\pm q} + \frac{1}{2} \mp \frac{1}{2} \right) \right] , \tag{4.93}$$

which is the Boltzmann collision integral for electron-LO phonon scattering. Here, $n_{ph}$ is the phonon population and the frequency differences $\Delta^{\pm}_{k,q}$ are given by

$$\hbar \Delta^{\pm}_{k,q} = \varepsilon_k - \left(\varepsilon_{k-q} \pm \hbar\omega_{LO}\right) . \tag{4.94}$$

The different terms in (4.93) describe the transition rates in and out of state $\boldsymbol{k}$ under absorption or emission of LO-phonons. The different terms describe the transitions $\boldsymbol{k} \to \boldsymbol{k} - \boldsymbol{q}$ and $\boldsymbol{k} - \boldsymbol{q} \to \boldsymbol{k}$ under emission (upper sign) or absorption (lower sign) of a phonon. The phonon population function $n_{ph,q}$ in general has to be computed self-consistently. However, it is often possible to simplify the problem by assuming that the phonons are in thermal equilibrium, so that

$$n_{ph,q} = \frac{1}{e^{\beta\hbar\omega_{LO}} - 1} , \tag{4.95}$$

i.e., the phonon distribution is described by a thermal Bose function. Examples of the numerical solution of (4.93) without or with the simultaneous inclusion of carrier-carrier scattering, (4.8), are shown in Figs. 4.3, 4, respectively.

For the interband polarization dynamics, the LO-phonon contribution has the form

$$\left.\frac{\partial p_k}{\partial t}\right|_{col} = -\sum_q \left(\Lambda^{e-p}_{kq} + \Lambda^{h-p}_{kq}\right) p_{k+q} , \tag{4.96}$$

where we again kept only the terms that are linear in the polarization. For $q = 0$,

$$\Lambda^{\alpha-p}_{k0} = \frac{1}{\hbar} \sum_q G_q^2 \Big\{ \left[(1-n_{\alpha,k-q}) n_{ph,q} + n_{\alpha,k-q}(1+n_{ph,q})\right] g(\Delta^{+}_{k,q}) + \left[(1-n_{\alpha,k-q})(1+n_{ph,q}) + n_{\alpha,k-q} n_{ph,q}\right] g(\Delta^{-}_{k,q}) \Big\} , \tag{4.97}$$

and for $q \neq 0$,

$$\Lambda^{\alpha-p}_{kq} = \frac{1}{\hbar} G_q^2 \Big\{ \left[(1-n_{\alpha,k}) n_{ph,q} + n_{\alpha,k}(1+n_{ph,q})\right] g(\Delta^{+}_{k,q}) + \left[(1-n_{\alpha,k})(1+n_{ph,q}) + n_{\alpha,k} n_{ph,q}\right] g(\Delta^{-}_{k,q}) \Big\} , \tag{4.98}$$

where $g(\Delta)$ is defined in (4.16).

## 4.6 Characteristic Relaxation Times

The notion of relaxation times is a very well established concept in atomic and molecular optics, and is also widely used in semiconductors. The basic concept can be understood, e.g., by analyzing a system that is coupled to incoherent reservoirs (heat baths). The system can then be de-excited through transfer of energy into the reservoir. As long as the reservoir is infinitely big and does not act back on the system, the relaxation of the system may be to a good approximation characterized by relaxation times that are basically determined by the coupling strength (coupling matrix element) of system and reservoir.

As we have seen in the previous sections, the relevant scattering processes in a semiconductor electron-hole plasma are mostly *not* of the reservoir coupling type. For example, for carrier-carrier scattering, the carriers are their own 'reservoir' and the occupation probabilities of the carrier states enter in a highly nonlinear fashion. Similarly, carrier-phonon scattering is also a process which depends nonlinearly on the carrier distribution functions.

In the simplest microscopic approximation all particle scattering processes lead to Boltzmann-type collision terms, with both, in- and out-scattering contributions. Hence, the characteristic *times* associated with these pocesses depend strongly on the situation under investigation, e.g., carrier relaxation due to scattering with phonon occurs at the time-scale of hundreds of femtoseconds if one considers a nonequilibrium carrier distribution excited energetically high in an otherwise empty band. However, the same scattering process yields relaxation times in the picosecond range if one considers situations where a quasi-equilibrium electron-hole plasma occupies the energetically lowest states.

With these precautions in mind, we concentrate the discussion in this section to typical semiconductor laser configurations, e.g. conventional III–V laser compounds at room temperature. For this situation, numerical evaluations of the electron-phonon Boltzmann equation yield scattering times on the order of picoseconds under normal operating conditions. Hence, in semiconductor lasers the carrier-carrier scattering is the dominant contributor to the dipole dephasing rate constant $\gamma_{\mathbf{k}}$.

In describing the carrier probabilities $n_{\alpha\mathbf{k}}$ themselves, the carrier-carrier scattering rates $\gamma_{\alpha\mathbf{k}}\{n\}$ dominate the response on subpicosecond time scales. Superimposed on this fast response is the relatively slow response resulting from radiative and nonradiative recombination and pumping processes along with the somewhat faster response that attempts to equilibrate the plasma and lattice temperatures via carrier-phonon scattering. These processes change the quasi-equilibrium Fermi-Dirac distributions to which the carrier probabilities are driven by the carrier-carrier scattering. Associated with these slower responses are $T_1$'s on the order of nanoseconds for the total carrier density and picoseconds for the temperature equilibration. Hence in general, semiconductor gain media involve a hierarchy of $T_1$ relaxation times.

If we are interested in phenomena that vary little in the carrier-carrier relaxation times, we may be able to adiabatically eliminate the corresponding transients by assuming that the carrier distributions are described by Fermi-Dirac distributions. This is again the quasi-equilibrium approximation and the slow transient response may be describeable using a single long $T_1$.

However, if we wish to study transient phenomena on a subpicosecond time scale, we might wonder how to define a "fast" $T_{1k}$. Such a $T_{1k}$ would be the lifetime of the inversion

$$d_k = n_{ek} + n_{hk} - 1 \tag{4.99}$$

as it shows up in the equation of motion

$$\frac{d}{dt} d_k = -\frac{d_k}{T_{1k}} + \text{other terms} \ . \tag{4.100}$$

Substituting (4.12, 99) into (4.100), we have

$$\left.\frac{d}{dt} d_k\right|_{col} = -\gamma_{ek}\{n\} n_{ek} - \gamma_{hk}\{n\} n_{hk} + \Sigma^{in}_{ek}\{n\} + \Sigma^{in}_{hk}\{n\} \ . \tag{4.101}$$

To write $d_k$ on the RHS of this equation, we introduce the probability "sum"

$$s_k = n_{ek} + (1 - n_{hk}) \ , \tag{4.102}$$

which complements the probability difference $d_k = n_{ek} - (1 - n_{hk})$. In terms of $d_k$ and $s_k$, we have that $n_{ek} = (d_k + s_k)/2$ and $n_{hk} = (d_k - s_k)/2 + 1$. Substituting these expressions into the equation of motion (4.101), we have

$$\begin{aligned} \left.\frac{d}{dt} d_k\right|_{col} = & -\frac{1}{2}(\gamma_{ek}\{n\} + \gamma_{hk}\{n\}) d_k - \frac{1}{2}(\gamma_{ek}\{n\} - \gamma_{hk}\{n\}) s_k \\ & - \gamma_{hk}\{n\} + \Sigma^{in}_{ek}\{n\} + \Sigma^{in}_{hk}\{n\} \ . \end{aligned} \tag{4.103}$$

Unless the electron and hole effective masses are equal, $\gamma_{ek}\{n\} \neq \gamma_{hk}\{n\}$, and we would not expect to be able to describe the fast response with a single $T_{1k}$. A similar situation is met in the case of the two-level atom problem with different upper- and lower-level decay constants. There, it is fairly well-known that the $T_1$ given by

$$T_1 = \frac{1}{2}\left(\frac{1}{\gamma_a} + \frac{1}{\gamma_b}\right) \ , \tag{4.104}$$

describes steady-state saturation correctly, although it fails to account for the transient response in general. Following a procedure similar to that used to derive (4.104), we consider the equation of motion for the sum term $s_k$. Using (4.12, 102) and dropping the $\{n\}$ for typographical simplicity, we have

$$\begin{aligned} \left.\frac{d}{dt} s_k\right|_{col} &= -\gamma_{ek} n_{ek} + \Sigma^{in}_{ek} + \gamma_{hk} n_{hk} - \Sigma^{in}_{hk} \\ &= -\frac{1}{2}\gamma_{ek}(d_k + s_k) + \frac{1}{2}\gamma_{hk}(d_k - s_k + 2) + \Sigma^{in}_{ek} - \Sigma^{in}_{hk} \\ &= -\gamma_k s_k - \frac{\gamma_{ek} - \gamma_{hk}}{2} d_k + \gamma_{hk} + \Sigma^{in}_{ek} - \Sigma^{in}_{hk} \ . \end{aligned} \tag{4.105}$$

In steady state ($\dot{s}_k|_{\text{col}} = 0$), this gives

$$s_k = \frac{1}{\gamma_k}\left(\frac{\gamma_{\text{h}k} - \gamma_{\text{e}k}}{2} d_k + \gamma_{\text{h}k} + \Sigma_{\text{e}k}^{\text{in}} - \Sigma_{\text{h}k}^{\text{in}}\right) \quad . \tag{4.106}$$

As a rough approximation to the near steady-state transient behavior of $d_k$, we substitute (4.106) into (4.103) and simplify to find

$$\frac{\mathrm{d}}{\mathrm{d}t} d_k = -\frac{d_k}{T_{1k}} + (\text{functions of } \gamma' s) \quad , \tag{4.107}$$

where the carrier-carrier scattering probability-difference decay time $T_{1k}$ is given approximately by

$$T_{1k} = \frac{1}{2}\left(\frac{1}{\gamma_{\text{e}k}} + \frac{1}{\gamma_{\text{h}k}}\right) \quad . \tag{4.108}$$

As noted above, in the limit of $\gamma_{\text{e}k} = \gamma_{\text{h}k} = \gamma_k$ , $T_{1k} = T_{2k}$. Equation (4.108) is similar to the two-level-atom (4.104), which is also only valid near steady state. The "steady state" considered here, however, is reached to all intents and purposes in a fraction of a picosecond. Hence, (4.108) could be useful in describing saturation involving hole burning by a single-mode field.

Finally, we wish to warn the reader that while relaxation rate approximations are widely used, it is important to remember that they are results of substantial approximations to the full Boltzmann equation and therefore have to be used with care. In general, the relaxation rates have to be obtained from fits to microscopic calculations or they might be used as fitting parameters in comparisons with experiments.

# 5. Bulk Band Structures

The ability to change the energy-level structure of the gain medium through material and structure design is one of the unique properties of semiconductor lasers. To take full advantage of this capability, one needs to be able to predict the band structure that results from a particular material and structure combination. In this chapter and the following, we show a procedure for performing band-structure calculations that are relevant to the laser physicist.

Section 5.1 begins with a review of the important concepts of lattice-periodicity, unit cells, Bloch functions, and energy bands. Section 5.2 describes the zone center ($k \cong 0$) states that are relevant for laser transitions taking place in a III–V compound like GaAs. To calculate the band structure that evolves from these states, we use the Kane or $\boldsymbol{k} \cdot \boldsymbol{p}$ theory, which is described in Sect. 5.3. Kane theory allows one to compute the band structure in the neigborhood of the zone center perturbatively, using the energy eigenvalues and the basic symmetries of the zone-center states. First Sect. 5.4 treats the conduction bands, which have spin-degenerate zone center eigenstates. Then, Sect. 5.5 describes the treatment of the valence bands which have degenerate $l = 1$ zone center eigenstates, and therefore requires the use of degenerate perturbation theory. Specializing to the top valence bands in typical III–V compounds with cubic crystal symmetry, Sect. 5.6 introduces the Luttinger Hamiltonian, whose diagonalization gives the heavy and light hole bands.

## 5.1 Bloch Theorem

An electron in a crystal sees a periodic potential due to the ions present at each lattice site. This potential is modified by more or less localized electrons that are originally bound to each atom making up the crystal. The net result may be approximated by a periodic effective potential $V_0$, such that an electronic energy eigenstate $|\phi_{n\boldsymbol{k}}\rangle$ in the solid obeys the time-independent Schrödinger equation

$$\left(\frac{p^2}{2m_0} + V_0\right) |\phi_{n\boldsymbol{k}}\rangle = \varepsilon_{n\boldsymbol{k}} |\phi_{n\boldsymbol{k}}\rangle \quad , \tag{5.1}$$

where $\boldsymbol{p}$ is the momentum operator, $m_0$ is the mass of the free electron, i.e., of the electron in vacuum, $n$ is the band index, and $\boldsymbol{k}$ is the electron wavevector. Translational symmetry in the lattice dictates that the energy eigenfunctions obey the *Bloch theorem*

$$\langle \boldsymbol{r} + \boldsymbol{R} | \phi_{n\boldsymbol{k}} \rangle = \mathrm{e}^{\mathrm{i}\boldsymbol{k}\cdot\boldsymbol{R}} \langle \boldsymbol{r} | \phi_{n\boldsymbol{k}} \rangle \quad , \tag{5.2}$$

where $\boldsymbol{r}+\boldsymbol{R}$ describes translation by a lattice vector. This condition is fulfilled when

$$\langle \boldsymbol{r} | \phi_{n\boldsymbol{k}} \rangle = \mathrm{e}^{\mathrm{i}\boldsymbol{k}\cdot\boldsymbol{r}} \langle \boldsymbol{r} | n\boldsymbol{k} \rangle \quad , \tag{5.3}$$

where $\langle \boldsymbol{r} | n\boldsymbol{k} \rangle$ is the *lattice periodic function* satisfying

$$\langle \boldsymbol{r} + \boldsymbol{R} | n\boldsymbol{k} \rangle = \langle \boldsymbol{r} | n\boldsymbol{k} \rangle \quad . \tag{5.4}$$

Substituting (5.3) into (5.1), we find the eigenvalue equation for the lattice periodic part of the wavefunction

$$\left( \frac{p^2}{2m_0} + V_0 + \frac{\hbar}{m_0} \boldsymbol{k}\cdot\boldsymbol{p} \right) | n\boldsymbol{k} \rangle = \left( \varepsilon_{nk} - \frac{\hbar^2 k^2}{2m_0} \right) | n\boldsymbol{k} \rangle \quad . \tag{5.5}$$

The eigenstates are orthonormal

$$\langle \phi_{m\boldsymbol{k}} | \phi_{n\boldsymbol{q}} \rangle = \delta_{m,n} \delta_{\boldsymbol{k},\boldsymbol{q}} \quad , \tag{5.6}$$

or equivalently,

$$\langle m\boldsymbol{k} | n\boldsymbol{k} \rangle = \delta_{m,n} \quad . \tag{5.7}$$

The first-principles calculation of energy bands and eigenfunctions in a solid is a specialized area in the field of condensed matter physics. First, one must develop an accurate model for the effective potential $V_0$. Then one solves (5.1), or equivalently (5.5), with the appropriate boundary conditions (5.2) or (5.4). The solution requires complicated numerical computation schemes, and a first-principles solution of all the $\boldsymbol{k}$ states involved in a laser transition can quickly become a prohibitively lengthy process. The $\boldsymbol{k}\cdot\boldsymbol{p}$ theory provides a shortcut through this process. This theory allows one to compute the energy eigenstates and the wavefunctions in the vicinity of any given $k$, and in particular in the vicinity of $k = 0$, which is the region most relevant for optical transitions in lasers. To use the method, we need to know the $k = 0$ eigenstates, which are discussed in the next section.

## 5.2 Electronic States at $\boldsymbol{k} = 0$

For a III–V compound like GaAs, the conduction and valence band states at $k = 0$ are summarized in Fig. 5.1. Without spin-orbit coupling, the $m_s = \pm 1/2$ states are uncoupled and degenerate. For typographical simplicity, we suppress the spin indices in working with $|l\ m_l\rangle$ and include a factor of 2 in the appropriate sums over states. The conduction-band $k = 0$ state has $s$-like

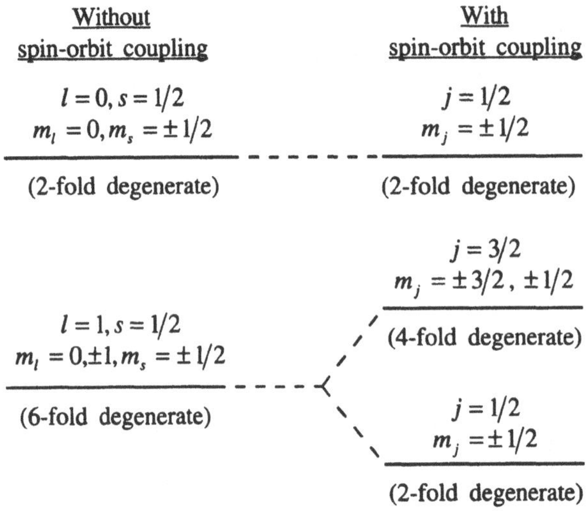


**Fig. 5.1.** The electron eigenstates at $k = 0$ that play a role in optical transitions. In the absence of spin-orbit coupling, the eigenstates are $|lsm_lm_s\rangle$, where $l$, $m_l$ and $s$, $m_s$ are the total and $z$-component quantum numbers for the orbital and spin angular momenta, respectively. With spin-orbit coupling, the eigenstates become $|ls; jm_j\rangle$, where $j$ and $m_j$ are the quantum numbers for the total (orbital plus spin) angular momentum

symmetry, with zero orbital angular momentum. We designate this state by $|S\rangle = |0\ 0\rangle$. The $k = 0$ states of the top valence band have $p$-like symmetry, which may be represented by the $l = 1$ states, $|1\ \pm 1\rangle$ and $|1\ 0\rangle$. In the next section, where we deal with the momentum-operator matrix elements, it is more convenient to work with the following combinations of orbital angular momentum eigenstates $|l\ m_l\rangle$

$$\begin{aligned} |X\rangle &= \frac{1}{\sqrt{2}} [|1\ -1\rangle + |1\ 1\rangle] \quad , \\ |Y\rangle &= \frac{\mathrm{i}}{\sqrt{2}} [|1\ -1\rangle - |1\ 1\rangle] \quad , \\ |Z\rangle &= |1\ 0\rangle \quad . \end{aligned} \tag{5.8}$$

In the coordinate representation, the corresponding eigenfunctions have the symmetry

$$\begin{aligned} \langle \boldsymbol{r}|X\rangle &\propto xf(r) \quad , \\ \langle \boldsymbol{r}|Y\rangle &\propto yf(r) \quad , \\ \langle \boldsymbol{r}|Z\rangle &\propto zf(r) \quad , \end{aligned} \tag{5.9}$$

where $f(r)$ is a spherically symmetric function.

The coupling between the electron spin and orbital angular momenta contributes a spin-orbit term in the Hamiltonian of the form

$$\frac{1}{2m_0^2c^2}(\boldsymbol{S} \times \nabla V) \cdot \boldsymbol{p} = \xi \boldsymbol{S} \cdot \boldsymbol{L} \quad , \tag{5.10}$$

where

$$\xi = \frac{1}{2m_0^2c^2 r}\frac{\mathrm{d}V}{\mathrm{d}r}$$

is the spin-orbit function [*Schiff* (1968)]. To obtain (5.10) we use

$$(\boldsymbol{S} \times \nabla V) \cdot \boldsymbol{p} = (\nabla V \times \boldsymbol{p}) \cdot \boldsymbol{S} \quad . \tag{5.11}$$

For a central potential

$$\nabla V(r) = \frac{\mathrm{d}V}{\mathrm{d}r}\frac{\boldsymbol{r}}{r} \quad , \tag{5.12}$$

so that

$$(\nabla V \times \boldsymbol{p}) \cdot \boldsymbol{S} = \frac{1}{r}\frac{\mathrm{d}V}{\mathrm{d}r}(\boldsymbol{r} \times \boldsymbol{p}) \cdot \boldsymbol{S} = \frac{1}{r}\frac{\mathrm{d}V}{\mathrm{d}r}\boldsymbol{L} \cdot \boldsymbol{S} \quad . \tag{5.13}$$

We assume that the spin-orbit interaction is small compared to the electrostatic interaction (*Russell-Saunders case* or *LS coupling scheme*). Furthermore, since

$$J^2 = (\boldsymbol{L} + \boldsymbol{S})^2 = L^2 + S^2 + 2\boldsymbol{L} \cdot \boldsymbol{S} \quad , \tag{5.14}$$

the new energy eigenstates are also eigenstates of the total angular momentum $\boldsymbol{J}$. These states are denoted as $|l\ s; j\ m_j\rangle$, and they can be expressed in terms of the old ones by

$$|l\ s; j\ m_j\rangle = \sum_{m_l, m_s} |l\ s\ m_l\ m_s\rangle\langle l\ s\ m_l\ m_s|l\ s; j\ m_j\rangle \quad , \tag{5.15}$$

where $\langle l\ s; m_l\ m_s|\,|l\ s; j\ m_j\rangle$ are the Clebsch-Gordan coefficients. For the conduction bands, the new states are simply

$$\left|0\ \frac{1}{2}; \frac{1}{2}\ \frac{1}{2}\right\rangle = \left|0\ \frac{1}{2}\ 0\ \frac{1}{2}\right\rangle$$

$$\left|0\ \frac{1}{2}; \frac{1}{2}\ -\frac{1}{2}\right\rangle = \left|0\ \frac{1}{2}\ 0\ -\frac{1}{2}\right\rangle \quad . \tag{5.16}$$

For the valence bands, $l = 1$ and $s = 1/2$ give $j = 3/2, 1/2$. The Clebsch-Gordan coefficients $\langle l\ s = \frac{1}{2}\ m_l\ m_s|l\ s = \frac{1}{2}; j\ m_j\rangle$ are given by the Table 5.1 (note that $m_l = m_j - m_s$).

Using these coefficients, we find the total angular momentum states for the heavy and light-hole valence bands of GaAs as

$$\left|\frac{3}{2}\ \frac{3}{2}\right\rangle = |1\uparrow\rangle \quad ,$$

$$\left|\frac{3}{2}\ \frac{1}{2}\right\rangle = -\sqrt{\frac{1}{3}}\,|1\downarrow\rangle + \sqrt{\frac{2}{3}}\,|0\uparrow\rangle \quad ,$$

**Table 5.1.** Clebsch-Gordan coefficients

| | $j = l + \frac{1}{2}$ | $j = l - \frac{1}{2}$ |
|---|---|---|
| $m_s = \frac{1}{2}$ | $\sqrt{\frac{l + m_j + 1/2}{2l+1}}$ | $-\sqrt{\frac{l - m_j + 1/2}{2l+1}}$ |
| $m_s = -\frac{1}{2}$ | $\sqrt{\frac{l - m_j + 1/2}{2l+1}}$ | $\sqrt{\frac{l + m_j + 1/2}{2l+1}}$ |

$$\begin{aligned}
\left|\frac{3}{2}\ -\frac{1}{2}\right\rangle &= \sqrt{\frac{2}{3}}\,|0\downarrow\rangle + \sqrt{\frac{1}{3}}\,|-1\uparrow\rangle \quad , \\
\left|\frac{3}{2}\ -\frac{3}{2}\right\rangle &= |-1\downarrow\rangle \quad , \\
\left|\frac{1}{2}\ \frac{1}{2}\right\rangle &= \sqrt{\frac{2}{3}}\,|1\downarrow\rangle + \sqrt{\frac{1}{3}}\,|0\uparrow\rangle \quad , \\
\left|\frac{1}{2}\ -\frac{1}{2}\right\rangle &= \sqrt{\frac{1}{3}}\,|0\downarrow\rangle - \sqrt{\frac{2}{3}}\,|-1\uparrow\rangle \quad .
\end{aligned} \tag{5.17}$$

Since in this discussion $l$ and $s$ are always 1 and 1/2, respectively, we use the following abbreviated notation for the states:

$$\begin{aligned}
&|l\ s; j\ m_j\rangle \rightarrow |j\ m_j\rangle \\
&|l\ s\ m_l\ m_s\rangle \rightarrow |m_l\ m_s\rangle \quad ,
\end{aligned}$$

with $\uparrow$ ($\downarrow$) denoting $m_s = 1/2(-1/2)$. In terms of $|X\uparrow\rangle = |X\rangle|\uparrow\rangle$, etc.,

$$\begin{aligned}
\left|\frac{3}{2}\ \frac{3}{2}\right\rangle &= -\sqrt{\frac{1}{2}}\,[|X\uparrow\rangle + \mathrm{i}\,|Y\uparrow\rangle] \quad , \\
\left|\frac{3}{2}\ \frac{1}{2}\right\rangle &= -\sqrt{\frac{1}{6}}\,[|X\downarrow\rangle + \mathrm{i}\,|Y\downarrow\rangle] + \sqrt{\frac{2}{3}}\,|Z\uparrow\rangle \quad , \\
\left|\frac{3}{2}\ -\frac{1}{2}\right\rangle &= \sqrt{\frac{1}{6}}\,[|X\uparrow\rangle - \mathrm{i}\,|Y\uparrow\rangle] + \sqrt{\frac{2}{3}}\,|Z\downarrow\rangle \quad , \\
\left|\frac{3}{2}\ -\frac{3}{2}\right\rangle &= \sqrt{\frac{1}{2}}\,[|X\downarrow\rangle - \mathrm{i}\,|Y\downarrow\rangle] \quad , \\
\left|\frac{1}{2}\ \frac{1}{2}\right\rangle &= \sqrt{\frac{1}{3}}\,[|X\downarrow\rangle + \mathrm{i}\,|Y\downarrow\rangle + |Z\uparrow\rangle] \quad , \\
\left|\frac{1}{2}\ -\frac{1}{2}\right\rangle &= \sqrt{\frac{1}{3}}\,[-|X\uparrow\rangle + \mathrm{i}\,|Y\uparrow\rangle + |Z\downarrow\rangle] \quad .
\end{aligned} \tag{5.18}$$

## 5.3 $\boldsymbol{k} \cdot \boldsymbol{p}$ Theory

Assuming that we know the band energies and eigenstates at the momentum value $\boldsymbol{k}_0$, we show in this section how the $\boldsymbol{k} \cdot \boldsymbol{p}$ theory allows one to compute the states in the vicinity of $\boldsymbol{k}_0$. Note that $|\phi_{n\boldsymbol{k}}\rangle$ cannot be expanded in terms of $|\phi_{m\boldsymbol{k}_0}\rangle$, because these functions are orthogonal to one another whenever $\boldsymbol{k}_0 \neq \boldsymbol{k}$. However, since the states for each $\boldsymbol{k}$ are complete, we may write

$$|n\boldsymbol{k}\rangle = \sum_m c_{nm\boldsymbol{k}} |m\boldsymbol{k}_0\rangle \quad . \tag{5.19}$$

If $\langle \boldsymbol{r}|n\boldsymbol{k}_0\rangle$ is periodic with respect to translation by a lattice vector, then $\langle \boldsymbol{r}|n\boldsymbol{k}\rangle$ is also periodic, so that the Bloch theorem is satisfied. The problem is then to use (5.5) to obtain the expansion coefficient $c_{nm\boldsymbol{k}}$. For this purpose we first rewrite (5.5) as

$$(H_0 + \lambda H_1)|n\boldsymbol{k}\rangle = W_{n\boldsymbol{k}}|n\boldsymbol{k}\rangle \quad , \tag{5.20}$$

where

$$H_0 = \frac{p^2}{2m_0} + V_0 \quad , \tag{5.21}$$

$$H_1 = \frac{\hbar}{m_0} \boldsymbol{k} \cdot \boldsymbol{p} \quad , \tag{5.22}$$

and

$$W_{n\boldsymbol{k}} = \varepsilon_{n\boldsymbol{k}} - \frac{\hbar^2 k^2}{2m_0} \quad . \tag{5.23}$$

We treat $H_1$ as a perturbation and use $\lambda$ to keep track of the order of the terms in the perturbation expansion. Later we then put $\lambda = 1$

In the following we discuss the analysis for the example of $k_0 = 0$, i.e., for the states around the center of the Brillouin zone ($\Gamma$ point). These states are the most relevant ones for the description of optical transitions near the semiconductor absorption edge. The zone-center eigenstates and energies

$$|m\ \boldsymbol{k} = 0\rangle \equiv |m\rangle \quad , \qquad \varepsilon_{m\boldsymbol{k}} \equiv \varepsilon_m \quad , \tag{5.24}$$

are solutions of the Schrödinger equation

$$H_0|m\rangle = \varepsilon_m|m\rangle \quad . \tag{5.25}$$

At first we ignore the spin-orbit coupling so that the states $|m\rangle$ are given by $|l\ m_l\rangle$. For the conduction band, we have the nondegenerate eigenstate $|S\rangle = |0\ 0\rangle$, and for the valence band, we have the degenerate eigenstates $|1\ \pm 1\rangle$ and $|1\ 0\rangle$, or $|X\rangle$, $|Y\rangle$ and $|Z\rangle$.

## 5.4 Conduction Bands

We now apply perturbation theory to solve the Schrödinger equation for the lattice periodic functions of a bulk semiconductor. First, we treat the case for bands with only spin-degenerate zone center eigenstates, such as the conduction band. Suppose we wish to find $|n\boldsymbol{k}\rangle$, which is the state that reduces to $|n\rangle$ at $k = 0$. First, we write $|n\boldsymbol{k}\rangle$ and $W_{n\boldsymbol{k}}$ as the expansions

$$|n\boldsymbol{k}\rangle = |n\boldsymbol{k}\rangle_0 + \lambda|n\boldsymbol{k}\rangle_1 + \lambda^2|n\boldsymbol{k}\rangle_2 + \cdots \quad , \tag{5.26}$$

$$W_{n\boldsymbol{k}} = W_{n\boldsymbol{k}}^{(0)} + \lambda W_{n\boldsymbol{k}}^{(1)} + \lambda^2 W_{n\boldsymbol{k}}^{(2)} + \cdots \quad . \tag{5.27}$$

Substituting (5.26, 27) into (5.20) and collecting terms with equal powers of $\lambda$, we get

$$\left(H_0 - W_{n\boldsymbol{k}}^{(0)}\right)|n\boldsymbol{k}\rangle_0 = 0 \quad , \tag{5.28}$$

$$\left(H_0 - W_{n\boldsymbol{k}}^{(0)}\right)|n\boldsymbol{k}\rangle_1 = \left(W_{n\boldsymbol{k}}^{(1)} - H_1\right)|n\boldsymbol{k}\rangle_0 \quad , \tag{5.29}$$

$$\left(H_0 - W_{n\boldsymbol{k}}^{(0)}\right)|n\boldsymbol{k}\rangle_2 = \left(W_{n\boldsymbol{k}}^{(1)} - H_1\right)|n\boldsymbol{k}\rangle_1 + W_{n\boldsymbol{k}}^{(2)}|n\boldsymbol{k}\rangle_0 \quad , \tag{5.30}$$

and so on. Equation (5.28) gives

$$W_{n\boldsymbol{k}}^{(0)} = \varepsilon_n \quad \text{and} \quad |n\boldsymbol{k}\rangle_0 = |n\rangle \quad . \tag{5.31}$$

Taking the scalar product of (5.29) with $\langle n|$, we find

$$W_{n\boldsymbol{k}}^{(1)} = \langle n|H_1|n\rangle = 0 \quad , \tag{5.32}$$

which vanishes because $H_0$ has inversion symmetry, so that $|n\rangle$ has definite parity. Consequently, all diagonal matrix elements of $\boldsymbol{p}$ vanish, since $\boldsymbol{p}$ changes sign under space inversion:

$$\langle n|\boldsymbol{p}|n\rangle = \int \mathrm{d}^3r\, \langle n|\boldsymbol{r}\rangle \frac{\hbar}{\mathrm{i}}\nabla\langle \boldsymbol{r}|n\rangle = 0 \quad , \tag{5.33}$$

and we have no first-order correction to the energy. If we take the scalar product of (5.29) with $\langle m|$, where $m \neq n$, then

$$\langle m|n\boldsymbol{k}\rangle_1 = \frac{\langle m|H_1|n\rangle}{\varepsilon_n - \varepsilon_m} \quad , \tag{5.34}$$

provided $\varepsilon_m \neq \varepsilon_n$. Then to first order in the perturbation, the eigenstate is

$$|n\boldsymbol{k}\rangle_1 = \sum_m |m\rangle\langle m|n\boldsymbol{k}\rangle_1 = \sum_{m\neq n} \frac{\langle m|H_1|n\rangle}{\varepsilon_n - \varepsilon_m}|m\rangle \quad . \tag{5.35}$$

Note, that we could add a term containing $|n\rangle$ to $|n\boldsymbol{k}\rangle_1$ and still satisfy (5.29). However, we choose not to do so in order to make the unperturbed eigenstate orthogonal to all the higher-order corrections, that is,

$$\langle n|n\boldsymbol{k}\rangle_j = 0 \tag{5.36}$$

for $j \neq 0$. Taking the scalar product of (5.30) with $\langle n|$ and using (5.34), we find

$$W_{nk}^{(2)} = \sum_{m \neq n} \frac{|\langle n|H_1|m\rangle|^2}{\varepsilon_n - \varepsilon_m} \quad . \tag{5.37}$$

From (5.23, 37), we have

$$\varepsilon_{nk} = \varepsilon_n + \frac{\hbar^2 k^2}{2m_0} + \frac{\hbar^2}{m_0^2} \sum_{m \neq n} \frac{|\langle n|\boldsymbol{k}\cdot\boldsymbol{p}|m\rangle|^2}{\varepsilon_n - \varepsilon_m} \quad . \tag{5.38}$$

Given that the eigenstates have definite parity, one can readily see that

$$\langle n|p_\alpha|m\rangle\langle m|p_\beta|n\rangle = 0 \tag{5.39}$$

for $\alpha \neq \beta$. Furthermore, for bulk III–V semiconductors,

$$|\langle n|p_x|m\rangle|^2 = |\langle n|p_y|m\rangle|^2 = |\langle n|p_z|m\rangle|^2 \quad , \tag{5.40}$$

because of the symmetry about the zone center. Then, we can write

$$\varepsilon_{nk} = \varepsilon_n + \frac{\hbar^2 k^2}{2m_{n,\text{eff}}} \quad , \tag{5.41}$$

where the effective mass $m_{n,\text{eff}}$ is defined by

$$\frac{1}{m_{n,\text{eff}}} = \frac{1}{m_0}\left(1 + \frac{2}{m_0} \sum_{m \neq n} \frac{|\langle n|p_x|m\rangle|^2}{\varepsilon_n - \varepsilon_m}\right) \quad . \tag{5.42}$$

Values of the effective masses for the lowest conduction bands are

| | $m_e/m_0$ |
|---|---|
| GaAs | 0.0665 |
| InAs | 0.027 |
| InP | 0.064 |
| GaP | 0.15 |
| AlP | 0.22 |

(5.43)

## 5.5 Valence Bands

For the valence bands, which have degenerate $l = 1$ zone center eigenstates, we need to use degenerate perturbation theory. Let us assume that the first $N$ eigenstates are $|1\rangle, |2\rangle, |3\rangle, \ldots, |N\rangle$, so that

$$\varepsilon_1 = \varepsilon_2 = \varepsilon_3 = \cdots = \varepsilon_N \quad . \tag{5.44}$$

The procedure involves finding $N$ new orthonormal eigenstates $|n\rangle'$ that evolve continuously into nondegenerate eigenstates as we leave the zone center. We write the new eigenstates as linear superpositions of the old ones, that is

$$|n\rangle' = \sum_{m=1}^{N} |m\rangle\langle m|n\rangle' \ , \tag{5.45}$$

where $1 \le n \le N$ and $\langle m|n\rangle' = 0$ for $m > N$. The scalar product of $\langle j|$ with (5.29) gives

$$\sum_{m=1}^{N} \langle j|H_1|m\rangle\langle m|n\rangle' - W_{nk}^{(1)}\langle j|n\rangle' = 0 \qquad \text{for} \quad 1 \le j \le N \tag{5.46}$$

and

$$\langle j|n\rangle' = \sum_{m=1}^{N} \frac{\langle j|H_1|m\rangle\langle m|n\rangle'}{\varepsilon_1 - \varepsilon_j} \qquad \text{for} \quad j > N \ . \tag{5.47}$$

Equation (5.46) provides us with $N$ coupled equations that can be solved for the $N$ new eigenstates. However, if $\langle j|H_1|m\rangle = 0$ for all $j$ and $m$ between 1 and $N$, then $W_{nk}^{(1)} = 0$ and we have to move to the next higher order in the perturbation. This turns out to be the case for every group of degenerate states, since all states within each group have the same orbital angular momentum. Repeating the above procedure with the second-order equation (5.30), we find

$$\sum_{l=1}^{N} \sum_{m>N} \frac{\langle j|H_1|m\rangle\langle m|H_1|l\rangle}{\varepsilon_1 - \varepsilon_m} \langle l|n\rangle' - W_{nk}^{(2)}\langle j|n\rangle' = 0 \ , \tag{5.48}$$

for $1 \le j \le N$. Equation (5.48) gives a set of $N$ coupled homogeneous equations that can be solved for $W_{nk}^{(2)}$ and $|n\rangle'$.

In matrix formalism, we have

$$\underline{H}\ \boldsymbol{A} = \varepsilon_{nk}\ \boldsymbol{A} \ , \tag{5.49}$$

where $\underline{H}$ is an $N \times N$ matrix with the matrix elements

$$H_{ij} = \left(\varepsilon_1 + \frac{\hbar^2 k^2}{2m_0}\right)\delta_{i,j} + \frac{\hbar^2}{m_0^2} \sum_{m>N} \frac{\boldsymbol{k}\cdot\boldsymbol{p}_{lm}\ \boldsymbol{k}\cdot\boldsymbol{p}_{mj}}{\varepsilon_1 - \varepsilon_m} \ . \tag{5.50}$$

Note that the degenerate $m = 1$ to $N$ states are excluded from the summation. In other words, the summation involves only the remote bands. The elements of the vector $\boldsymbol{A}$ are the probability amplitudes

$$A_j = \langle j|n\rangle' \ . \tag{5.51}$$

For nontrivial solutions,

$$\det(\underline{H} - \varepsilon_{nk}\underline{I}) = 0 \ , \tag{5.52}$$

where $\underline{I}$ is the identity matrix. The solutions of the resulting secular equation yield $\varepsilon_{nk}$ for $n = 1$ to $N$. The transformation matrix $\underline{S}$ that diagonalizes $\underline{H}$ gives the new eigenstates at $k = 0$ according to

$$\underline{S}^{-1}\underline{H}\ \underline{S} = \underline{E} \ , \tag{5.53}$$

or

$$S_{mn} = \langle m|n\rangle' \ , \tag{5.54}$$

and the elements of the matrix $\underline{E}$ are

$$\varepsilon_{nm} = \varepsilon_{n\boldsymbol{k}} \, \delta_{n,m} \ . \tag{5.55}$$

## 5.6 Luttinger Hamiltonian

Specializing our discussion to the two top valence bands, we first ignore the spin of the electron so that the $k = 0$ eigenstates have $l = 1$ and are threefold degenerate, with $m_l = 0, \pm 1$. To second order in the perturbation, the energy of the $n$th band is

$$\varepsilon_{n\boldsymbol{k}} = \varepsilon_n + \frac{\hbar^2 k^2}{2m_0} + W_{n\boldsymbol{k}}^{(2)} \ , \tag{5.56}$$

where $W_{n\boldsymbol{k}}^{(2)}$ is given by (5.48). Hence we have the set of three equations

$$\sum_{l=1}^{3} \left[ \sum_{m>3} \frac{\langle j|H_1|m\rangle\langle m|H_1|l\rangle}{\varepsilon_1 - \varepsilon_m} + \left( \varepsilon_1 + \frac{\hbar^2 k^2}{2m_0} - \varepsilon_{n\boldsymbol{k}} \right) \delta_{j,l} \right] \langle l|n\boldsymbol{k}\rangle = 0 \ , \tag{5.57}$$

where $l = 1$ states are excluded in the $m$ summation, or in other words, the summation involves only the remote bands. Nontrivial solutions of this set of $N$ coupled homogeneous equations occur only if

$$\det(\underline{H} - \varepsilon_{n\boldsymbol{k}} \underline{I}) = 0 \ , \tag{5.58}$$

where $\underline{H}$ is a $3 \times 3$ matrix, whose elements are

$$H_{jl} = \left( \varepsilon_1 + \frac{\hbar^2 k^2}{2m_0} \right) \delta_{j,l} + \sum_{m>3} \frac{\langle j|H_1|m\rangle\langle m|H_1|l\rangle}{\varepsilon_1 - \varepsilon_m} \ . \tag{5.59}$$

The computation of the matrix elements is straightforward. For example, if we order the basis states so that $|j\rangle = 1$, 2 and 3 are $X$, $Y$ and $Z$, then

$$H_{11} \equiv \langle X|H|X\rangle = \varepsilon_1 + \frac{\hbar^2 k^2}{2m_0} + \sum_{m>3} \frac{|\langle X|H_1|m\rangle|^2}{\varepsilon_1 - \varepsilon_m} \ . \tag{5.60}$$

Using (5.9), we can convince ourselves that

$$\frac{m_0^2}{\hbar^2} |\langle X|H_1|m\rangle|^2 = |\langle X|p_x|m\rangle|^2 k_x^2 + |\langle X|p_y|m\rangle|^2 k_y^2 + |\langle X|p_z|m\rangle|^2 k_z^2 \ , \tag{5.61}$$

so that

$$H_{11} = \varepsilon_1 + \sum_{j=x,y,z} \left( \frac{\hbar^2}{2m_0} + \frac{\hbar^2}{m_0^2} \sum_{m>3} \frac{|\langle X|p_j|m\rangle|^2}{\varepsilon_1 - \varepsilon_m} \right) k_j^2 \ . \tag{5.62}$$

Due to the symmetry of III–V compound semiconductors at $k = 0$,

$$|\langle X|p_y|m\rangle|^2 = |\langle X|p_z|m\rangle|^2 \quad , \tag{5.63}$$

which reduces (5.62) to

$$H_{11} = \varepsilon_1 + Ak_x^2 + B\left(k_y^2 + k_z^2\right) \quad , \tag{5.64}$$

where

$$A = \frac{\hbar^2}{2m_0} + \frac{\hbar^2}{m_0^2}\sum_{j>3}\frac{|\langle X|p_x|j\rangle|^2}{\varepsilon_1 - \varepsilon_j} \quad , \tag{5.65}$$

$$B = \frac{\hbar^2}{2m_0} + \frac{\hbar^2}{m_0^2}\sum_{j>3}\frac{|\langle X|p_y|j\rangle|^2}{\varepsilon_1 - \varepsilon_j} \quad . \tag{5.66}$$

The same procedure can be used for the remaining matrix elements, resulting in

$$H = \begin{pmatrix} \varepsilon_1 + Ak_x^2 + B\left(k_y^2 + k_z^2\right) & Ck_xk_y & Ck_xk_z \\ Ck_xk_y & \varepsilon_1 + Ak_y^2 + B\left(k_x^2 + k_z^2\right) & Ck_yk_z \\ Ck_xk_z & Ck_yk_z & \varepsilon_1 + Ak_z^2 + B\left(k_x^2 + k_y^2\right) \end{pmatrix} , \tag{5.67}$$

where

$$C = \frac{\hbar^2}{m_0^2}\sum_{j>3}\frac{\langle X|p_x|j\rangle\langle j|p_y|Y\rangle + \langle X|p_y|j\rangle\langle j|p_x|Y\rangle}{\varepsilon_1 - \varepsilon_j} \quad . \tag{5.68}$$

As discussed in Sect. 5.2, to include the effects of spin-orbit coupling, we need to use the eigenstates of the total (orbit and spin) angular momentum $|j, m_j\rangle$. The spin-orbit coupling removes the degeneracy between states with different total angular momenta. For the $j = 3/2$ and $j = 1/2$ states the energy separation is $\Delta \simeq 9\langle\xi\rangle/8$, where $\langle\xi\rangle$ is the expectation value of the spin-orbit function, usually treated as a parameter obtained from experiment. The spin-orbit energy in GaAs is 0.34 eV. As a consequence of this large energy splitting very often only the lower energy $j = \pm 3/2$ states are directly involved in optical transitions.

So, with the addition of spin, and taking into account only the $j = 3/2$ states, $\underline{H}$ in (5.58) becomes a $4 \times 4$ matrix, where the matrix elements are computed using (5.67). For example, if we arrange the basis states so that

$$|1\rangle = \left|\frac{3}{2}\;\frac{3}{2}\right\rangle \quad ,$$

$$|2\rangle = \left|\frac{3}{2}\;-\frac{1}{2}\right\rangle \quad ,$$

$$|3\rangle = \left|\frac{3}{2}\;\frac{1}{2}\right\rangle \quad ,$$

$$|4\rangle = \left|\frac{3}{2} - \frac{3}{2}\right\rangle \ , \tag{5.69}$$

then

$$\begin{aligned} H_{11} &\equiv \left\langle \frac{3}{2}\,\frac{3}{2}\right| H \left|\frac{3}{2}\,\frac{3}{2}\right\rangle \\ &= \frac{1}{2}[\langle X\uparrow|H|X\uparrow\rangle + \langle Y\uparrow|H|Y\uparrow\rangle \\ &\quad + \mathrm{i}\langle X\uparrow|H|Y\uparrow\rangle - \mathrm{i}\langle Y\uparrow|H|X\uparrow\rangle] \end{aligned} \tag{5.70}$$

$$= \varepsilon_1 + \frac{A}{2}\left(k_x^2 + k_y^2\right) + \frac{B}{2}\left(k_x^2 + k_y^2 + 2k_z^2\right) \ . \tag{5.71}$$

Here we use the fact that the momentum operator $\boldsymbol{p}$ does not couple states with different spin orientations, so that matrix elements like $\langle X\uparrow|H|X\downarrow\rangle$ vanish. In practice, instead of computing the matrix elements $\langle n|\boldsymbol{p}|m\rangle$ from first principles, one replaces them with experimentally determined parameters called *Luttinger parameters.* For the material systems discussed in this book there are three Luttinger parameters

$$\gamma_1 = -\frac{2m_0}{3\hbar^2}(A + 2B) \ , \tag{5.72}$$

$$\gamma_2 = -\frac{m_0}{3\hbar^2}(A - B) \ , \tag{5.73}$$

$$\gamma_3 = -\frac{m_0}{3\hbar^2}C \ . \tag{5.74}$$

These parameters are determined experimentally using various techniques sensitive to the band structure. The Luttinger parameters for many semiconductor materials are listed in books like Landolt-Börnstein. Typical values for some III–V semiconductors are

| | $\gamma_1$ | $\gamma_2$ | $\gamma_3$ |
|---|---|---|---|
| GaAs | 6.85 | 2.1 | 2.9 |
| InAs | 19.67 | 8.37 | 9.29 |
| InP | 6.35 | 2.08 | 2.76 |
| GaP | 4.2 | 0.98 | 1.66 |
| AlP | 3.47 | 0.06 | 1.15 |

(5.75)

In terms of the Luttinger parameters,

$$H_{11} = \varepsilon_1 - \frac{\hbar^2 k_z^2}{2m_0}(\gamma_1 - 2\gamma_2) - \frac{\hbar^2\left(k_x^2 + k_y^2\right)}{2m_0}(\gamma_1 + \gamma_2) \ . \tag{5.76}$$

The band structure calculated using the Luttinger parameters is the hole band structure because the experiments performed to measure the Luttinger parameters kept track of the hole in an otherwise filled valence band. Changing to the convention where the hole energy is positive, (5.76) becomes

$$H_{\mathrm{hh}} \equiv H_{11} = \frac{\hbar^2 k_z^2}{2m_0}(\gamma_1 - 2\gamma_2) + \frac{\hbar^2\left(k_x^2 + k_y^2\right)}{2m_0}(\gamma_1 + \gamma_2) \ , \tag{5.77}$$

where the zero-energy reference for $H_{\mathrm{hh}}$ is usually defined such that $\varepsilon_1 = 0$. Repeating the above calculation for the other matrix elements, we find the Luttinger Hamiltonian

$$H = \begin{pmatrix} H_{\mathrm{hh}} & -c & -b & 0 \\ -c^* & H_{\mathrm{lh}} & 0 & b \\ -b^* & 0 & H_{\mathrm{lh}} & -c \\ 0 & b^* & -c^* & H_{\mathrm{hh}} \end{pmatrix} , \tag{5.78}$$

where

$$\begin{aligned} H_{\mathrm{lh}} &= \frac{\hbar^2 k_z^2}{2m_0}(\gamma_1 + 2\gamma_2) + \frac{\hbar^2\left(k_x^2 + k_y^2\right)}{2m_0}(\gamma_1 - \gamma_2) \ , \\ c &= \frac{\sqrt{3}\hbar^2}{2m_0}\left[\gamma_2\left(k_x^2 - k_y^2\right) - 2\mathrm{i}\gamma_3 k_x k_y\right] \ , \\ b &= \frac{\sqrt{3}\hbar^2}{m_0}\gamma_3 k_z\left(k_x - \mathrm{i}k_y\right) \ . \end{aligned} \tag{5.79}$$

Notice that we find several compounds in the list (5.75) for which

$$\gamma_1 > \gamma_2 \simeq \gamma_3 \ . \tag{5.80}$$

For these systems, as long as we are interested in band properties in the vicinity of $k = 0$, we can introduce the so-called axial approximation. This approximation involves replacing the Luttinger parameters $\gamma_2$ and $\gamma_3$ in the function $c$ of (5.79) by an effective Luttinger parameter

$$\overline{\gamma} = \frac{\gamma_2 + \gamma_3}{2} \ . \tag{5.81}$$

The function c then simplifies to

$$c \simeq \frac{\sqrt{3}\,\hbar^2\overline{\gamma}}{2m_0}\left(k_x - ik_y\right)^2 \ . \tag{5.82}$$

In the axial approximation, the Luttinger Hamiltonian (5.78) can be transformed into a block diagonal form by a unitary transformation [*Broido* and *Sham* (1985)]

$$H' = UHU^\dagger \ , \tag{5.83}$$

where

$$U = \begin{pmatrix} v^* & 0 & 0 & -v \\ 0 & w^* & -w & 0 \\ 0 & w^* & w & 0 \\ v^* & 0 & 0 & v \end{pmatrix} , \tag{5.84}$$

$$v = \frac{1}{\sqrt{2}}\mathrm{e}^{\mathrm{i}(3\pi/4 - 3\xi/2)} \ , \tag{5.85}$$

$$w = \frac{1}{\sqrt{2}}\mathrm{e}^{\mathrm{i}(-\pi/4 + \xi/2)} \ , \tag{5.86}$$

and

$$\xi = \mathrm{atan}\left(\frac{k_y}{k_x}\right) \quad . \tag{5.87}$$

Evaluating (5.83), we find the block-diagonal Luttinger Hamiltonian

$$H' = \begin{pmatrix} H^U & 0 \\ 0 & H^L \end{pmatrix} \quad , \tag{5.88}$$

where

$$H^U = \begin{pmatrix} H_{\mathrm{hh}} & R \\ R^* & H_{\mathrm{lh}} \end{pmatrix} \quad , \tag{5.89}$$

$$H^L = \begin{pmatrix} H_{\mathrm{lh}} & R \\ R^* & H_{\mathrm{hh}} \end{pmatrix} \quad , \tag{5.90}$$

and

$$R = |c| - \mathrm{i}\,|b| \quad . \tag{5.91}$$

The block-diagonal basis is

$$\begin{aligned}
|1\rangle &= v^* \left|\frac{3}{2}\;\frac{3}{2}\right\rangle - v\left|\frac{3}{2}\;-\frac{3}{2}\right\rangle \quad , \\
|2\rangle &= w^* \left|\frac{3}{2}\;-\frac{1}{2}\right\rangle - w\left|\frac{3}{2}\;\frac{1}{2}\right\rangle \quad , \\
|3\rangle &= w^* \left|\frac{3}{2}\;-\frac{1}{2}\right\rangle + w\left|\frac{3}{2}\;\frac{1}{2}\right\rangle \quad , \\
|4\rangle &= v^* \left|\frac{3}{2}\;\frac{3}{2}\right\rangle + v\left|\frac{3}{2}\;-\frac{3}{2}\right\rangle \quad ,
\end{aligned} \tag{5.92}$$

or

$$\begin{aligned}
|1\rangle &= -v^*\left(\frac{|X\uparrow\rangle + \mathrm{i}\,|Y\uparrow\rangle}{\sqrt{2}}\right) - v\left(\frac{|X\downarrow\rangle + \mathrm{i}\,|Y\downarrow\rangle}{\sqrt{2}}\right) \quad , \\
|2\rangle &= w^*\left(\frac{|X\uparrow\rangle - \mathrm{i}\,|Y\uparrow\rangle}{\sqrt{6}} + \sqrt{\frac{2}{3}}\,|Z\downarrow\rangle\right) \\
&\quad + w\left(\frac{|X\downarrow\rangle + \mathrm{i}\,|Y\downarrow\rangle}{\sqrt{6}} - \sqrt{\frac{2}{3}}\,|Z\uparrow\rangle\right) \quad , \\
|3\rangle &= w^*\left(\frac{|X\uparrow\rangle - \mathrm{i}\,|Y\uparrow\rangle}{\sqrt{6}} + \sqrt{\frac{2}{3}}\,|Z\downarrow\rangle\right) \\
&\quad - w\left(\frac{|X\downarrow\rangle + \mathrm{i}\,|Y\downarrow\rangle}{\sqrt{6}} - \sqrt{\frac{2}{3}}\,|Z\uparrow\rangle\right) \quad , \\
|4\rangle &= -v^*\left(\frac{|X\uparrow\rangle + \mathrm{i}\,|Y\uparrow\rangle}{\sqrt{2}}\right) + v\left(\frac{|X\downarrow\rangle + \mathrm{i}\,|Y\downarrow\rangle}{\sqrt{2}}\right) \quad .
\end{aligned} \tag{5.93}$$

The axial approximation removes the small anisotropy of the band structure for different directions in the $k_x$, $k_y$ plane, resulting in cylindrical symmetry around the $k_z$ axis [*Altarelli* (1985)]. To completely diagonalize the block-diagonal Luttinger Hamiltonian (5.88), we only need to diagonalize each block individually. This gives the eigenvalues

$$\begin{aligned} \varepsilon_{lh,k} &= \frac{H_{\mathrm{hh}}+H_{\mathrm{lh}}}{2} + \sqrt{\left(\frac{H_{\mathrm{hh}}-H_{\mathrm{lh}}}{2}\right)^2 + |c|^2 + |b|^2} \\ &= \frac{\hbar^2 k^2}{2m_0}(\gamma_1 + 2\gamma_2) \ , \end{aligned} \tag{5.94}$$

$$\begin{aligned} \varepsilon_{hh,k} &= \frac{H_{\mathrm{hh}}+H_{\mathrm{lh}}}{2} - \sqrt{\left(\frac{H_{\mathrm{hh}}-H_{\mathrm{lh}}}{2}\right)^2 + |c|^2 + |b|^2} \\ &= \frac{\hbar^2 k^2}{2m_0}(\gamma_1 - 2\gamma_2) \ , \end{aligned} \tag{5.95}$$

where we use $\overline{\gamma}$ and $\gamma_2$ interchangably. Since the upper and lower blocks give the same results, the band structure is composed of two twice-degenerate isotropic valence bands with effective masses

$$m_{\mathrm{hh}} \equiv \frac{m_0}{(\gamma_1 - 2\gamma_2)} \ , \tag{5.96}$$

$$m_{\mathrm{lh}} \equiv \frac{m_0}{(\gamma_1 + 2\gamma_2)} \ . \tag{5.97}$$

Using the Luttinger parameters in (5.75), we obtain

| | $m_{\mathrm{hh}}/m_0$ | $m_{\mathrm{lh}}/m_0$ |
|---|---|---|
| GaAs | 0.377 | 0.091 |
| InAs | 0.302 | 0.028 |
| InP | 0.095 | 0.095 |
| GaP | 0.446 | 0.162 |
| AlP | 0.299 | 0.279 |

(5.98)

Figure 5.2 shows the GaAs band structure in the wavenumber region of interest for optical laser transitions.

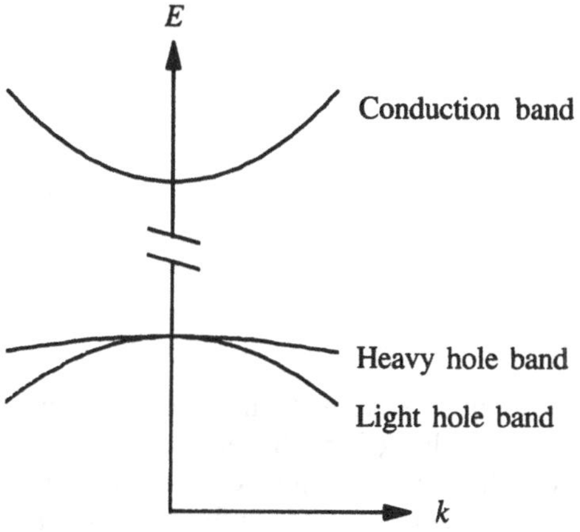


**Fig. 5.2.** The GaAs energy bands involved in optical transitions. The effective masses for the conduction, heavy-hole and light-hole bands are $0.067m_0$, $0.377m_0$ and $0.09m_0$, respectively

# 6. Quantum Wells

We continue the development of the previous chapter to include the band-structure modifications in quantum wells resulting from the quantum-confinement geometry. Section 6.1 shows how the envelope approximation method incorporates confinement effects into the $\boldsymbol{k} \cdot \boldsymbol{p}$ theory. The influence of quantum confinement on the valence band structure can be quite significant mixing especially the top two bulk semiconductor valence bands, i.e. the heavy-hole and light-hole bands. We show in Sect. 6.2 how this mixing is treated in the context of the Luttinger Hamiltonian. Section 6.3 introduces the concept of elastically strained systems and shows how strain effects may be incorporated into the band-structure calculations. In order to compute gain and refractive index, we need the dipole matrix elements, which we derive in Sect. 6.4. Up to that point, the hole band-structure calculations are based on the bulk-material $4 \times 4$ Luttinger Hamiltonian, which ignores the effects of the additional split-off hole states with total angular momentum $j = 1/2$. Section 6.5 describes how these states can be included in the band-structure calculations. Reasons for doing so involves laser compounds based on phosphides and nitrides, where the spin-orbit energies are smaller than those of the arsenides. The nitride based compounds exist in the cubic and hexagonal crystal structures. Section 6.6 shows the modifications of the Luttinger Hamiltonian which are necessary in order to be applicable to the hexagonal geometry.

## 6.1 Envelope Approximation Method

Figure 6.1 schematically shows the effective potential influencing the carriers in a quantum-well structure. For an electron in such a structure, the time-independent Schrödinger equation is

$$\left( \frac{p^2}{2m_0} + V_0 + V_{\mathrm{con}} \right) |\phi_\lambda^{\mathrm{QW}}\rangle = \varepsilon_\lambda |\phi_\lambda^{\mathrm{QW}}\rangle \ , \tag{6.1}$$

where $V_0$ is the periodic potential due to the lattice, $V_{\mathrm{con}}$ is the confinement potential due to the epitaxially grown heterostructure, and $\lambda$ represents the combination of quantum numbers to be specified later for identifying the quantum-well states. We assume that the potential $V_{\mathrm{con}}$ is sufficiently small in comparison to $V_0$ and varies sufficiently little within a unit cell, so that the new eigenstates may be approximated as the linear superposition of the bulk-material eigenstates

$$|\phi_\lambda^{\mathrm{QW}}\rangle = \sum_{n,\boldsymbol{k}} |\phi_{n,\boldsymbol{k}}\rangle\langle\phi_{n,\boldsymbol{k}}|\phi_\lambda^{\mathrm{QW}}\rangle \ . \tag{6.2}$$

Here $|\phi_{n,\boldsymbol{k}}\rangle$ is the bulk eigenstate satisfying (5.1) and $\langle\phi_{n,\boldsymbol{k}}|\phi_\lambda^{\mathrm{QW}}\rangle$ is the projection of the quantum-well eigenstate $|\phi_\lambda^{\mathrm{QW}}\rangle$ on the bulk eigenstate $|\phi_{n,\boldsymbol{k}}\rangle$. In the coordinate representation,

$$\begin{aligned}\langle\boldsymbol{r}|\phi_\lambda^{\mathrm{QW}}\rangle &= \sum_{n,\boldsymbol{k}} \langle\boldsymbol{r}|\phi_{n,\boldsymbol{k}}\rangle\langle\phi_{n,\boldsymbol{k}}|\phi_\lambda^{\mathrm{QW}}\rangle \\ &= \sum_{n,\boldsymbol{k}} \mathrm{e}^{\mathrm{i}\boldsymbol{k}\cdot\boldsymbol{r}}\langle\boldsymbol{r}|n,\boldsymbol{k}\rangle\langle\phi_{n,\boldsymbol{k}}|\phi_\lambda^{\mathrm{QW}}\rangle \ ,\end{aligned} \tag{6.3}$$

where we used the Bloch theorem and $|n,\boldsymbol{k}\rangle$ is a lattice periodic eigenstate of the bulk-material. Expanding $|n,\boldsymbol{k}\rangle$ in terms of the eigenstates at $k=0$, we find

$$\begin{aligned}\langle\boldsymbol{r}|\phi_\lambda^{\mathrm{QW}}\rangle &= \sum_{m,n,\boldsymbol{k}} \mathrm{e}^{\mathrm{i}\boldsymbol{k}\cdot\boldsymbol{r}}\langle\boldsymbol{r}|m\rangle\langle m|n,\boldsymbol{k}\rangle\langle\phi_{n,\boldsymbol{k}}|\phi_\lambda^{\mathrm{QW}}\rangle \\ &= \sum_m \left(\sum_{\boldsymbol{k}} \mathrm{e}^{\mathrm{i}\boldsymbol{k}\cdot\boldsymbol{r}} W_{\lambda m\boldsymbol{k}}\right)\langle\boldsymbol{r}|m\rangle \ ,\end{aligned} \tag{6.4}$$

where

$$W_{\lambda m\boldsymbol{k}} = \sum_n \langle m|n,\boldsymbol{k}\rangle\langle\phi_{n,\boldsymbol{k}}|\phi_\lambda^{\mathrm{QW}}\rangle \ . \tag{6.5}$$

We note that (6.3) may be written as

$$\langle\boldsymbol{r}|\phi_\lambda^{\mathrm{QW}}\rangle = \sum_m W_{\lambda m}(\boldsymbol{r})\langle\boldsymbol{r}|m\rangle \ , \tag{6.6}$$

where

$$W_{\lambda m}(\boldsymbol{r}) = \sum_{\boldsymbol{k}} \mathrm{e}^{\mathrm{i}\boldsymbol{k}\cdot\boldsymbol{r}} W_{\lambda m\boldsymbol{k}} \ . \tag{6.7}$$

Using (6.3) in (6.1), multiplying the result by $\exp(-\mathrm{i}\boldsymbol{k}'\cdot\boldsymbol{r})\langle\boldsymbol{r}|m\rangle$ and integrating over the volume of the crystal, we find

$$\begin{aligned}&\sum_{n,\boldsymbol{k}} W_{\lambda n\boldsymbol{k}} \frac{1}{V}\int_V \mathrm{d}^3r\, \mathrm{e}^{\mathrm{i}(\boldsymbol{k}-\boldsymbol{k}')\cdot\boldsymbol{r}}\langle m|\boldsymbol{r}\rangle \\ &\times\left[\frac{\hbar^2k^2}{2m_0} + \varepsilon_n - \varepsilon_\lambda - \frac{\mathrm{i}\hbar^2}{m_0}\boldsymbol{k}\cdot\nabla + V_{\mathrm{con}}(z)\right]\langle\boldsymbol{r}|n\rangle = 0 \ .\end{aligned} \tag{6.8}$$

As indicated in Fig. 6.1, the quantum-well structure gives rise to two length scales: one for coarse spatial variations on the scale of the heterostructure and the other for fine spatial variations on the scale of the lattice unit cell. It is sometimes useful to differentiate between these two length scales,

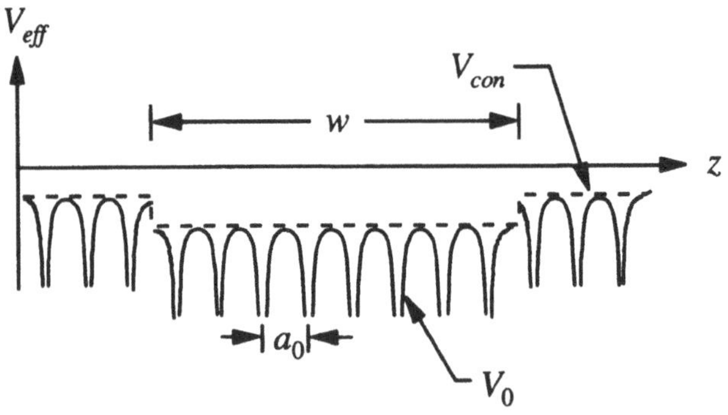


**Fig. 6.1.** Effective potential for a quantum-well structure showing the two length scales: a coarse one characterized by the quantum-well width $w$, and a fine one characterized by the lattice constant $a$

and we do so in the following by using upper-case letters (e.g., $\boldsymbol{R}$, $Z$, and $\boldsymbol{R}_\perp$) for the coarse variations and lower-case letters (e.g., $\boldsymbol{r}$, $x$, $y$, and $z$) for the fine variations. We make use of these two scales when evaluating the spatial integrals in (6.8). First, we write $\boldsymbol{r} = \boldsymbol{R}_l + \boldsymbol{\rho}$, where $\boldsymbol{R}_l$ is a lattice vector and $\boldsymbol{\rho}$ lies within the unit cell. Then a typical integral can be evaluated as

$$\begin{aligned}
&\frac{1}{V}\int_V \mathrm{d}^3 r\, \mathrm{e}^{\mathrm{i}(\boldsymbol{k}-\boldsymbol{k}')\cdot\boldsymbol{r}}\langle m|\boldsymbol{r}\rangle\langle\boldsymbol{r}|n\rangle \\
&\quad = \frac{1}{N}\sum_{\nu=1}^{N} \mathrm{e}^{\mathrm{i}(\boldsymbol{k}-\boldsymbol{k}')\cdot\boldsymbol{R}_\nu}\frac{1}{v}\int_v \mathrm{d}^3\rho\, \mathrm{e}^{\mathrm{i}(\boldsymbol{k}-\boldsymbol{k}')\cdot\boldsymbol{\rho}}\langle m|\boldsymbol{\rho}\rangle\langle\boldsymbol{\rho}|n\rangle \\
&\quad = \delta_{\boldsymbol{k},\boldsymbol{k}'}\frac{1}{v}\int_v \mathrm{d}^3\rho\,\langle m|\boldsymbol{\rho}\rangle\langle\boldsymbol{\rho}|n\rangle = \delta_{\boldsymbol{k},\boldsymbol{k}'}\delta_{m,n} \quad .
\end{aligned} \tag{6.9}$$

Here we note that $\langle \boldsymbol{R}_\nu - \boldsymbol{\rho}|n\rangle = \langle\boldsymbol{\rho}|n\rangle$ and the crystal volume, $V = Nv$ [compare (2.102, 103)]. Similarly,

$$\begin{aligned}
&\frac{1}{V}\int_V \mathrm{d}^3 r\, \mathrm{e}^{\mathrm{i}(\boldsymbol{k}-\boldsymbol{k}')\cdot\boldsymbol{r}}\langle m|\boldsymbol{r}\rangle\left(-\frac{\mathrm{i}\hbar^2}{m_0}\boldsymbol{k}\cdot\nabla\right)\langle\boldsymbol{r}|n\rangle \\
&\quad = \delta_{\boldsymbol{k},\boldsymbol{k}'}\frac{\hbar}{m_0}\boldsymbol{k}\cdot\boldsymbol{p}_{mn} \quad ,
\end{aligned} \tag{6.10}$$

and

$$\begin{aligned}
&\frac{1}{V}\int_V \mathrm{d}^3 r\, \mathrm{e}^{\mathrm{i}(\boldsymbol{k}-\boldsymbol{k}')\cdot\boldsymbol{r}}\langle m|\boldsymbol{r}\rangle V_{\mathrm{con}}(z)\langle\boldsymbol{r}|n\rangle \\
&\quad \simeq \frac{1}{N}\sum_{\nu=1}^{N} \mathrm{e}^{\mathrm{i}(\boldsymbol{k}-\boldsymbol{k}')\cdot\boldsymbol{R}_\nu} V_{\mathrm{con}}(Z^\nu)\frac{1}{v}\int_v \mathrm{d}^3\rho\, \mathrm{e}^{\mathrm{i}(\boldsymbol{k}-\boldsymbol{k}')\cdot\boldsymbol{\rho}}\langle m|\boldsymbol{r}\rangle\langle\boldsymbol{\rho}|n\rangle \\
&\quad \simeq V^{\mathrm{con}}_{\boldsymbol{k}-\boldsymbol{k}'}\delta_{m,n} \quad .
\end{aligned} \tag{6.11}$$

Here we abbreviated $\boldsymbol{p}_{mn} = \langle m|\boldsymbol{p}|n\rangle$ and

$$V_{\boldsymbol{k}-\boldsymbol{k}'}^{\mathrm{con}} = \frac{1}{N}\sum_{\nu=1}^{N} \mathrm{e}^{\mathrm{i}(\boldsymbol{k}-\boldsymbol{k}')\cdot \boldsymbol{R}_\nu} V_{\mathrm{con}}(Z^\nu) \quad . \tag{6.12}$$

In the evaluation of (6.11), we assumed that the external potential varies little over a unit cell so that $V_{\mathrm{con}}(z) \simeq V_{\mathrm{con}}(Z^\nu)$ and $\exp[\mathrm{i}(\boldsymbol{k}-\boldsymbol{k}')\cdot\boldsymbol{\rho}] \simeq 1$ which is a good approximation in the case when the electronic state mixing is such that $|\boldsymbol{k}-\boldsymbol{k}'|$ is much smaller than the reciprocal lattice vector. With these results, (6.8) becomes

$$\left(\frac{\hbar^2k^2}{2m_0} + \varepsilon_n\right) W_{\lambda nk} + \frac{\hbar}{m_0}\sum_m \boldsymbol{k}\cdot\boldsymbol{p}_{nm} W_{\lambda mk} + \sum_{k'} V_{\boldsymbol{k}-\boldsymbol{k}'}^{\mathrm{con}} W_{\lambda nk'} = \varepsilon_\lambda W_{\lambda nk} \quad . \tag{6.13}$$

This equation shows that within the current set of approximations the confinement potential does not mix states from different bands.

To solve the set of equations given by (6.13), we again use perturbation theory. First we consider the bands that are nondegenerate at $k = 0$, except for the spin degeneracy. An example is the $s$-like conduction band. We label that band with the subscript $n$. Equation (6.13) for all other bands may be approximated by

$$\left(\frac{\hbar^2k^2}{2m_0} + \varepsilon_j\right) W_{\lambda jk} + \frac{\hbar}{m_0}\boldsymbol{k}\cdot\boldsymbol{p}_{jn} W_{\lambda nk} = \varepsilon_\lambda W_{\lambda jk} \quad , \tag{6.14}$$

where we ignore the effects of $V_{\mathrm{con}}$ and consider only the coupling to the $n$th band. Approximating $\varepsilon_\lambda \simeq \varepsilon_n + \hbar^2k^2/(2m_0)$, we have

$$W_{\lambda jk} = \frac{\hbar}{m_0}\frac{\boldsymbol{k}\cdot\boldsymbol{p}_{jn}}{\varepsilon_n - \varepsilon_j} W_{\lambda nk} \quad . \tag{6.15}$$

Substituting (6.15) into (6.13) for the $n$th band gives

$$\left(\varepsilon_n + \frac{\hbar^2k^2}{2m_0} + \frac{\hbar^2}{m_0^2}\sum_{m\neq n}\frac{|\boldsymbol{k}\cdot\boldsymbol{p}_{mn}|^2}{\varepsilon_n - \varepsilon_m}\right) W_{\lambda nk} + \sum_{k'} V_{\boldsymbol{k}-\boldsymbol{k}'}^{\mathrm{con}} W_{\lambda nk'} = \varepsilon_\lambda W_{\lambda nk} \quad , \tag{6.16}$$

which can be written in the form

$$\left(\varepsilon_n + \frac{\hbar^2k^2}{2m_n}\right) W_{\lambda nk} + \sum_{k'} V_{\boldsymbol{k}-\boldsymbol{k}'}^{\mathrm{con}} W_{\lambda nk'} = \varepsilon_\lambda W_{\lambda nk} \quad , \tag{6.17}$$

where $m_n$ is the effective mass defined in (5.42). The Fourier transform of (6.17) is

$$\left[-\frac{\hbar^2\nabla^2}{2m_n} + V_{\mathrm{con}}(Z)\right] W_{\lambda n}(\boldsymbol{R}) = \varepsilon_{\lambda n} W_{\lambda n}(\boldsymbol{R}) \quad , \tag{6.18}$$

which we recognize as the equation for a particle in a one-dimensional potential $V_{\mathrm{con}}(Z)$. $W_{\lambda n}(\boldsymbol{R})$, which varies slowly compared to $\langle \boldsymbol{r}|m\rangle$, is the *envelope function* introduced (6.4–7).

Equation (6.18) is separable so that

$$W_{\lambda n}(\boldsymbol{R}) = A_{n_Z n}(Z) B_{\boldsymbol{k}_\perp n}(\boldsymbol{R}_\perp) \quad , \tag{6.19}$$

where $\lambda \to n_Z \boldsymbol{k}_\perp$ and $\boldsymbol{R}_\perp$ is a vector which lies in the plane of the heterostructure. Note, that the function $A(Z)$ is the generalization of the function $\zeta(z)$ of Sect. 1.7. This $Z$-dependent parts of the envelope functions obey

$$\left[-\frac{\mathrm{d}}{\mathrm{d}Z}\frac{\hbar^2}{2m_n}\frac{\mathrm{d}}{\mathrm{d}Z} + V_{\mathrm{con}}(Z)\right] A_{n_Z n}(Z) = \varepsilon_{n_Z n} A_{n_Z n}(Z) \quad , \tag{6.20}$$

i.e., the motion in the direction perpendicular to the well is that of a particle with mass $m_n$ in a one-dimensional square well potential. Since the effective mass in the $Z$-direction of a quantum-well heterostructure depends on space, we must symmetrize the second derivative as shown in (6.20) to assure that the Hamiltonian is Hermitian. For the case of a single quantum-well, (6.20) reduces to (1.39). Furthermore,

$$-\frac{\hbar^2}{2m_n}\nabla_\perp^2 B_{\boldsymbol{k}_\perp n}(\boldsymbol{R}_\perp) = \varepsilon_{\boldsymbol{k}_\perp n} B_{\boldsymbol{k}_\perp n}(\boldsymbol{R}_\perp) \quad , \tag{6.21}$$

which describes free-particle motion in the plane of the well. The total energy is

$$\varepsilon_{n_Z \boldsymbol{k}_\perp n} = \varepsilon_n + \varepsilon_{n_Z n} + \varepsilon_{\boldsymbol{k}_\perp n} \quad , \tag{6.22}$$

where

$$\varepsilon_{\boldsymbol{k}_\perp n} = \frac{\hbar^2 k_\perp^2}{2m_n} \quad , \tag{6.23}$$

and the solution for $\varepsilon_{n_Z n}$ is discussed in Sect. 1.7. Hence, the energy of the quantum-confined electron has contributions from the bulk structure ($\varepsilon_n$), the quantum confinement ($\varepsilon_{n_Z n}$), and the free motion in the $x$-$y$ plane ($\varepsilon_{\boldsymbol{k}_\perp n}$). Instead of an infinite number of states in every direction, the $z$-direction has only a finite number of states, which are the bound solutions of (6.20).

## 6.2 Band Mixing

The effects of quantum confinement are much more interesting for bands that are degenerate at $k = 0$, such as the top two valence bands in GaAs and similar materials. Let us label these degenerate bands with the subscripts 1 to $N$. Repeating the steps discussed in the previous section for the nondegenerate case gives now

$$W_{\lambda j \boldsymbol{k}} = \frac{\hbar}{m_0(\varepsilon_1 - \varepsilon_j)} \sum_{n=1}^{N} \boldsymbol{k} \cdot \boldsymbol{p}_{jn} W_{\lambda n \boldsymbol{k}} \quad , \tag{6.24}$$

where $j > N$. Substituting (6.24) into. (6.13) for the $N$ degenerate bands yields

$$\left(\varepsilon_1 + \frac{\hbar^2 k^2}{2m_0}\right) W_{\lambda n \boldsymbol{k}} + \frac{\hbar^2}{m_0^2} \sum_{m=1}^{N} \sum_{j>N} \frac{\boldsymbol{k}\cdot\boldsymbol{p}_{nj}\boldsymbol{k}\cdot\boldsymbol{p}_{jm}}{\varepsilon_1 - \varepsilon_j} W_{\lambda m \boldsymbol{k}'}$$
$$+ \sum_{\boldsymbol{k}'} V^{\mathrm{con}}_{\boldsymbol{k}-\boldsymbol{k}'} W_{\lambda n \boldsymbol{k}'} = \varepsilon_\lambda W_{\lambda n \boldsymbol{k}} \quad , \tag{6.25}$$

where we assume that the degenerate states have equal parity, so that the matrix element of $\boldsymbol{p}$ between any two degenerate states vanishes. With the exception of the terms containing $V^{\mathrm{con}}_{\boldsymbol{k}-\boldsymbol{k}'}$, (6.25) is similar to (5.49) and (5.50). If we repeat the degenerate perturbation theory outlined in Chap. 5, we get for the holes

$$\begin{aligned}
H_{\mathrm{hh}} W_{\lambda 1 \boldsymbol{k}} + R W_{\lambda 2 \boldsymbol{k}} + \sum_{\boldsymbol{k}'} V^{\mathrm{con}}_{\boldsymbol{k}-\boldsymbol{k}'} W_{\lambda 1 \boldsymbol{k}'} &= \varepsilon_\lambda W_{\lambda 1 \boldsymbol{k}} \quad , \\
H_{\mathrm{lh}} W_{\lambda 2 \boldsymbol{k}} + R^* W_{\lambda 1 \boldsymbol{k}} + \sum_{\boldsymbol{k}'} V^{\mathrm{con}}_{\boldsymbol{k}-\boldsymbol{k}'} W_{\lambda 2 \boldsymbol{k}'} &= \varepsilon_\lambda W_{\lambda 2 \boldsymbol{k}} \quad , \\
H_{\mathrm{lh}} W_{\lambda 3 \boldsymbol{k}} + R W_{\lambda 4 \boldsymbol{k}} + \sum_{\boldsymbol{k}'} V^{\mathrm{con}}_{\boldsymbol{k}-\boldsymbol{k}'} W_{\lambda 3 \boldsymbol{k}'} &= \varepsilon_\lambda W_{\lambda 3 \boldsymbol{k}} \quad , \\
H_{\mathrm{hh}} W_{\lambda 4 \boldsymbol{k}} + R^* W_{\lambda 3 \boldsymbol{k}} + \sum_{\boldsymbol{k}'} V^{\mathrm{con}}_{\boldsymbol{k}-\boldsymbol{k}'} W_{\lambda 4 \boldsymbol{k}'} &= \varepsilon_\lambda W_{\lambda 4 \boldsymbol{k}} \quad ,
\end{aligned} \tag{6.26}$$

where we used the states given by (6.5) and $H_{\mathrm{hh}}$, $H_{\mathrm{lh}}$, and $R$ are defined in (5.77, 79, 91), respectively. The symmetrized Fourier transform of (6.26) is

$$\left[-\frac{\partial}{\partial Z}\frac{\hbar^2}{2m_{\mathrm{hh}Z}}\frac{\partial}{\partial Z} - \frac{\hbar^2}{2m_{\mathrm{hh}\perp}}\left(\frac{\partial^2}{\partial X^2} + \frac{\partial^2}{\partial Y^2}\right) + V_{\mathrm{con}}(Z)\right] W_{\lambda n}(\boldsymbol{R})$$
$$+\frac{\sqrt{3}\hbar^2 k_\perp}{2m_0}\left(\gamma_2 k_\perp - 2\gamma_3 \frac{\partial}{\partial Z}\right) W_{\lambda m}(\boldsymbol{R}) = \varepsilon_{\lambda 1} W_{\lambda n}(\boldsymbol{R}) \quad , \tag{6.27}$$

for $n, m = 1, 2$ and $4, 3$ and

$$\left[-\frac{\partial}{\partial Z}\frac{\hbar^2}{2m_{\mathrm{lh}Z}}\frac{\partial}{\partial Z} - \frac{\hbar^2}{2m_{\mathrm{lh}\perp}}\left(\frac{\partial^2}{\partial X^2} + \frac{\partial^2}{\partial Y^2}\right) + V_{\mathrm{con}}(Z)\right] W_{\lambda n}(\boldsymbol{R})$$
$$+\frac{\sqrt{3}\hbar^2 k_\perp}{2m_0}\left(\gamma_2 k_\perp - 2\gamma_3 \frac{\partial}{\partial Z}\right) W_{\lambda m}(\boldsymbol{R}) = \varepsilon_{\lambda 2} W_{\lambda n}(\boldsymbol{R}) \quad , \tag{6.28}$$

for $n, m = 2, 1$ and $3, 4$. Similar to the situation for the bulk-material, the solutions to (6.27, 28) are doubly degenerate and we only need to consider either $n = 1, 2$ or $3, 4$. We identify $n = 1$ and $4$ as hh (for heavy hole) and $n = 2$ and $3$ as lh (for light hole), so that

$$\begin{aligned}
m_{\mathrm{hh}z} &= \frac{m_0}{\gamma_1 - 2\gamma_2} \quad , \\
m_{\mathrm{lh}z} &= \frac{m_0}{\gamma_1 + 2\gamma_2} \quad , \\
m_{\mathrm{hh}\perp} &= \frac{m_0}{\gamma_1 + \gamma_2} \quad , \\
m_{\mathrm{lh}\perp} &= \frac{m_0}{\gamma_1 - \gamma_2} \quad .
\end{aligned} \tag{6.29}$$

As in the nondegenerate case, the quantum-well states are products of two-dimensional $(X, Y)$ free-particle eigenstates with the one-dimensional $(Z)$ square-well eigenstates of (6.18). The quantum numbers are $\lambda \to (n_Z, \boldsymbol{k}_\perp)$, where $n_Z$ is between 1 and $N_{Z,\mathrm{hh}}$ for the heavy holes and $u_z$ is between 1 and $N_{Z,\mathrm{lh}}$ for the light holes, respectively. $N_{Z,\mathrm{hh}}$ and $N_{Z,\mathrm{lh}}$ are the numbers of solutions to the $Z$ part of (6.27, 28). A difference between the present situation and the case of isotropic bulk semiconductors is that the effective mass in the $Z$-direction deviates from that in the transverse direction for both heavy and light holes. In fact, according to (6.29), the states with the heavier effective mass in the $Z$-direction have the lighter effective mass in the transverse direction. This is commonly referred to as *mass reversal.* The convention is to use the terms *light* and *heavy* holes according to the respective mass in $Z$-direction. Figure 6.2a illustrates the result of mass reversal. At $k_\perp = 0$, the hole energies are the eigenvalues of (6.27), with masses $m_n = m_{\mathrm{hh}Z}$ and $m_{\mathrm{lh}Z}$ for the heavy and light holes, respectively. Because $m_{\mathrm{hh}Z} > m_{\mathrm{lh}Z}$, the heavy-hole state is below the light-hole state in energy at $k_\perp = 0$, as shown in the figure. If we ignore the nondiagonal terms in (6.27, 28), then the band dispersions are given by (6.23), with $m_n = m_{\mathrm{hh}\perp}$ and $m_{\mathrm{lh}\perp}$ for the heavy and light holes, respectively. Since $m_{\mathrm{hh}\perp} < m_{\mathrm{lh}\perp}$, the heavy-hole band has a higher curvature than that of the light hole, and the hole bands intersect as shown in Fig. 6.2a.

Of course, the band-crossings are unphysical and are removed when we solve the full equations (6.27, 28). To do that, we first write these equations in matrix form,

$$\underline{H} W_{n\boldsymbol{k}_\perp} = \varepsilon_{n\boldsymbol{k}_\perp} W_{n\boldsymbol{k}_\perp} \quad , \tag{6.30}$$

where $n$ is the band index and $\underline{H}$ is a $N_Z \times N_Z$ matrix with $N_Z = N_{Z,\mathrm{hh}} + N_{Z,\mathrm{lh}}$. The diagonal matrix elements are

$$H_{ii} = -\frac{\partial}{\partial Z}\frac{\hbar^2}{2m_{nZ}}\frac{\partial}{\partial Z} + \frac{\hbar^2 k_\perp^2}{2m_{n\perp}} \tag{6.31}$$

and the off-diagonal elements are

$$H_{ij} = \frac{\sqrt{3}\hbar^2 k_\perp}{2m_0}\left(\gamma_2 k_\perp - 2\gamma_3 \frac{\partial}{\partial Z}\right) \quad . \tag{6.32}$$

The matrix elements, which involve integrals with the eigenfunctions of (6.27) are usually evaluated numerically. The diagonalization of (6.30) then gives the hole band structure (see, for example, the solid curves in Fig. 6.2b). The removal of the band crossings by the nondiagonal elements, (6.32), also leads to nonparabolic hole energy dispersions (compare solid and dashed curves in Fig. 6.2b).

The state mixing due to the nondiagonal terms in the Hamiltonian $\underline{H}$ involves both confined and unconfined states of the quantum well. Figure 6.3 explains the differences between the confined and unconfined states. We label

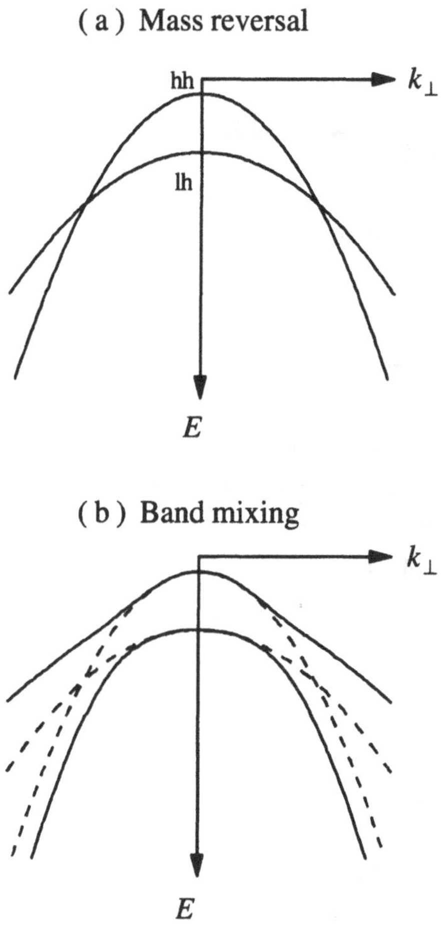


**Fig. 6.2a, b.** (**a**) Heavy (hh) and light (lh) hole bands in the absence of band mixing. (**b**) Solution of the full hole Hamiltonian (*solid curves*) showing the removal of the degeneracies that are present without band mixing (*dashed curves*)

as confined states those states which are described by the bounded solutions of (6.27, 28). The eigenfunctions of confined states are typically localized within the quantum well. The unconfined states are the states whose energies exceed the quantum-well confinement potential. The eigenfunctions of the unconfined states extend into the barrier regions.

The band-structure calculation then involves solving (6.30) for a basis consisting of confined and unconfined states. To do so, we first evaluate the matrix elements, (6.31, 32), and then diagonalize $\underline{H}$, (6.30), which gives the hole band structure and eigenfunctions. Figure 6.4 shows the results for a 4 nm $In_{0.2}Ga_{0.8}As$ quantum well between 8 nm GaAs barriers. For this example, the confined states are the $n = 1$ and 2 heavy hole states. The unconfined states consist of three heavy hole and two light hole states. The effects of mixing between confined and unconfined states are clearly shown by the noticeable

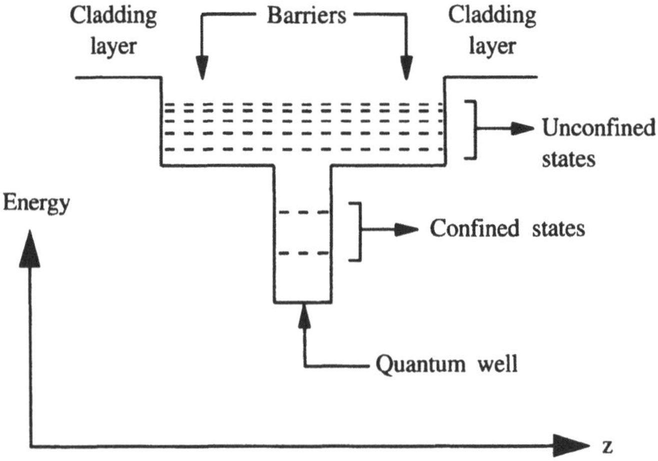


**Fig. 6.3.** Confined and unconfined states of a quantum-well structure

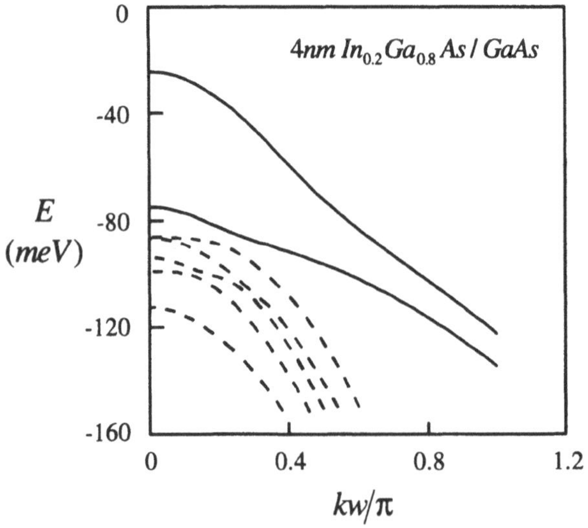


**Fig. 6.4.** Hole band structure for a 4 nm $In_{0.2}Ga_{0.8}As$/GaAs quantum-well structure, showing the effects of coupling between confined (*solid curves*) and unconfined states (*dashed curves*)

change in slopes in the two energetically lowest hole bands (solid curves) at $kw/\pi > 0.3$. In this case the mixing is between the confined heavy hole states and the unconfined light hole states.

## 6.3 Strained Quantum Wells

Semiconductor heterostructures can be grown epitaxially also with materials that are not perfectly lattice matched, provided this mismatch is not too large. A too large mismatch may prevent epitaxial growth altogether or lead to fractures, island formation, and other usually undesirable defects. However, under proper conditions, one may obtain a stable structure in which the materials are under elastic strain. One situation of practical interest is the case of relatively thick barriers and thin quantum wells grown on a substrate of barrier material. Ideally, in this case the well material assumes the lattice constant $a_b$ of the barrier material. Hence the well material is under strain, which is compressive (tensile) if the lattice constant of the barrier is smaller (larger) than that of the well material. Strained-layer quantum wells are interesting for applications in semiconductor lasers, because they allow both, a wider range of material combinations, and a certain amount of band structure and gain engineering. The simplest example is a frequency shift of the gain spectrum, but gain increases are also possible.

To appreciate the band-structure engineering aspect, we discuss in this section the simplest modifications of quantum-well band structures caused by elastic strain. For this purpose, we modify the analysis of the previous sections of this chapter to include the most important strain effects. In particular, we repeat the $\boldsymbol{k} \cdot \boldsymbol{p}$ analysis, here with the perturbation due to strain added to that due to the $\boldsymbol{k} \cdot \boldsymbol{p}$ term. This theory is originally due to *Pikus* and *Bir* (1960), and we more or less follow their original work. Rather than presenting the general theory of band structures in strained semiconductors, we restrict our discussion to the ideal case of a single quantum well with cubic symmetry and bulk lattice constant $a_w$, which grows under elastic strain in the $x$-$y$ plane and assumes the lattice constant $a_b$ of the barrier material. Under these conditions the strain tensor is

$$e = \begin{pmatrix} e_{xx} & 0 & 0 \\ 0 & e_{yy} & 0 \\ 0 & 0 & e_{zz} \end{pmatrix} , \tag{6.33}$$

where the in-plane components are given by the lattice mismatch parameters

$$e_{xx} = e_{yy} = \frac{a_b - a_w}{a_w} \equiv e_0 \quad . \tag{6.34}$$

As a consequence of the structural geometry, all strain components

$$e_{\alpha\beta} = 0 \text{ for } \alpha \neq \beta \quad , \tag{6.35}$$

so that the only other nonvanishing component is $e_{zz}$. To determine $e_{zz}$ we make use of the fact that there is no net force acting pendicular to the quantum-well plane. We can write

$$Z_z = C_{12}e_{xx} + C_{12}e_{yy} + C_{11}e_{zz} \tag{6.36}$$

where $Z$ denotes the force acting in $z$-direction and the index of $Z$ is the normal of the plane, which in this case is also in the $z$-direction [*Kittel* (1971), Chap. 4]. The quantities $C_{ij}$ are the elastic moduli or elastic stiffness constants. Without going into detail, we just note that cubic symmetry requires that only three, i.e., $C_{11}$, $C_{12}$, and $C_{44}$, of the generally possible twenty-one $C_{ij}$ are unequal to zero. The values of these material dependent constants can be found, e.g., in *Landolt-Börnstein* (1982)

| | $a$(Å) | $C_{11}$(GPa) | $C_{12}$(GPa) | $C_{44}$(GPa) |
|---|---|---|---|---|
| GaAs | 5.653 | 11.88 | 5.38 | 5.94 |
| AlAs | 5.660 | 12.5 | 5.3 | 5.4 |
| InAs | 6.058 | 8.33 | 4.53 | 3.96 |
| InP | 5.869 | 10.22 | 5.76 | 4.6 |
| GaP | 5.451 | 14.1 | 6.2 | 7.0 |
| AlP | 5.451 | 13.2 | 6.3 | 6.2 |

(6.37)

where the MKS units for $C_{ij}$ is giga Pascal (GPa). For the condition that $Z_z = 0$, (6.34, 36) yield

$$e_{zz} = -2\frac{C_{12}}{C_{11}}e_0 \quad . \tag{6.38}$$

Knowing the components of the strain tensor, we now proceed with the analysis of the band-structure modifications. Clearly, in the strained material an equation like (5.5) holds, where however all space variables are now in the strained system. On the other hand, the material parameters are given in the unstrained coordinate system. Therefore, we need to transform (5.5) from the strained coordinate system back to the unstrained system. Since we always assume small amounts of strain, we can expand the functions of the new variables in the basis of the old, unstrained variables. In terms of the vector $\boldsymbol{r}$ in the unstrained lattice, the component $\alpha$ of the vector $\boldsymbol{r}'$ of the strained lattice is

$$r'_\alpha = r_\alpha + \sum_\beta e_{\alpha\beta} r_\beta \quad . \tag{6.39}$$

Ignoring all terms of order $\mathcal{O}(e^2)$ or higher, we can write

$$\begin{aligned} \frac{\partial}{\partial r'_\alpha} &= \sum_\beta \frac{\partial r_\beta}{\partial r'_\alpha}\frac{\partial}{\partial r_\beta} \\ &= \sum_\beta \frac{1}{\delta_{\alpha,\beta} + e_{\alpha\beta}}\frac{\partial}{\partial r_\beta} \\ &\simeq \sum_\beta (\delta_{\alpha,\beta} - e_{\alpha\beta})\frac{\partial}{\partial r_\beta} \quad , \end{aligned} \tag{6.40}$$

and correspondingly

$$\frac{\partial^2}{\partial r_\alpha'^2} \simeq \frac{\partial^2}{\partial r_\alpha^2} - 2\sum_\beta e_{\alpha\beta}\frac{\partial^2}{\partial r_\alpha \partial r_\beta} \quad , \tag{6.41}$$

$$k_\alpha p_\alpha' \simeq k_\alpha p_\alpha - \sum_\beta k_\alpha e_{\alpha\beta} p_\beta \quad , \tag{6.42}$$

and

$$V_0(r') \simeq V_0(r) + \sum_{\alpha\beta} V_{\alpha\beta} e_{\alpha\beta} \quad , \tag{6.43}$$

where $V_{\alpha\beta}$ is the derivative of the lattice periodic potential $V_0(r')$ with respect to $e_{\alpha\beta}$.

Inserting these expansions into (5.5), we obtain

$$\left[ -\frac{\hbar^2\nabla^2}{2m_0} + V_0(r) + \frac{\hbar}{m_0}\boldsymbol{k}\cdot\boldsymbol{p} + \sum_{\alpha\beta} S^{\alpha\beta} e_{\alpha\beta} \right] |n\boldsymbol{k}\rangle$$
$$= \left( \varepsilon_{n\boldsymbol{k}} - \frac{\hbar^2 k^2}{2m_0} \right) |n\boldsymbol{k}\rangle \quad , \tag{6.44}$$

where

$$S^{\alpha\beta} = \frac{\hbar}{m_0} k_\alpha p_\beta + \frac{\hbar^2}{m_0} e_{\alpha\beta} \frac{\partial^2}{\partial r_\alpha \partial r_\beta} + V_{\alpha\beta} \quad . \tag{6.45}$$

The comparison shows that (6.45) and (5.5) differ only by the additional term proportional to the strain tensor elements. Hence, we can now repeat all steps of the previous sections including the additional strain terms. For example, in the $\boldsymbol{k}\cdot\boldsymbol{p}$ theory of Sect. 5.3 we obtain

$$\sum_j W_{ij}(\boldsymbol{k}) \left\{ \left[ \frac{\hbar^2 k^2}{2m_0} + E_j(0) - E_i(\boldsymbol{k}) \right] \delta_{i,j} + \frac{\hbar}{m_0} \boldsymbol{k}\cdot\boldsymbol{p}_{ij} + S_{ij} \right\} = 0 \tag{6.46}$$

where

$$S_{ij} = \sum_{\alpha\beta} S_{ij}^{\alpha\beta} e_{\alpha\beta} \quad \text{and} \quad S_{ij}^{\alpha\beta} = \frac{1}{V} \int \mathrm{d}^3 r\, u_i^*(0,\boldsymbol{r}) S^{\alpha\beta} u_j(0,\boldsymbol{r}) \quad . \tag{6.47}$$

In the subsequent perturbation theory we keep terms of order $k$, $k^2$, as in Sect. 5.3, but now we include also terms linear in the strain tensor, $\mathcal{O}(e)$. We ignore all contributions containing products of $k$ and strain tensor, or higher orders. This way, the effective $\boldsymbol{k}\cdot\boldsymbol{p}$ Hamiltonian is generalized as

$$H_{ij} = E_0 \delta_{i,j} + \frac{\hbar}{m_0} \boldsymbol{k}\cdot\boldsymbol{p}_{ij} + \sum_{\alpha,\beta=x,y,z} \left( D_{ij}^{\alpha\beta} k_\alpha k_\beta + S_{ij}^{\alpha\beta} e_{\alpha\beta} \right) \quad . \tag{6.48}$$

Equation (6.48) shows a one-to-one correspondence between the terms proportional to $k_\alpha k_\beta$ and to $e_{\alpha\beta}$. Hence, we can proceed to generalize the Luttinger Hamiltonian simply by adding the proper $e_{\alpha\beta}$ terms to the corresponding $k_\alpha k_\beta$ terms. As in the unstrained case, where we introduced the empirical

Luttinger parameters, we do not attempt to explicitly compute the matrix elements entering the strain part of the Hamiltonian. Instead, we introduce new parameters, the so-called *hydrostatic* and *shear deformation potentials*, which are obtained from experiments.

The strain induced addition to the term $b$, (5.79), vanishes for the present example,

$$b_{\text{strain}} \propto (e_{zx} - \mathrm{i}e_{zy}) = 0 \quad , \tag{6.49}$$

since $e_{zx} = e_{zy} = 0$, see (6.35). The correction to the term $c$, (5.79), vanishes as well,

$$c_{\text{strain}} \propto c_1(e_{xx} - e_{yy}) - c_2 e_{xy} = 0 \quad , \tag{6.50}$$

since $e_{xx} = e_{yy}$ and $e_{xy} = 0$. Hence, we have only strain corrections to the diagonal terms in the Luttinger Hamiltonian

$$H_{\text{strain}} = \begin{pmatrix} H_{\text{hh}}^{\text{strain}} & 0 & 0 & 0 \\ 0 & H_{\text{lh}}^{\text{strain}} & 0 & 0 \\ 0 & 0 & H_{\text{lh}}^{\text{strain}} & 0 \\ 0 & 0 & 0 & H_{\text{hh}}^{\text{strain}} \end{pmatrix} . \tag{6.51}$$

Using the one-to-one correspondence between the terms proportional to $k_\alpha k_\beta$ and to $e_{\alpha\beta}$, and replacing the parameter combinations $\hbar^2\gamma_1/2m_0$ and $\hbar^2\gamma_2/2m_0$ in (5.77, 79) by $a_1$ and $a_2$, respectively, we write the strain corrections $H_{\text{hh}}^{\text{strain}}$ and $H_{\text{lh}}^{\text{strain}}$ in complete symmetry to the respective unstrained parts:

$$H_{\text{hh}}^{\text{strain}} = -e_{zz}(a_1 - 2a_2) - (e_{xx} + e_{yy})(a_1 + a_2) \tag{6.52}$$

and

$$H_{\text{lh}}^{\text{strain}} = -e_{zz}(a_1 + 2a_2) - (e_{xx} + e_{yy})(a_1 - a_2) \quad . \tag{6.53}$$

Reordering the terms and using (6.34, 38) we obtain

$$\begin{aligned} H_{\text{hh}}^{\text{strain}} &= -a_1(e_{xx} + e_{yy} + e_{zz}) - a_2(e_{xx} + e_{yy} - 2e_{zz}) \\ &= -2a_1 e_0 \frac{C_{11} - C_{12}}{C_{11}} - 2a_2 e_0 \frac{C_{11} + 2C_{12}}{C_{11}} \\ &= -\delta\varepsilon_H - \frac{\delta\varepsilon_s}{2} \quad , \end{aligned} \tag{6.54}$$

where

$$\delta\varepsilon_H = 2a_1 e_0 \frac{C_{11} - C_{12}}{C_{11}} \tag{6.55}$$

and

$$\delta\varepsilon_s = 2a_2 e_0 \frac{C_{11} + 2C_{12}}{C_{11}} \quad . \tag{6.56}$$

Similarly, we obtain

$$H_{\mathrm{lh}}^{\mathrm{strain}} = -\delta\varepsilon_H + \frac{\delta\varepsilon_{\mathrm{s}}}{2} \quad . \tag{6.57}$$

Representative values for the deformation potentials are listed below:

| | $a_1$(eV) | $a_2$(eV) |
|---|---|---|
| GaAs | −7.1 | −1.7 |
| AlAs | −5.64 | −1.5 |
| InAs | −5.9 | −1.8 |
| InP | −6.35 | −2.0 |
| GaP | −9.3 | −1.5 |
| AlP | −5.54 | −1.6 |

(6.58)

To compute the strain-induced energy changes, we diagonalize the total Hamiltonian

$$H_{total} = H' + H_{\mathrm{strain}} \quad , \tag{6.59}$$

where $H'$ and $H_{\mathrm{strain}}$ are given in (5.88, 6.51), respectively. For the case of bulk-material we obtain

$$\begin{aligned} \varepsilon_{1(2)} &= \frac{H_{\mathrm{hh}} + H_{\mathrm{lh}}}{2} + \delta\varepsilon_H \pm \sqrt{\left(\frac{H_{\mathrm{hh}} - H_{\mathrm{lh}} - \delta\varepsilon_{\mathrm{s}}}{2}\right)^2 + |c|^2 + |b|^2} \\ &= \frac{\hbar^2 k^2}{2m_0}(\gamma_1 \pm \gamma_2) + \delta\varepsilon_H \pm \frac{\delta\varepsilon_{\mathrm{s}}}{2} + \mathcal{O}(k_\alpha, k_\beta, \alpha \neq \beta) \quad . \end{aligned} \tag{6.60}$$

This result shows that $\delta\varepsilon_H$ shifts all valence bands by the same amount. Hence, it can be considered a strain-induced bandgap shift. The term $\delta\varepsilon/2$, however, enters the heavy- and light-hole energies with opposite sign. Using the definitions (5.96, 5.97), we obtain from (6.60)

$$\varepsilon_{\mathrm{hh}} = \frac{\hbar^2 k^2}{2m_{\mathrm{hh}}} + \delta\varepsilon_H - \frac{\delta\varepsilon_{\mathrm{s}}}{2} \tag{6.61}$$

$$\varepsilon_{\mathrm{hh}} = \frac{\hbar^2 k^2}{2m_{\mathrm{lh}}} + \delta\varepsilon_H + \frac{\delta\varepsilon_{\mathrm{s}}}{2} \quad , \tag{6.62}$$

showing the occurrence of a strain-induced heavy-hole light-hole splitting by the amount $\delta\varepsilon_{\mathrm{s}}$.

Figure 6.5 indicates that the strain-induced energy shifts can be very different for compressive and tensile strain. The figure shows the heavy and light hole energy levels at $k = 0$. For compressive strain, the hydrostatic shift increases the hole energies of the heavy and light holes by an equal amount, $\varepsilon_H$. However, the shear shift decreases the heavy hole energy and increases the light hole energy by $\varepsilon_S/2$. This leads to a greater separation between heavy and light hole states, as shown in the figure. For tensile strain, the hydrostatic and shear energy shifts reverse in sign from the compressively strained case. The usual result is a reduction in the heavy and light hole energy separation (Fig. 6.5).

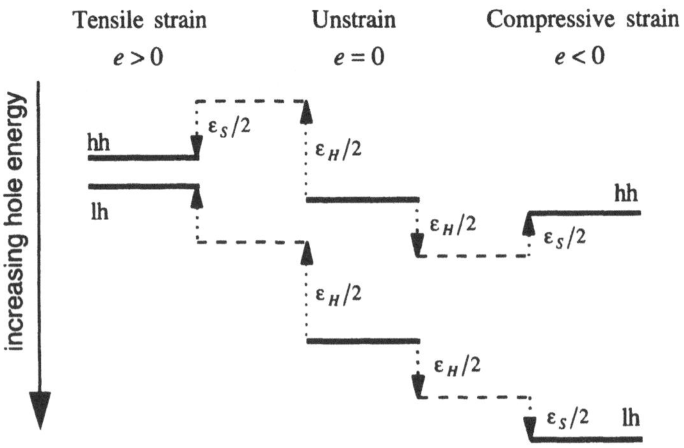


**Fig. 6.5.** Strain-induced energy shifts for heavy and light hole states

The relative energy placement of the heavy and light hole states at $k = 0$ plays an important role in the energy dispersions of the hole bands. A good example for illustrating this an $In_xGa_{1-x}As$ quantum well between InP barriers. The structure is unstrained for $x \simeq 0.53$, and under tensile (compressive) strain for smaller (larger) indium concentration. For a well width of 2 nm, the confined states consist of a heavy hole and a light hole state. In the unstrained structure, the bandgap energies for the heavy and light holes are equal. As discussed earlier, the heavy hole has the lowest energy as shown in Fig. 6.6 (middle). The lowest energy band then starts out from $k = 0$ with a curvature corresponding to $m_{hh\perp}$, until mixing with the light hole states changes it to a smaller curvature. For the compressively strained case (bottom figure) with $x = 0.73$, the light hole bandgap is shifted further up in energy. This shifts the mixing of heavy and light hole states to higher $k$, resulting in a higher band curvature over a greater range of $k > 0$ for the lowest energy hole band. The situation is reversed for the tensile strained case (top figure) with $x = 0.33$. Here the light hole bandgap energy is sufficiently reduced so that the light hole state has lower energy at $k = 0$. The lowest energy band curvature then corresponds to that for the higher mass, $m_{lh\perp}$.

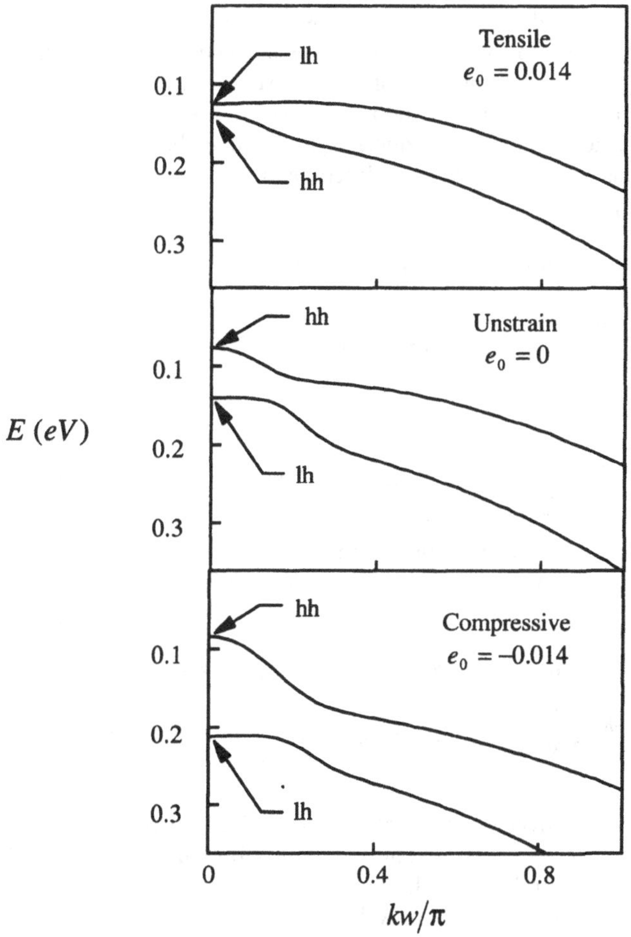


**Fig. 6.6.** Band structure for 2 nm $In_xGa_{1-x}As/InP$ under tensile strain (**top**), no strain (**middle**), and compressive strain (**bottom**) conditions

## 6.4 Dipole Matrix Elements

In order to use the computed band structures in the calculation of gain and refractive index, we have to replace the kinetic energy terms in the microscopic expressions by the respective single-particle energies computed from (6.30). In addition, we have to take into consideration that the dipole matrix elements also change. We therefore compute the dipole matrix elements for light which propagates along the $x$-$y$ plane of the quantum well, as is usually the case in edge emitting semiconductor lasers. Here we can have two possible polarization directions of the light: in the case of a TM mode the polarization is parallel to the $z$-direction, and for a TE mode the polarization is in the $x$-$y$

plane. To have a well defined configuration we assume TE polarization along the $x$-direction. In addition to the energy dispersions, the calculation of gain requires the dipole matrix elements. To proceed with the derivation of the dipole matrix elements for TE and TM polarizations, we first summarize our results for the wavefunctions of the conduction and valence bands.

The wavefunctions for the conduction electrons are

$$\langle \boldsymbol{r}|\phi^{\mathrm{e}}_{n\boldsymbol{k}_\perp m_s}\rangle = \mathrm{e}^{\mathrm{i}\boldsymbol{k}_\perp \cdot \boldsymbol{R}_\perp} A_{en}(Z)\langle \boldsymbol{r}|Sm_s\rangle \quad . \tag{6.63}$$

Correspondingly, the wavefunctions for the hole states are

$$\langle \boldsymbol{r}|\phi^{U}_{n\boldsymbol{k}_\perp}\rangle = \mathrm{e}^{\mathrm{i}\boldsymbol{k}_\perp \cdot \boldsymbol{R}_\perp} \sum_{m=1}^{2} \sum_{n_m=1}^{N_m} C^{U}_{n_m,n} A_{m,n_m}(Z)\langle \boldsymbol{r}|m\rangle \tag{6.64}$$

for the upper block, and

$$\langle \boldsymbol{r}|\phi^{L}_{n\boldsymbol{k}_\perp}\rangle = \mathrm{e}^{\mathrm{i}\boldsymbol{k}_\perp \cdot \boldsymbol{R}_\perp} \sum_{m=3}^{4} \sum_{n_m=1}^{N_m} C^{L}_{n_m,n} A_{m,n_m}(z)\langle \boldsymbol{r}|m\rangle \tag{6.65}$$

for the lower block of the Luttinger Hamiltonian (5.88).

For the TM mode we have to evaluate the matrix element

$$\mu_{\mathrm{TM}} = \langle \phi^{\mathrm{e}}_{n\boldsymbol{k}_\perp m_s}|ez|\phi^{h}_{n\boldsymbol{k}_\perp}\rangle \tag{6.66}$$

where $h = U$ or $L$. Substituting (6.63, 64) or (6.65) into (6.66) gives

$$\mu^{\mathrm{TM}}_{ij} = \sum_{m}\sum_{n_m} C^{h}_{n_m,j}\frac{1}{N}\sum_{\nu} A_{ei}(Z_\nu)A_{m,n_m}(Z_\nu)\frac{1}{v}\int_v \mathrm{d}^3\rho\,\langle Sm_s|\boldsymbol{\rho}\rangle ez\langle \boldsymbol{r}|m\rangle \ , \tag{6.67}$$

where we split the integral over all space into the sum over unit cells $\nu$ and the integral within the unit cells, (2.102). The $m$-summation runs over $m = 1, 2$ for $h = U$, and $m = 3, 4$ for $h = L$. The $n_m$ summation is over the number of heavy or light hole confined quantum-well states.

We first note that because of the symmetry between $m_s = \pm 1/2$ electron states, and the symmetry (except for phase factors) between the upper and lower eigenfunctions corresponding to the heavy (or light) holes, we have

$$\begin{aligned} |\langle \phi^{\mathrm{e}}_{n\boldsymbol{k}_\perp\uparrow}|ez|\phi^{U}_{n\boldsymbol{k}_\perp}\rangle|^2 &= |\langle \phi^{\mathrm{e}}_{n\boldsymbol{k}_\perp\downarrow}|ez|\phi^{U}_{n\boldsymbol{k}_\perp}\rangle|^2 \\ &= |\langle \phi^{\mathrm{e}}_{n\boldsymbol{k}_\perp\uparrow}|ez|\phi^{L}_{n\boldsymbol{k}_\perp}\rangle|^2 = |\langle \phi^{\mathrm{e}}_{n\boldsymbol{k}_\perp\downarrow}|ez|\phi^{L}_{n\boldsymbol{k}_\perp}\rangle|^2 \quad . \end{aligned} \tag{6.68}$$

Therefore, in terms of the gain calculations, we can limit the dipole matrix element derivation to electron states involving $m_s = 1/2$ and hole states in the upper block. We obtain,

$$\mu^{\mathrm{TM}}_{ij} = -w\sqrt{\frac{2}{3}}\sum_{n_2} C^{U}_{n_2,j}\frac{1}{N}\sum_{\nu} A_{ei}(Z_\nu)A_{2,n_2}(Z_\nu)\langle S\uparrow|ez|Z\uparrow\rangle \ , \tag{6.69}$$

and the absolute square is

$$|\mu_{ij}^{\mathrm{TM}}|^2 = \frac{1}{3}\left|\sum_{n_2} C^U_{n_2,j}\langle A_{ei}|A_{2,n_2}\rangle\right|^2 |\langle S\uparrow|ez|Z\uparrow\rangle|^2 \quad , \tag{6.70}$$

where

$$\langle A_{ei}|A_{2,n_2}\rangle = \frac{1}{N}\sum_{\nu} A_{ei}(Z_\nu)A_{2,n_2}(Z_\nu) \tag{6.71}$$

and we used (5.86) to see that

$$|w|^2 = \frac{1}{2} \quad . \tag{6.72}$$

For the TE mode we have to evaluate

$$\mu_{\mathrm{TE}} = \langle \phi^e_{n\boldsymbol{k}_\perp m_s}|ex|\phi^h_{n\boldsymbol{k}_\perp}\rangle \quad . \tag{6.73}$$

Using the same arguments as before, we can convince ourselves that for the case of conduction band spin $\uparrow$ the only nonvanishing matrix element is $\langle S\uparrow|X\uparrow\rangle$. Furthermore, we use the results from Chap. 5 to get

$$\begin{aligned}\mu_{ij}^{\mathrm{TE}} &= \frac{1}{N}\sum_{\nu} A_{ei}(Z_\nu)\Bigg[-\sqrt{\frac{1}{2}}v^*\sum_{n_1} C^U_{n_1,j}A_{1,n_1}(Z_\nu) \\ &\quad +\sqrt{\frac{1}{6}}w^*\sum_{n_2} C^U_{n_2,j}A_{2,n_2}(Z_\nu)\Bigg]\langle S\uparrow|ex|X\uparrow\rangle \quad .\end{aligned} \tag{6.74}$$

The absolute square is

$$\begin{aligned}&|\mu_{ij}^{\mathrm{TE}}|^2 \\ &= \frac{|\langle S\uparrow|ex|Z\uparrow\rangle|^2}{4}\Bigg[\left|\sum_{n_1} C^U_{n_1,j}\langle A_{ei}|A_{1,n_1}\rangle\right|^2 + \frac{1}{3}\left|\sum_{n_2} C^U_{n_2,j}\langle A_{ei}|A_{2,n_2}\rangle\right|^2 \\ &\quad + \frac{2}{\sqrt{3}}\left(\sum_{n_1} C^U_{n_1,j}\langle A_{ei}|A_{1,n_1}\rangle\right)\left(\sum_{n_2} C^U_{n_2,j}\langle A_{ei}|A_{2,n_2}\rangle\right)\cos(2\phi)\Bigg] \quad .\end{aligned} \tag{6.75}$$

Here we used

$$v^*w = -\tfrac{1}{2}\,\mathrm{e}^{2\mathrm{i}\phi} \tag{6.76}$$

where $\phi$ is the angle between $\boldsymbol{k}$ and $\boldsymbol{x}$.

Figure 6.7 shows typical quantum-well TE and TM dipole matrix elements as a function of carrier momentum. To identify the transitions we use the following notation: we label each band according to the quantum numbers that are valid at the zone center, $k = 0$. The conduction bands are labeled $en$, where $n = 1, 2, 3, \ldots$ are the square-well bound state quantum numbers. Similarly, since there is usually little band mixing around $k = 0$, the valence bands can be denoted as hh1, hh2, ... and lh1, lh2, ... where hh and lh stand for heavy hole and light hole, respectively. We emphasize that away from the

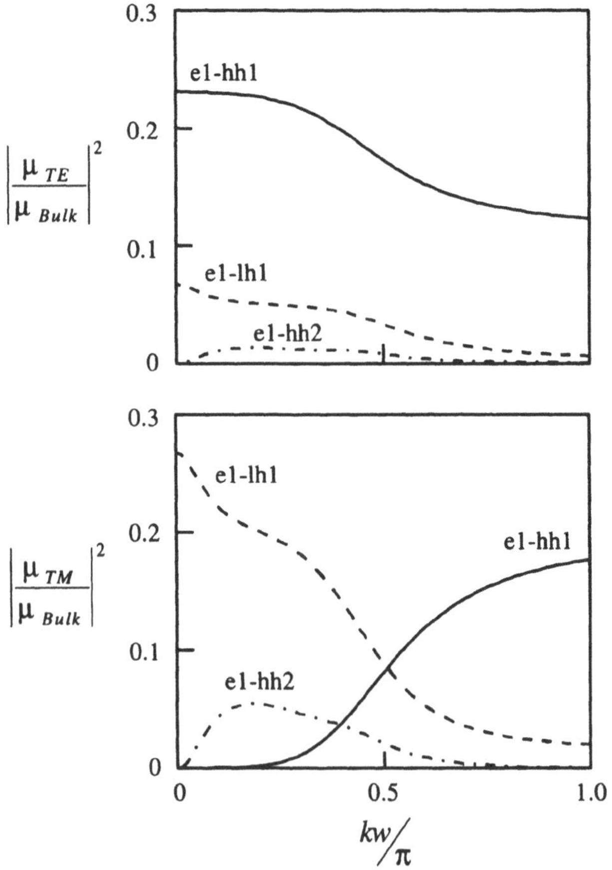


**Fig. 6.7.** TE (**top**) and TM (**bottom**) dipole-transition matrix elements versus transverse carrier momentum for 4 nm $In_{0.1}Ga_{0.9}As/Al_{0.1}Ga_{0.9}As$

zone center, the valence bands are mixtures of heavy and light hole states with possibly different square-well quantum numbers. This is evident, e.g., by the changes in the band curvatures for $k > 0$, as seen in Figs. 6.2, 4.

The behavior shown in Fig. 6.7 can be understood from symmetry arguments. From the derivations resulting in (6.69) and (6.75), we note that because of the symmetry of the electron and hole lattice periodic functions, the TM dipole matrix element only couples the electron state to the light hole components of a state, whereas the TE dipole matrix element has contributions from both electron to heavy hole and electron to light hole transitions. This property shows up in Fig. 6.7 (bottom) where the TM e1–hh1 dipole matrix element is negligible for small $k$ values. This matrix element increases at higher $k$ because of the mixing of heavy hole and light hole states in the hh1 band. The figure also reflects the fact that the dipole matrix elements involve inner products of envelope functions. Since the envelope functions are the orthonormal eigenfunctions of a square well potential, a dipole matrix element that couples states with different square well quantum numbers is negligible around the zone center, thus explaining the smaller values for the e1–hh2 curves in Fig. 6.7. Again, because of state mixing these matrix elements grow with increasing $k$.

## 6.5 6 × 6 Luttinger Hamiltonian

In this section, we describe the changes to the hole band-structure problem when the $j = 1/2$ states are taken to account. Reasons for including these states involve laser compounds based on phosphides and nitrides, where the spin-orbit energies are smaller than those of the arsenides (e.g., 0.11 eV and 0.017 eV for InP and GaN, respectively, compared to 0.34 eV in GaAs). The derivation of the Luttinger Hamiltonian is similar to what is described in the previous chapter. The end result is again a block diagonal Hamiltonian:

$$H = \begin{pmatrix} H^U & 0 \\ 0 & H^L \end{pmatrix} \tag{6.77}$$

where

$$H^u = \begin{pmatrix} H_{\text{hh}} & H_{\text{hh,lh}} & H_{\text{hh,so}} \\ H_{\text{lh,hh}} & H_{\text{lh}} & H_{\text{lh,so}} \\ H_{\text{so,hh}} & H_{\text{so,lh}} & H_{\text{so,so}} \end{pmatrix} , \tag{6.78}$$

and $H^L = (H^u)^\dagger$. The diagonal matrix elements for the submatrices $H_{\text{hh}}$, $H_{\text{lh}}$, and $H_{\text{ch}}$ are

$$H_{\text{hh}n,\text{hh}n} = -E^{\text{conf}}_{\text{hh},n} - \frac{\hbar^2 k^2}{2m_0}[\gamma_1 + \gamma_2] , \tag{6.79}$$

$$H_{\text{lh}n,\text{lh}n} = -E^{\text{conf}}_{\text{lh},n} - \frac{\hbar^2 k^2}{2m_0}(\gamma_1 - \gamma_2) , \tag{6.80}$$

$$H_{\text{so}n,\text{so}n} = -E^{\text{conf}}_{\text{so},n} - \frac{\hbar^2 k^2}{2m_0}\gamma_1 , \tag{6.81}$$

respectively, and the off-diagonal elements are

$$H_{\alpha n,\alpha m} = 0 . \tag{6.82}$$

The couplings among heavy, light and split-off holes are described by the matrix elements for the submatrices $H_{\text{hh,lh}}$, $H_{\text{hh,ch}}$, and $H_{\text{lh,ch}}$:

$$H_{\text{hh}n,\text{lh}m} = \frac{\sqrt{3}\hbar^2 k_\perp}{2m_0} \int_{-\infty}^{\infty} \mathrm{d}z\, u_{\text{hh},n}(z) \left(\overline{\gamma} k_\perp - 2\gamma_3 \frac{\mathrm{d}}{\mathrm{d}z}\right) u_{\text{lh},m}(z) , \tag{6.83}$$

$$H_{\text{hh}n,\text{so}m} = \sqrt{\frac{3}{2}} \frac{\hbar^2 k_\perp}{m_0} \int_{-\infty}^{\infty} \mathrm{d}z\, u_{\text{hh},n}(z) \left(\overline{\gamma} k_\perp + \gamma_3 \frac{\mathrm{d}}{\mathrm{d}z}\right) u_{\text{so},m}(z) , \tag{6.84}$$

$$H_{\mathrm{lh}n,\mathrm{so}m} = \int_{-\infty}^{\infty} \mathrm{d}z\, u_{\mathrm{lh},n}(z) \times \left[ \frac{\varepsilon_{\mathrm{s}}}{\sqrt{2}} + \frac{\hbar^2 \gamma_2}{\sqrt{2} m_0} \left( k_\perp^2 + 2\frac{\mathrm{d}^2}{\mathrm{d}z^2} \right) + \frac{3\hbar^2 \gamma_3 k_\perp}{m_0} \frac{\mathrm{d}}{\mathrm{d}z} \right] u_{\mathrm{so},m}(z) \, . \tag{6.85}$$

The diagonalization of (6.77) gives doubly degenerate hole bands, one from the upper block and the other from the lower block. The eigenfunctions from the upper and lower block are similar except for phase factors. Figure 6.8 shows an example of the contibutions of the spin-orbit states to the band structure of a 4nm $In_{0.5}Ga_{0.5}P/(Al_{0.5}Ga_{0.5})_{0.5}In_{0.5}P$ quantum well. The lowest energy hole band keeps its purely heavy hole character at $k = 0$. However, the second hole band is now a mixture of the light hole and split-off states at $k = 0$.

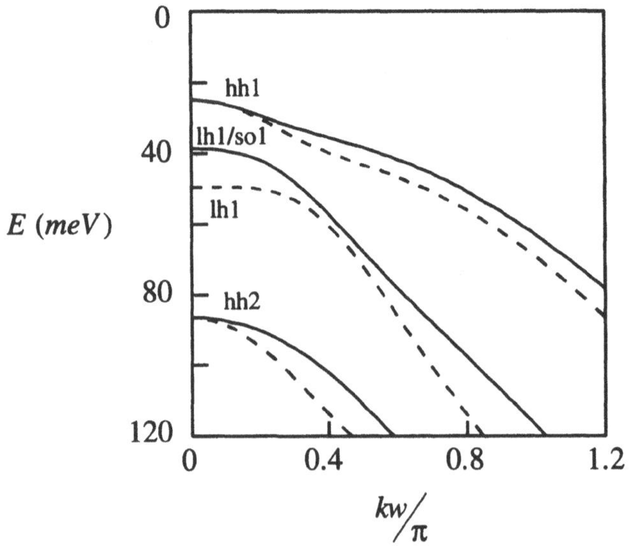


**Fig. 6.8.** Hole band structure for 4 nm $In_{0.5}Ga_{0.5}P/(Al_{0.5}Ga_{0.5})_{0.5}In_{0.5}P$ quantum-well structure with (*solid curves*) and without (*dashed curves*) split-off states taken into account

The effects of mixing between the $j = 3/2$ and $j = 1/2$ states are more significant in the optical dipole matrix elements. As in the previous sections, we can use the spatial symmetry of the bulk-material states to identify the electric field polarizations parallel (TM) and perpendicular (TE) to the growth direction. For the corresponding dipole matrix elements we obtain

$$|\mu_{ij}^{\mathrm{TM}}|^2 = \frac{|\wp_z|^2}{\sqrt{3}} \left| \sum_{n_2=1}^{N_2} C_{n_2,j}^U \langle A_{ei} | A_{2,n_2} \rangle - \frac{1}{\sqrt{2}} \sum_{n_3=1}^{N_3} C_{n_3,j}^U \langle A_{ei} | A_{3,n_3} \rangle \right|^2 \tag{6.86}$$

$$|\mu_{ij}^{\mathrm{TE}}|^2 = |\wp_x|^2 \left( \frac{1}{\sqrt{6}} \left| \frac{1}{\sqrt{2}} \sum_{n_2=1}^{N_2} C_{n_2,j}^U \langle A_{ei}|A_{2,n_2}\rangle + \sum_{n_3=1}^{N_3} C_{n_3,j}^U \langle A_{ei}|A_{3,n_3}\rangle \right|^2 + \frac{1}{4} \sum_{n_1=1}^{N_1} |C_{n_1,j}^U \langle A_{ei}|A_{1,n_1}\rangle|^2 \right) . \tag{6.87}$$

Figure 6.9 shows the quantum-well dipole matrix elements for transitions involving the electron band e1 and the two lowest energy hole bands hh1 and lh1/so1.

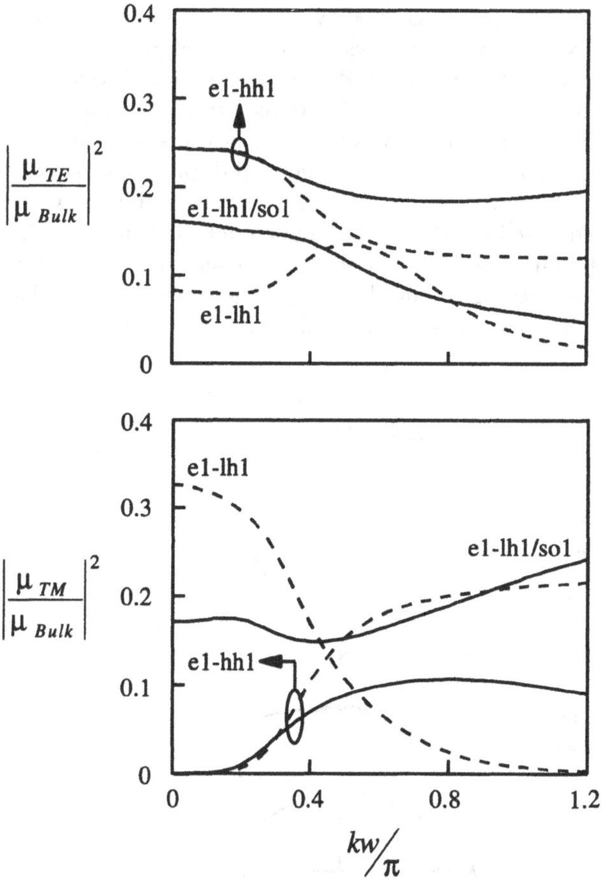


**Fig. 6.9.** TE (**top**) and TM (**bottom**) dipole-transition matrix elements versus transverse carrier momentum for 4 nm $In_{0.5}Ga_{0.5}P/(Al_{0.5}Ga_{0.5})_{0.5}In_{0.5}P$ quantum-well structure. Only the transitions involving the two lowest-energy hole bands are shown. The *solid* and *dashed curves* are computed with and without the split-off states, respectively

## 6.6 Wurtzite Crystal

A very interesting and significant recent development in semiconductor lasers is the demonstration of lasing in wide bandgap group-III nitride compounds. There are many applications for compact light sources emitting in the spectral region of blue-green to uv wavelengths. In several aspects the group-III nitride compounds are quite different from those of conventional near-infrared lasers. For example, the spin-orbit splitting is significantly smaller so that the effects of the $j = 1/2$ states are magnified. Also, these compounds typically grow in the hexagonal wurtzite crystal structure.

In this section we discuss a $\boldsymbol{k} \cdot \boldsymbol{p}$ treatment for wurtzite structures. We consider quantum wells of AlGaN or InGaN alloys, pseudomorphically grown along the $c$ axis of the hexagonal wurtzite crystal structure. Figure 6.10 compares the bulk cubic and wurtzite structures. We see several basic differences. They are: (a) the asymmetry between the directions parallel ($z$) and perpendicular ($\perp$) to the $c$ axis, (b) the removal of the heavy and light hole degeneracy at the $\Gamma$ point (zone center, $k = 0$) due to the spin-orbit interaction, (c) the small spin-orbit splitting leading to significant mixing between $j = 1/2$ and $j = 3/2$ states, and (d) the nonparabolicity of the hole bands in the $\perp$ direction due to this mixing. For the conduction band, the zone center states are the degenerate states $|S\uparrow\rangle$ and $|S\downarrow\rangle$. For the valence bands, the important zone center states are the six $p$-like hole states with total angular momenta $j = 3/2$ and $1/2$.

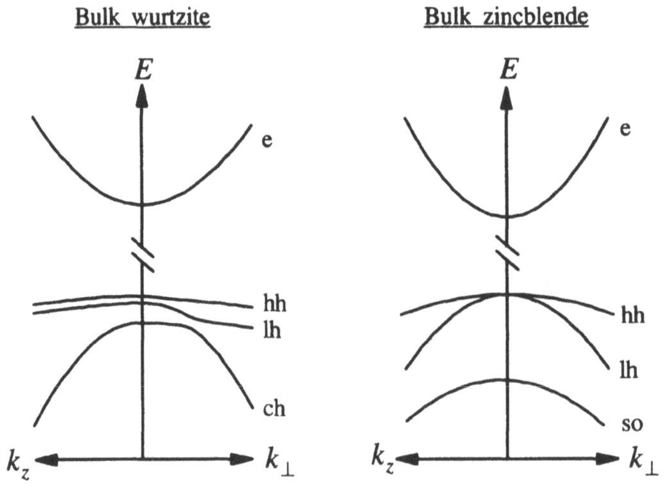


**Fig. 6.10.** Bulk-material band structures for wurtzite (**left figure**) and zincblende (**right figure**) semiconductors. The conduction band (e) is shown together with the top three hole bands. Following the usual convention, the hole bands are labeled heavy-hole (hh), light-hole (lh) and crystal-field split-hole (ch) for the wurtzite structure. For the zincblende structure, the spin-orbit split-hole (so) band replaces the ch band

Other bulk-material properties important to the quantum-well band structure include the effective masses for the electron (e), heavy hole (hh), light hole (lh) and crystal-split hole (ch), parallel and perpendicular to the $c$-axis. The hole effective masses may be expressed in terms of parameters usually denoted as $A_1$, $A_2$ and $A_5$. The energy separation between the hole states at $k = 0$ are described by the two energy splittings $\Delta_1$ and $\Delta_2$. The parameters relating to strain effects are the lattice constants $a$ ($\perp$ direction) and $c$ ($z$ direction), the elastic constants $C_{11}$, $C_{12}$, $C_{13}$ and $C_{33}$, the conduction band deformation potentials $a_{cz}$ and $a_{c\perp}$, and the valence band deformation potentials $D_1$, $D_2$, $D_3$ and $D_4$. Table 6.1 lists the typical values for these parameters

**Table 6.1.** Material parameters for AlN, GaN and InN used in quantum-well band-structure calculations

| | AlN | GaN | InN |
|---|---|---|---|
| $m_{e,z}/m_0$ | 0.31 | 0.18 | 0.11 |
| $m_{e,\perp}/m_0$ | 0.32 | 0.20 | 0.11 |
| $A_1$ | −4.00 | −6.67 | −9.09 |
| $A_2$ | −0.23 | −0.50 | −0.63 |
| $A_5$ | −2.07 | −3.12 | −4.36 |
| $E_g$ (eV) | 6.20 | 3.44 | 1.89 |
| $\Delta_1$ (eV) | −0.221 | 0.019 | 0.025 |
| $\Delta_2$ (eV) | 0.0043 | 0.0063 | 0.0003 |
| $d_{13}$ ($10^{-12}$ mV$^{-1}$) | −2 | −1.7 | −1.1 |
| $a$ (Å) | 3.081 | 3.150 | 3.494 |
| $c$ (Å) | 4.948 | 5.142 | 5.669 |
| $C_{11}$ (GPa) | 398 | 396 | 271 |
| $C_{12}$ (GPa) | 140 | 144 | 124 |
| $C_{13}$ (GPa) | 108 | 103 | 92 |
| $C_{33}$ (GPa) | 373 | 405 | 224 |
| $(a_{cz} - D_1)$ (eV) | −4.21 | −6.11 | −4.05 |
| $(a_{c\perp} - D_2)$ (eV) | −12.04 | −9.62 | −6.67 |
| $D_3$ (eV) | 9.06 | 5.76 | 4.92 |
| $D_4$ (eV) | −4.05 | −3.04 | −1.79 |

For the bandgap energies of the composite materials, typical empirical formulas are:

$$E_g = 1.89x + 3.44(1 - x) - 1.02x(1 - x) \tag{6.88}$$

for $In_xGa_{1-x}N$, and

$$E_g = 6.28x + 3.44(1-x) - 0.98x(1-x) \tag{6.89}$$

for $Al_xGa_{1-x}N$.

In materials like the group-III nitrides the strain induces an electric field because of an effective displacement between the core electrons and the ions (piezoelectric effects). It is convenient to incorporate this piezoelectric effect in the band-structure calculations instead of including it into the electron many-body Hamiltonian, (3.6). For this purpose the energy eigenfunctions are now written as superpositions of the subband states,

$$\psi^{e}_{n\boldsymbol{k}\uparrow}(\boldsymbol{r}) = e^{i\boldsymbol{k}\cdot\boldsymbol{r}_\perp} \sum_{m=1}^{N_e} C^{\uparrow}_{m,n} u_{e,m}(z)\langle \boldsymbol{r}|S\uparrow\rangle \ , \tag{6.90}$$

$$\psi^{e}_{n\boldsymbol{k}\downarrow}(\boldsymbol{r}) = e^{i\boldsymbol{k}\cdot\boldsymbol{r}_\perp} \sum_{m=1}^{N_e} C^{\downarrow}_{m,n} u_{e,m}(z)\langle \boldsymbol{r}|S\downarrow\rangle \ , \tag{6.91}$$

for the electrons, and

$$\phi^{U}_{n,\boldsymbol{k}}(\boldsymbol{r}) = e^{i\boldsymbol{k}\cdot\boldsymbol{r}_\perp} \sum_{m=1}^{3}\sum_{n_m=1}^{N_m} A^{U}_{n_m,n} u_{m,n_m}(z)\langle \boldsymbol{r}|m\rangle \ , \tag{6.92}$$

$$\phi^{L}_{n,\boldsymbol{k}}(\boldsymbol{r}) = e^{i\boldsymbol{k}\cdot\boldsymbol{r}_\perp} \sum_{m=4}^{6}\sum_{n_m=1}^{N_m} A^{L}_{n_m,n} u_{m,n_m}(z)\langle \boldsymbol{r}|m\rangle \ , \tag{6.93}$$

for the holes. The basis $\langle \boldsymbol{r}|m\rangle$ for the holes are the bulk-material hole eigenfunctions at zone center in the absence of spin-orbit coupling and $u_{\alpha,m}(z)$ is an envelope function. These eigenfunctions may be expressed in terms of the orbital angular momentum $|l,m_l\rangle$ and spin eigenstates $|\uparrow(\downarrow)\rangle$:

$$|1\rangle = \alpha^*|1,1\rangle|\uparrow\rangle + \alpha|1,-1\rangle|\downarrow\rangle \ , \tag{6.94}$$

$$|2\rangle = \beta|1,-1\rangle|\uparrow\rangle + \beta^*|1,1\rangle|\downarrow\rangle \ , \tag{6.95}$$

$$|3\rangle = \beta^*|1,0\rangle|\uparrow\rangle + \beta|1,0\rangle|\downarrow\rangle \ , \tag{6.96}$$

$$|4\rangle = \alpha^*|1,1\rangle|\uparrow\rangle - \alpha|1,-1\rangle|\downarrow\rangle \ , \tag{6.97}$$

$$|5\rangle = \beta|1,-1\rangle|\uparrow\rangle - \beta^*|1,1\rangle|\downarrow\rangle \ , \tag{6.98}$$

$$|6\rangle = -\beta^*|1,0\rangle|\uparrow\rangle + \beta|1,0\rangle|\downarrow\rangle \ , \tag{6.99}$$

where $\sqrt{2}\alpha = \exp[i(3\pi/4 + 3\phi/2)]$, $\sqrt{2}\beta = \exp[i(\pi/4 + \phi/2)]$, and $\tan\phi = k_x/k_y$.

The influence of strain on the energy levels may be determined with the help of Fig. 6.11, which shows the bandedge shifts for the electron and holes in the presence of compressive strain. (The sign of the shifts are reversed for tensile strain.) The shifts may be grouped into

$$\varepsilon_e = -2e\left(a_{c\perp} - \frac{C_{13}}{C_{33}}a_{cz}\right) \ , \tag{6.100}$$

$$\varepsilon_h = -2e\left[D_2 + D_4 - \frac{C_{13}}{C_{33}}(D_1 + D_3)\right] \ , \tag{6.101}$$

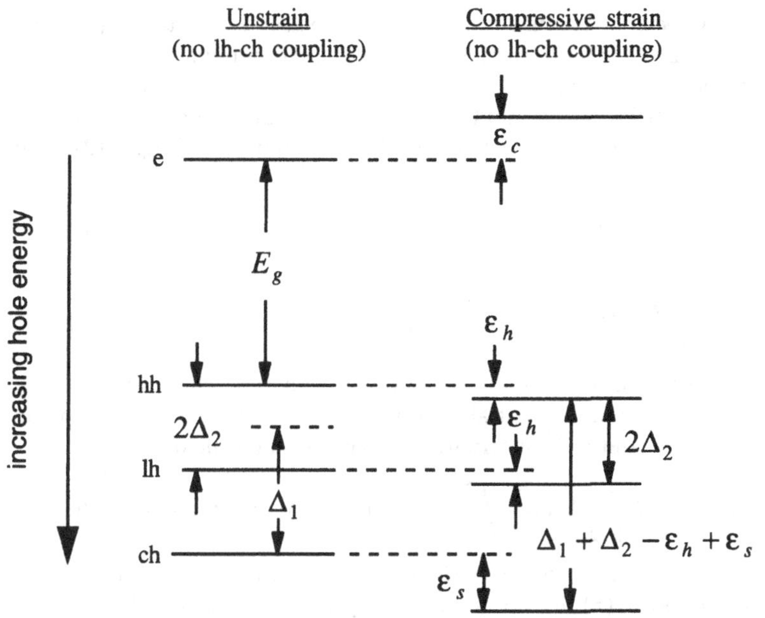


**Fig. 6.11.** Energy shifts at zone center due to strain in the wurtzite quantum well. The level shifts are drawn for the case of compressive strain, ignoring coupling the lh and ch states. The shifts are in the opposite direction for tensile strain

$$\varepsilon_s = -2e\left(D_2 - \frac{C_{13}}{C_{33}}D_1\right) , \tag{6.102}$$

where $e_0 = (a_b - a_q)/a_q$ and $a_{q(b)}$ is the lattice constant in the quantum-well (barrier) regions.

Using a basis consisting of the following $2N_e$ eigenfunctions,

$$\xi_{n\uparrow}(\boldsymbol{r}) = u_{e,n}(z)\langle \boldsymbol{r}|S\uparrow\rangle \tag{6.103}$$

$$\xi_{n\downarrow}(\boldsymbol{r}) = u_{e,n}(z)\langle \boldsymbol{r}|S\downarrow\rangle \tag{6.104}$$

where $1 \le n \le N_e$, we get the matrix equation

$$\begin{pmatrix} H^e & 0 \\ 0 & H^e \end{pmatrix}\begin{pmatrix} C_n^\uparrow \\ C_n^\downarrow \end{pmatrix} = E_{en\perp}\begin{pmatrix} C_n^\uparrow \\ C_n^\downarrow \end{pmatrix} , \tag{6.105}$$

where the elements for the $N_e \times N_e$ matrix $H_e$ are

$$H^e_{n,m} = \left(E^{\mathrm{conf}}_{e,n} + \frac{\hbar^2 k^2}{2m_{e\perp}}\right)\delta_{n,m} + eE_p(N)\int_{-\infty}^{\infty} \mathrm{d}z\, u_{e,n}(z)\, z\, u_{e,m}(z) , \tag{6.106}$$

and the elements of the column submatrix $C_n$ are the amplitudes $C^\uparrow_{m,n} = C^\downarrow_{m,n} \equiv C_{m,n}$ in (6.90, 91). The off-diagonal matrix elements in $H^e$ are due to the presence of a strain-induced electric field along the $c$ axis, that is, sometimes appreciable because of the large piezoelectric constant $d_{31}$ in the wurtzite structure. For finite carrier densities the strain induced electric field is reduced due to screening. The resulting effective field has to be computed from the coupled Schrödinger and Poisson equations taking into consideration the spatial separation in the electron and hole distributions (Fig. 6.12). The net electric field is

$$E_p(N) = E_p(0) + E_{\mathrm{scr}}(N) \tag{6.107}$$

where

$$E_p(0) = \frac{2d_{31}e}{\varepsilon_b}\left(C_{11} + C_{12} - \frac{2C_{13}^2}{C_{33}}\right) \tag{6.108}$$

is the piezoelectric field in the absence of carriers. The effective screening field may be estimated by spatially averaging the electric field due to the combined electron and hole distributions over the quantum well:

$$E_{\mathrm{scr}}(N) = \frac{1}{w}\int_{-w/2}^{w/2} \mathrm{d}z_0 \frac{eN}{2\varepsilon_b}\int_{-\infty}^{\infty} \mathrm{d}z \left[|u_e(z)|^2 - |u_h(z)|^2\right]\frac{z - z_0}{|z - z_0|}\,, \tag{6.109}$$

where $N$ is the two-dimensional carrier density, and the spatial carrier distributions $u_e(z)$ and $u_h(z)$ are approximated by the lowest order electron and hole subband envelope functions at zone center. The screening field $E_{\mathrm{scr}}(N)$ is determined by iterating (6.107) and the solution to (6.105) until convergence is reached.

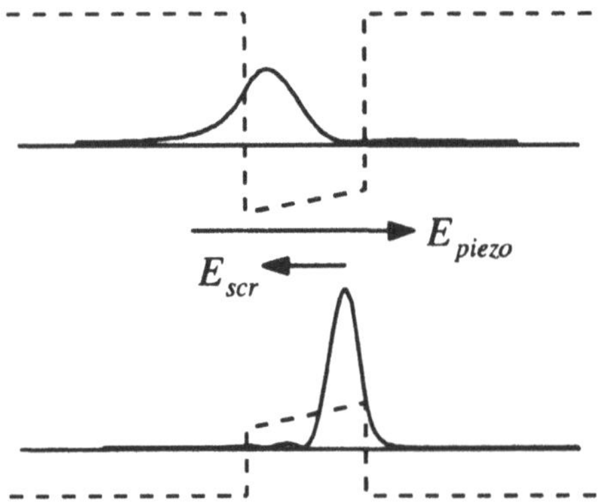


**Fig. 6.12.** Piezoelectric effect in a strained wurtzite quantum well schematically shown as *dashed line*. The piezoelectric field $E_{\mathrm{piezo}}$ causes a spatial separation of electron (**top**) and hole (**bottom**) distributions. The charge separation creates a screening field $E_{\mathrm{scr}}$, which leads to the carrier density dependence of $E_{\mathrm{piezo}}$

As a consequence of the carrier density dependent screening of the piezoelectric field we have now a density dependent band structure. In principle one could include this density dependence in the many-body Hamiltonian and deal with a zero density band structure, as in the cases without the piezoelectric effect. Even though this would be a more systematic approach we use the simpler treatment presented here since this avoids modifications in our many-body theory.

For the holes, the basis consists of the $2N_h = 2(N_{\text{hh}}+N_{\text{lh}}+N_{\text{ch}})$ functions, where $N_{\text{hh}}$, $N_{\text{lh}}$, and $N_{\text{ch}}$ are the number of solutions for $\alpha =$ hh, lh and *ch*, respectively. These functions are

$$\zeta_{n,m}(\boldsymbol{r}) = u_{\alpha,n}(z)\langle \boldsymbol{r}|m\rangle \tag{6.110}$$

where $\alpha =$ hh, lh and *ch* for $m = 1$ or 4, 2 or 5, and 3 or 6, respectively. The eigenvalue equation in this representation is

$$\begin{pmatrix} H^U & 0 \\ 0 & H^L \end{pmatrix} \begin{pmatrix} A_n^U \\ A_n^L \end{pmatrix} = E_{hn\perp} \begin{pmatrix} A_n^U \\ A_n^L \end{pmatrix} \tag{6.111}$$

where $H^L_{i,j} = (H^u)^*_{i,j}$, and the elements of the column submatrix $A_n^{U(L)}$ are the amplitudes $A^{U(L)}_{n_m,n}$ in (6.92, 93). The $N_h \times N_h$ matrix $H^U$ may be written in the form

$$H^u = \begin{pmatrix} H_{\text{hh}} & H_{\text{hh,lh}} & H_{\text{hh,ch}} \\ H_{\text{lh,hh}} & H_{\text{lh}} & H_{\text{lh,ch}} \\ H_{\text{ch,hh}} & H_{\text{ch,lh}} & H_{\text{ch,ch}} \end{pmatrix} . \tag{6.112}$$

The diagonal matrix elements for the submatrices $H_{\text{hh}}$, $H_{\text{lh}}$, and $H_{\text{ch}}$ are

$$H_{\text{hh}n,\text{hh}n} = -E^{\text{conf}}_{\text{hh},n} + \frac{\hbar^2 k^2}{4m_0}[A_1 + A_2] \tag{6.113}$$

$$H_{\text{lh}n,\text{lh}n} = -E^{\text{conf}}_{\text{lh},n} - 2\Delta_2 + \frac{\hbar^2 k^2}{4m_0}(A_1 + A_2) \tag{6.114}$$

$$H_{\text{ch}n,\text{ch}n} = -E^{\text{conf}}_{\text{ch},n} - \Delta_1 - \Delta_2 + \varepsilon_h - \varepsilon_s + \frac{\hbar^2 k^2}{2m_0} A_2 \tag{6.115}$$

respectively, and the off-diagonal elements are

$$H_{\alpha n,\alpha m} = eE_p(N) \int_{-\infty}^{\infty} \mathrm{d}z\, u_{\alpha,n}(z)\, z\, u_{\alpha,m}(z) \ . \tag{6.116}$$

The couplings among heavy, light and crystal-split holes are described by the matrix elements for the submatrices $H_{\text{hh,lh}}$, $H_{\text{hh,ch}}$, and $H_{\text{lh,ch}}$:

$$H_{\text{hh}n,\text{lh}m} = \frac{\hbar^2 k^2}{2m_0} A_5 \int_{-\infty}^{\infty} \mathrm{d}z\, u_{\text{hh},n}(z) u_{\text{lh},m}(z) \ , \tag{6.117}$$

$$H_{\text{hh}n,\text{ch}m} = -\frac{\hbar^2 k}{\sqrt{2}m_0}(A_2 - A_1 + 4A_5) \int_{-\infty}^{\infty} \mathrm{d}z\, u_{\text{hh},n}(z) \frac{\mathrm{d}}{\mathrm{d}z} u_{\text{ch},m}(z) \ , \tag{6.118}$$

$$\begin{aligned} H_{\text{lh}n,\text{ch}m} = {} & \sqrt{2}\Delta_2 \int_{-\infty}^{\infty} \mathrm{d}z\, u_{\text{lh},n}(z) u_{\text{ch},m}(z) \\ & - \frac{\hbar^2 k}{\sqrt{2}m_0}(A_2 - A_1 + 4A_5) \int_{-\infty}^{\infty} \mathrm{d}z\, u_{\text{lh},n}(z) \frac{\mathrm{d}}{\mathrm{d}z} u_{\text{ch},m}(z) \ . \end{aligned} \tag{6.119}$$

The diagonalization of (6.111) gives doubly degenerate hole bands, one from the upper block and the other from the lower block. The eigenfunctions

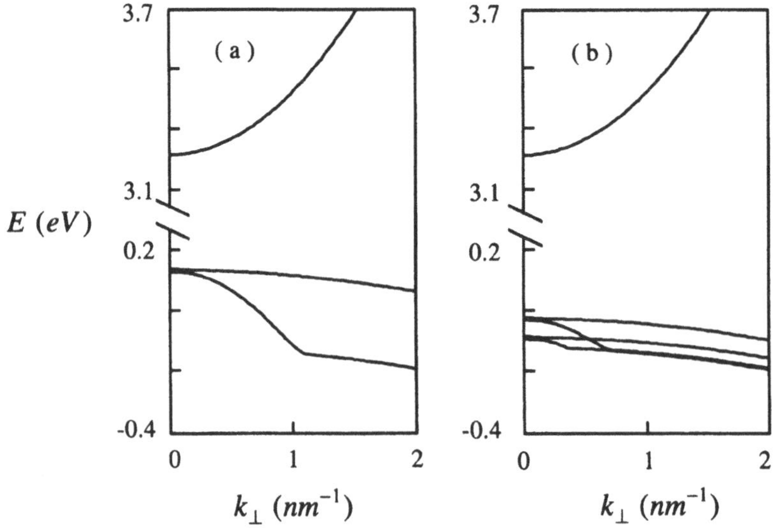


**Fig. 6.13.** Band structure for strained 2 nm $In_{0.2}Ga_{0.8}N$ quantum well between GaN barrier layers, (**a**) at low carrier density where the piezoelectric field is unscreened, and (**b**) at high density where the piezoelectric effect is strongly screened

from the upper and lower block are similar except for phase factors. Figure 6.13 shows examples of the quantum-well band structure obtained using the calculation procedure outlined above.

Finally, to compute the optical dipole matrix elements we make use of the wavefunctions, (6.90–93). From the spatial symmetry of the bulk-material states, the dipole matrix elements for the electric field polarizations parallel (TM) and perpendicular (TE) to the growth direction are

$$|\mu_{i,j}^{\mathrm{TM}}|^2 = \frac{|\mu_{\mathrm{Bulk},z}|^2}{2}\left|\sum_{n=1}^{N_e}\sum_{m=1}^{N_{\mathrm{ch}}} C_{n,i}A_{n,j}^{U(L)}\int_{-\infty}^{\infty}\mathrm{d}z\, u_{e,n}(z)u_{\mathrm{ch},m}(z)\right|^2 \quad (6.120)$$

$$|\mu_{i,j}^{\mathrm{TE}}|^2 = \frac{|\mu_{\mathrm{Bulk},x}|^2}{4}\left[\left|\sum_{n=1}^{N_e}\sum_{n_1=1}^{N_1} C_{n,i}A_{n_1,1}^{U(L)}\int_{-\infty}^{\infty}\mathrm{d}z\, u_{e,n}(z)u_{\mathrm{ch},m}(z)\right|^2 \right.$$
$$\left. \times\left|\sum_{n=1}^{N_e}\sum_{n_2=1}^{N_2} C_{n,i}A_{n_2,2}^{U(L)}\int_{-\infty}^{\infty}\mathrm{d}z\, u_{en}(z)u_{n_2,2}(z)\right|^2\right] \quad (6.121)$$

if $j$ is an upper (lower) block state. The bulk dipole matrix elements are

$$|\mu_{\mathrm{Bulk},z}|^2 = \frac{\hbar^2}{2m_0E_{g0}}\left(\frac{m_0}{m_{ez}}-1\right)\left(1+\frac{\Delta_1+\Delta_2}{E_{g0}}\right) , \quad (6.122)$$

$$|\mu_{\mathrm{Bulk},x}|^2 = \frac{\hbar^2}{2m_0E_{g0}}\left(\frac{m_0}{m_{e\perp}}-1\right)\left(1+\frac{\Delta_1+\Delta_2}{E_{g0}}\right) . \quad (6.123)$$

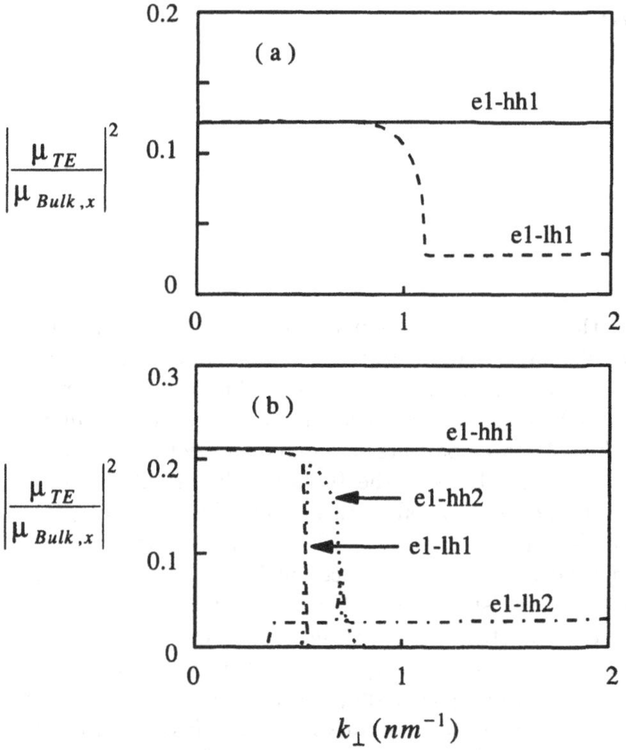


**Fig. 6.14.** TE dipole transition matrix elements versus transverse carrier momentum for wurtzite structure of Fig. 6.13. The limits of low (**a**) and high (**b**) carrier densities are shown

Figure 6.14 shows the quantum-well dipole matrix elements for transitions involving some of the low energy electron and hole states. Note that unlike the near-infrared III–V compounds where the spin-orbit splitting is large, both heavy and light hole contribute strongly to the TE mode in the wurtzite structure. This is because the strong mixing between the $j = 3/2$ and $1/2$ states lead to hh and lh states that have basically $x$ or $y$ symmetry. The $z$-like state at zone center occurs only with the higher lying crystal-field split state.

# 7. Applications

This chapter combines the gain theory of Chaps. 3, 4, with the band-structure calculation techniques of Chaps. 5, 6, to study some important material combinations which often serve as gain media in semiconductor lasers. First of all, the widely used laser heterostructure consisting of a GaAs quantum well between lattice matched AlGaAs barriers is treated in Sect. 7.1. Strained quantum-well structures on the basis of the InGaAs–AlGaAs material system are studied in Sect. 7.2. Here, we illustrate some of the advantages of the strain-induced band-structure changes, such as reduced carrier density for transparency, and polarization selection. Section 7.3 describes results for another strained quantum-well structure, InGaAs–InP, which allows us to obtain some systematic understanding of strain effects. In this context, it is particularly interesting that one can vary the quantum well continuously from tensile to compressive strain by changing the InGaAs composition. Section 7.4 describes and analyzes results for InGaP–InAlGaP quantum wells, which are interesting for applications requiring optical emission at red wavelengths below 700 nm. This section also discusses carrier leakage, which is an important loss mechanism for short wavelength or high temperature operation of InGaP lasers. Section 7.5 describes the wide bandgap II–VI CdZnSe quantum-well structure, where strong Coulomb interactions lead to significant contributions of the many-body effects discussed in Chaps. 3, 4. We end the chapter with a discussion of wide bandgap group-III nitride structures. In addition to exhibiting strong Coulomb effects, this material system deviates noticeably from the more conventional near-infrared laser materials because of the strong influence of the split-off $j = 1/2$ states. Furthermore, the typical crystal symmetry for group-III nitride laser structures is hexagonal instead of cubic as in most other semiconductor laser materials.

Whenever possible, we make connections to experiments. In particular, Sects. 7.2, 4, 5 contain comparisons between theory and experiment that illustrate the importance of band structure and Coulomb effects.

## 7.1 GaAs–AlGaAs Quantum Wells

Semiconductor lasers using lattice matched GaAs–AlGaAs quantum wells typically emit between 780 to 870 nm. Countless devices are made with this material combination, including the first microcavity lasers and laser arrays. The material growth technology is well established and one has considerable data on device performance and reliability. The relevant bulk-material parameters may be found in Chap. 6. For the bandgap of $Al_xGa_{1-x}As$, the composition dependence of the bandgap energy can be fitted by the empirical formulas,

$$\varepsilon_{g0} = 1.519 + 1.247x - 0.0005405\frac{T^2}{T + 204\,\mathrm{K}} \tag{7.1}$$

for $x < 0.45$, and

$$\varepsilon_{g0} = 1.519 + 1.247x + 1.147(x - 0.45)^2 - 0.0005405\frac{T^2}{T + 210\,\mathrm{K}} \tag{7.2}$$

for $x \geq 0.45$. The offset ratio between conduction and valence band is usually taken to be 67 : 33. Figure 7.1 shows two computed GaAs–$Al_{0.2}Ga_{0.8}As$ quantum-well band structures using the theory of Chap. 6. Since the spin-orbit splitting is around 340 meV in GaAs, the $j = 1/2$ valence bands are well removed from the lowest energy $j = 3/2$ bands. Hence, it is sufficient to use the $4 \times 4$ Luttinger Hamiltonian to determine the hole band structure. Following the convention introduced in Chap. 6, we label each band according to the quantum numbers that are valid at the zone center, $k = 0$. The conduction bands are labeled $en$, where $n = 1, 2, 3, \ldots$ are the square-well

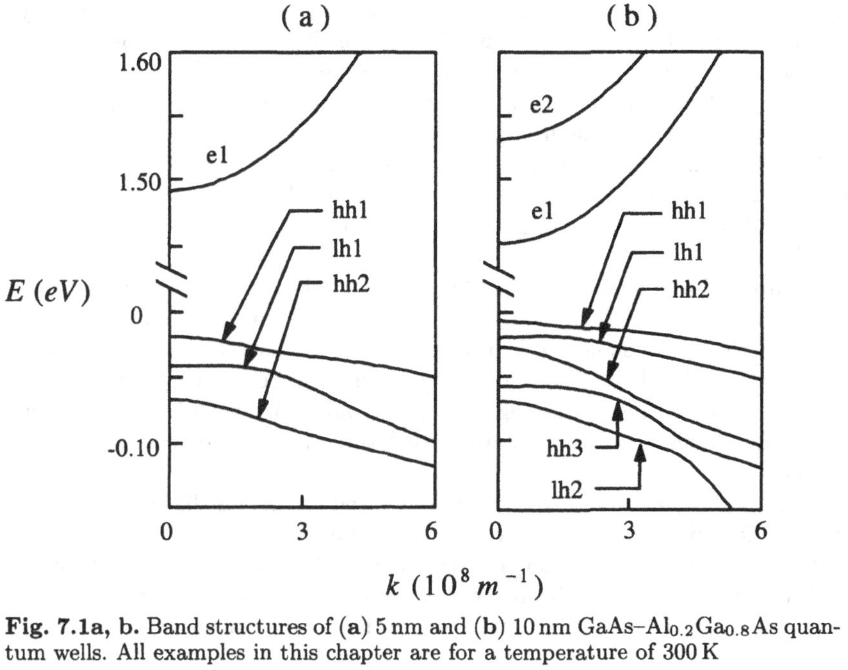


**Fig. 7.1a, b.** Band structures of (**a**) 5 nm and (**b**) 10 nm GaAs–$Al_{0.2}Ga_{0.8}As$ quantum wells. All examples in this chapter are for a temperature of 300 K

bound state quantum numbers. Similarly, the valence bands are denoted as hh1, hh2,... and lh1, lh2,... where hh and lh stand for heavy hole and light hole, respectively. We emphasize that away from the zone center, the valence bands are mixtures of heavy and light hole states with possibly different square-well quantum numbers. This is evident, e.g., by the changes in the band curvatures for $k > 0$.

For narrow well widths, there are only few confined well states, and the relevant band structure is relatively simple. For example, Fig. 7.1a shows the computed band structure of a 5 nm GaAs quantum well with only one conduction and three valence bands. For a wider quantum well the band structure becomes more complicated because of the increase in the number of confined states. For a 10 nm quantum well, Figure 7.1b shows that the number of conduction and valence bands increases to two and five, respectively, with the valence bands originating from three heavy-hole and two light-hole states at $k = 0$.

Besides the dispersion of the energy bands, the band-structure calculation also gives the electron and hole eigenfunctions, which we use to compute the dipole transition matrix elements. Figure 7.2 shows the results for the 5 nm quantum well of Fig. 7.1a. We see that the curves exhibit a rather strong $k$ dependence. In terms of the gain, the important values of the dipole matrix elements are those around the zone center ($\Gamma$ point) because the inversion is highest in this region of the Brillouin zone. Since there is usually little band mixing around $k = 0$, one can get a feeling for the significance of the individual transitions to the dipole matrix elements already from symmetry arguments. In Chap. 6 we show that because of the symmetry of the electron and hole lattice periodic functions, the TM dipole matrix element only couples the electron state to the light hole state, whereas the TE dipole matrix element has contributions from both electron to heavy hole and electron to light hole transitions. This property shows up in Fig. 7.2b where the TM e1–hh1 dipole matrix element is negligible for small $k$. Its value increases at higher $k$ because of the mixing of heavy hole and light hole states in the hh1 band.

Chapter 6 also shows that the dipole matrix elements involve inner products of envelope functions. Since the envelope functions are the orthonormal eigenfunctions of a square well potential, a dipole matrix element that couples states with different square well quantum numbers is negligible around the zone center, thus explaining the smaller values for the e1–hh2 curves in the figure. Again, because of state mixing these matrix elements grow with increasing $k$.

Our discussions in Chaps. 2–4 are presented for the idealized case of two-band bulk and quantum well structures and ideal two-dimensional quantum confinement. However, by applying the results of Chaps. 5, 6, we can compute realistic band structures. To use these band-structure results in gain calculations and thus avoid the restriction to a perfectly two-dimensional structure,

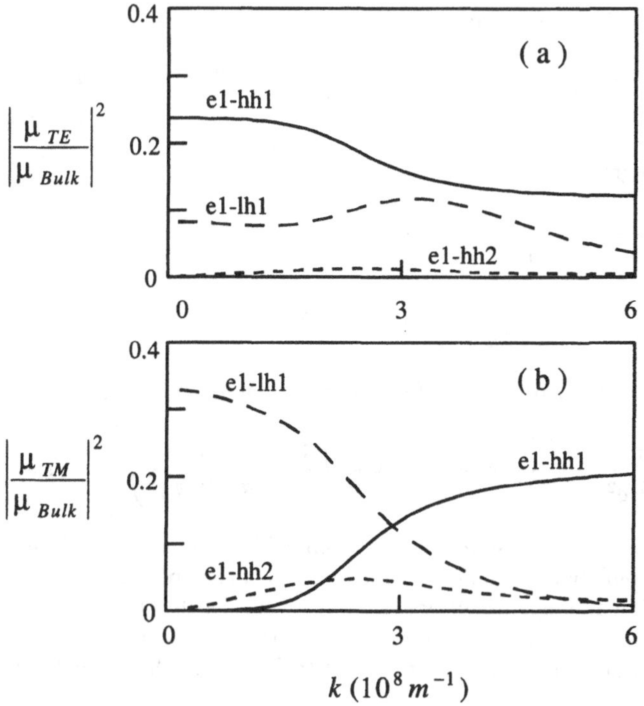


**Fig. 7.2.** Absolute square of the dipole matrix elements for TE (**a**) and TM (**b**) polarization versus $k$ for the band structure shown in Fig. 7.1a. The transitions are e1–hh1 (*solid line*), e1–lh1 (*long dashed line*) and e1–hh2 (*short dashed line*). $\mu_{\text{Bulk}}$ is the bulk GaAs dipole matrix element for the electron to heavy hole transition

we introduce a form factor that accounts for the deviations from the ideal case. For a finite quantum-well width, the Coulomb potential becomes

$$V_q = f_q \frac{e^2}{2\epsilon_b A q} \quad , \tag{7.3}$$

where the form factor is

$$f_q = \int_{-\infty}^{\infty} dz \int_{-\infty}^{\infty} dz' \, |u(z)|^2 |u(z')|^2 \, e^{-q|z-z'|} \quad . \tag{7.4}$$

To obtain an analytic expression for the integrals we approximate the envelope functions with those of an infinitely deep well extended from 0 to $w$

$$u_n(z) = \sqrt{\frac{2}{w}} \sin\left(\frac{n\pi z}{w}\right) \quad . \tag{7.5}$$

Then, the form factor becomes

$$f_q = \frac{2}{w^2} \int_0^w dz \sin^2\left(\frac{n\pi z}{w}\right) F(z) \quad , \tag{7.6}$$

where $F(z)$ involves the $z'$ integral

$$F(z) = \mathrm{e}^{-qz} \int_0^z \mathrm{d}z' \sin^2\left(\frac{n\pi z'}{w}\right) \mathrm{e}^{qz'} + \mathrm{e}^{qz} \int_z^w \mathrm{d}z' \sin^2\left(\frac{n\pi z'}{w}\right) \mathrm{e}^{-qz'} . \tag{7.7}$$

Performing the integrations in (7.7) gives

$$\begin{aligned} F(z) &= \frac{1}{2q} - \frac{q}{q^2 + \left(\frac{2n\pi}{w}\right)^2} \cos\left(\frac{2n\pi z}{w}\right) \\ &\quad - \frac{q}{2}\left(\frac{1}{q^2} - \frac{1}{q^2 + \left(\frac{2n\pi}{w}\right)^2}\right)\left(\mathrm{e}^{-qz} + \mathrm{e}^{-q|w-z|z}\right) . \end{aligned} \tag{7.8}$$

Using the result (7.8) in (7.6), and performing the $z$ integration leads to

$$f_q = \frac{2}{w}\left[\frac{1}{q} + \frac{1}{2}\frac{q}{q^2 + \left(\frac{2\pi}{w}\right)^2} + (\mathrm{e}^{-qw} - 1)\frac{1}{w}\left(\frac{1}{q} - \frac{q}{q^2 + \left(\frac{2\pi}{w}\right)^2}\right)^2\right] . \tag{7.9}$$

Equation (7.9) captures the essence of the well width dependence of the quantum-well Coulomb potential. Some improvement of the above result may be achieved by using the eigenfunctions for finite well depth. However, the resulting expression for $f_q$ becomes significantly more complicated, and the differences in the final results are not significant for the cases that have been tested.

With the more accurate band structures and a well width dependence in the Coulomb potential, we are now in the position to study the properties of a realistic gain structure under different experimental conditions. For example, we can investigate the influence of the well width on the emission properties of quantum-well systems. We have already demonstrated in Fig. 7.1 that the band structure can vary noticeably with the quantum-well width. Comparison of Figs. 7.3, 4 now shows the corresponding changes in the gain/absorption spectra. These spectra are computed using the many-body gain theory of Chap. 4. First, we examine the spectra of the 5 nm GaAs–$Al_{0.2}Ga_{0.8}As$ system. We note that the TM gain spectra are shifted toward higher frequencies than the corresponding TE gain spectra, and a higher carrier density is necessary to reach transparency. These results are consequences of the property that there are no TM transitions to the energetically lower heavy hole states as discussed in the previous paragraph.

Figure 7.4 shows that a wider quantum-well width leads to broader gain spectra. Furthermore the change in the shape of the gain spectrum with carrier density is larger in wider wells than in the narrower ones. At a carrier density of $N_{2\mathrm{d}} \leq 4 \times 10^{12}\,\mathrm{cm}^{-2}$, the TE and TM gain of the 10 nm quantum well has a simple shape because the small carrier density only causes appreciable population of the e1, hh1 and lh1 bands. At the peak gain frequency, the main contributions to the gain result from transitions involving the recombination of an electron from the e1 band and a hole from the hh1

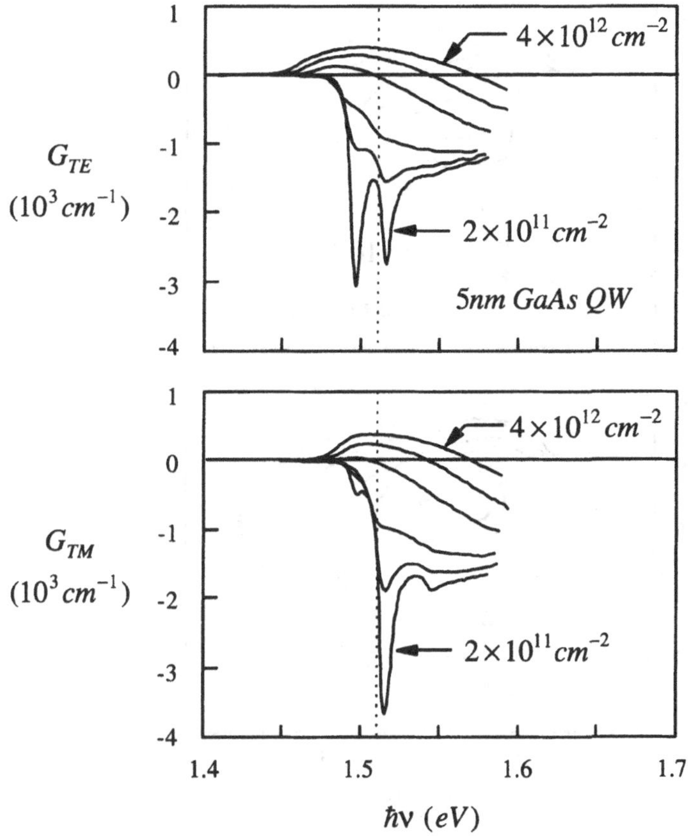


**Fig. 7.3.** TE (**top**) and TM (**bottom**) gain spectra for 5 nm GaAs–$Al_{0.2}Ga_{0.8}As$ quantum well for carrier densities, $N_{2d} = 2 \times 10^{11}, 6 \times 10^{11}, 10^{12}, 2 \times 10^{12}, 3 \times 10^{12}$ and $4 \times 10^{12}\,cm^{-2}$ (from bottom to top). The *dotted line* shows the energy of the e1–hh1 bandedge

or lh1 band. The transitions involving the different valence bands are essentially indistinguishable because their energetic separation is not much greater than the transition broadening (effective dephasing). When the carrier density is sufficiently increased such that an inversion is established between the conduction band, e2, and the valence bands, hh2 and lh2, a secondary peak appears in the gain curve, as seen in the TE and TM gain spectra in Fig. 7.4 at $N_{2d} = 6 \times 10^{12}$ and $8 \times 10^{12}\,cm^{-2}$, respectively. These peaks are due to the combined contributions of transitions originating from e1 and e2. They may eventually become larger than the original peaks, as shown in the TE gain spectra for $N_{2d} = 8 \times 10^{12}\,cm^{-2}$.

One feature of the general approach described in Chap. 4 is that it correctly predicts the appearance of the exciton resonance at low carrier densities. Such resonances are clearly visible in the low density spectra of

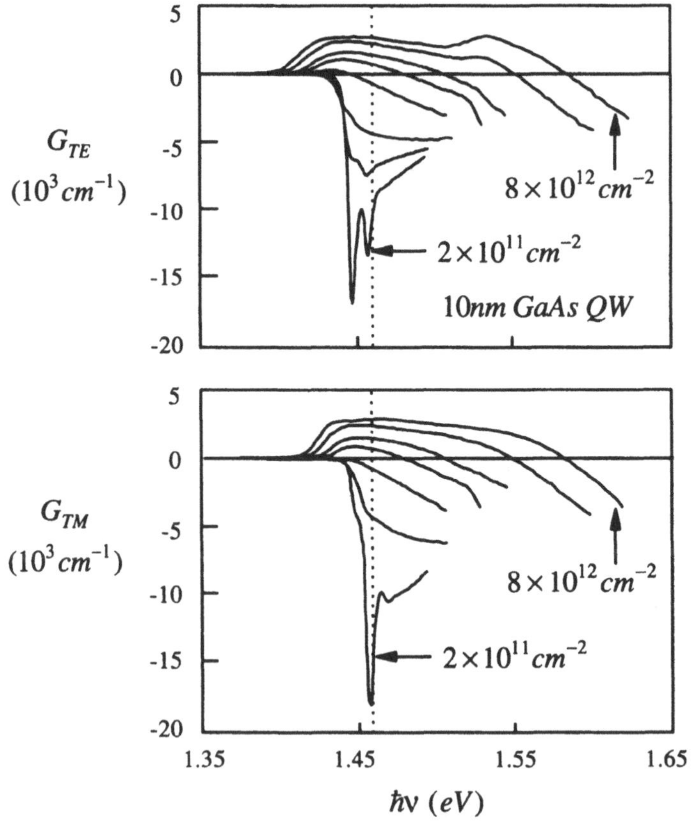


**Fig. 7.4.** TE (**top**) and TM (**bottom**) gain spectra for 10 nm GaAs–$Al_{0.2}Ga_{0.8}As$ quantum well for carrier densities, $N_{2d} = 2\times10^{11}, 6\times10^{11}, 10^{12}, 2\times10^{12}, 3\times10^{12}, 4\times10^{12}, 6\times10^{12}$ and $8\times10^{12}\,cm^{-2}$ (from bottom to top). The *dotted line* shows the energy of the el–hh1 bandedge

Figs. 7.3, 4. The lowest energy absorption resonance in the TE spectrum is due to the el–hh1 exciton. The dotted lines in the figures indicate the el–hh1 bandedges. The energy differences between the resonances and the corresponding bandedges allow us to determine the el-hh1 exciton binding energies of 15 and 12 meV for the 5 and 10 nm quantum wells, respectively. These values illustrate the dependence of the exciton binding energy on the quantum-well width. In terms of the three-dimensional exciton binding energy $\varepsilon_R^{3d}$ we have for the 5 nm well the binding energy of $3.4\varepsilon_R^{3d}$, while the wider well gives a binding energy of $2.9\varepsilon_R^{3d}$. In the limit of an ideal two-dimensional well with infinitely high potential barriers the binding energy approaches four times that of the corresponding bulk material.

Another feature of the excitonic resonances is that their energetic positions are basically independent of carrier density. This is alluded to in Chap. 4

and is a result of the proper cancellation of the respective influences of the diagonal and nondiagonal Coulomb contributions.

The increased band-structure complexity in wider quantum wells also results in additional structures in the spectra of the antiguiding or linewidth enhancement factor, $\alpha$. Figure 7.5 plots the $\alpha$ spectra for the 5 nm (solid curves) and 10 nm (dashed curves) quantum wells. The densities ($3 \times 10^{12}$ and $6 \times 10^{12}\,\mathrm{cm}^{-2}$ for the 5 and 6 nm quantum wells, respectively) are choosen to give peak gains values in the range of $2 \times 10^3$–$3 \times 10^3\,\mathrm{cm}^{-1}$. Both cases show a flattening of the spectra in the energy range where gain exists.

The value of $\alpha$ has important implications on the fundamental linewidth of quantum-well lasers. As $\alpha$ approaches zero, the refractive index is increasingly decoupled from changes in gain (e.g., due to spontaneous emission), resulting

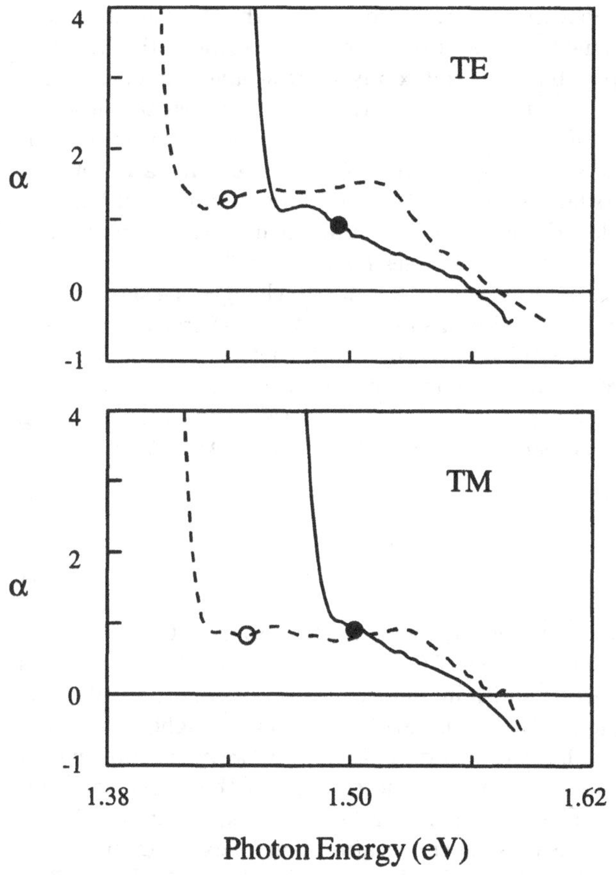


**Fig. 7.5.** Spectra of linewidth enhancement factor for 5 nm (*solid curve*) and 10 nm (*dashed curve*) GaAs–$Al_{0.2}Ga_{0.8}As$ for TE (**top**) and TM (**bottom**) polarization. The densities are $3 \times 10^{12}$ and $6 \times 10^{12}\,\mathrm{cm}^{-2}$ for the 5 and 10 nm quantum wells, respectively. The *dots* indicate the values at the gain peaks

in a reduction of the spontaneous emission contribution to the laser linewidth. The coupling between refractive index and gain also leads to filamentation or self-focusing, which limits the scalability of semiconductor lasers to higher output power. Specifically, small $\alpha$ values allow broad-area semiconductor lasers to operate more efficiently far above threshold. Whether we can make use of these advantages depends on $\alpha$ at the peak gain frequency, which is where a laser usually operates. The dots in the curves in Fig. 7.5 indicates the position of the gain peaks.

## 7.2 InGaAs–AlGaAs Strained Quantum Wells

As described in Chap. 6, quantum wells may be grown with materials having different lattice constants. Under suitable conditions, the thin quantum-well material grows in a strained configuration in order to become lattice matched to the barriers. The possibility to epitaxially grow strained quantum wells makes it possible to use a wide variety of material combinations that leads to laser emission over wavelength ranges that are desirable for applications. An extensively investigated strained quantum-well structure is InGaAs between AlGaAs barriers. Lasers fabricated with this structure typically emit between 910 and 980 nm. Wavelengths as long as 1.07 μm having been reported. Such lasers are of interest, e.g., as pump sources for fiber amplifiers.

In addition to the lasing wavelength, strain also changes the shape of the valence bands. To see this for the InGaAs–AlGaAs structure, we first note that the lattice constant for InGaAs is larger than that for AlGaAs. As a result, the InGaAs quantum well is under compressive strain. To compute the band structure, we use (7.1, 2) for the bandgap of AlGaAs and the following empirical formulas for the bandgap of InGaAs: $In_xGa_{1-x}As$ at 77 K,

$$\varepsilon_{g0} = 1.508 - 1.47x + 0.375x^2 \quad , \tag{7.10}$$

and at 300 K

$$\varepsilon_{g0} = 1.43 - 1.53x + 0.45x^2 \quad . \tag{7.11}$$

The bandgap energies at the other temperatures are assumed to be those obtained by linear interpolation. Figure 7.6 shows the calculated band structures for $In_xGa_{1-x}As$–$Al_{0.2}Ga_{0.8}As$ with indium concentrations of 0.1 and 0.2. According to Fig. 6.5, compressive strain causes the heavy and light hole square-well states to shift such that the energy separation between the hh1 and lh1 bands at $k = 0$ increases. This moves the crossing of the heavy and light hole bands to higher $k$ values (compare Fig. 6.2). Consequently, the highest valence band hh1 maintains its smaller curvature over a larger region, as may be seen by comparing the solid and dashed curves in Fig. 7.6a. Figure 7.6b shows that increasing the indium concentration to 0.2 further increases the average hh1 band curvature.

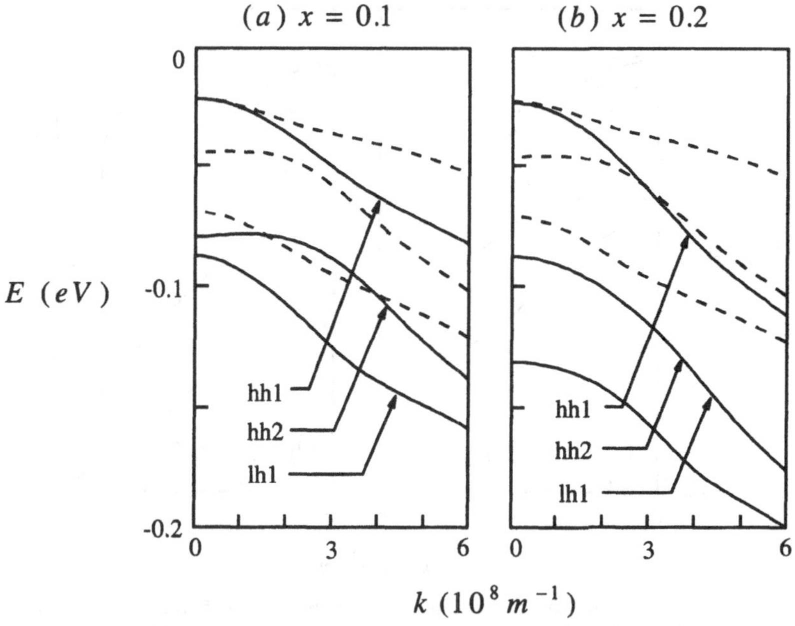


**Fig. 7.6a, b.** Band structures for 5 nm $In_xGa_{1-x}As$–$Al_{0.2}Ga_{0.8}As$ strained quantum wells with (**a**) $x = 0.1$ and (**b**) $x = 0.2$. For comparison, the *dashed curves* depict the valence bands for 5 nm GaAs–$Al_{0.2}Ga_{0.8}As$

A consequence of the changes in the band structure due to compressive strain is a reduction in the carrier density at transparency. As discussed in the earlier chapters, increasing a band curvature (or equivalently for a parabolic band, reducing its effective mass) makes it easier to populate the states around the band extremum. Figure 7.7 shows that for a given carrier density, the hole chemical potential lies further inside the valence bands for higher indium concentration.

One of the problems in creating gain in a semiconductor medium is to obtain a sufficent hole population. Here, the introduction of strain helps alleviate the problem by increasing the curvature of the lowest energy hole band, which is equivalent to reducing its density of states. This is seen in Fig. 7.8, where we plot the peak gain versus carrier density for the two cases of 5 nm $In_{0.2}Ga_{0.8}As$–$Al_{0.2}Ga_{0.8}As$ and 5 nm GaAs–$Al_{0.2}Ga_{0.8}As$. The curves also show that until the onset of rollover, compressive strain gives rise to a greater differential gain $dG/dN$.

Another result of compressive strain is the possibility to introduce polarization discrimination. Compressively strained structures usually have significantly smaller TM than TE gain (Fig. 7.9). This is again a consequence of the TM dipole transition occurring only between the electron and light-hole states. As a consequence of the increased splitting between the hh1 and lh1

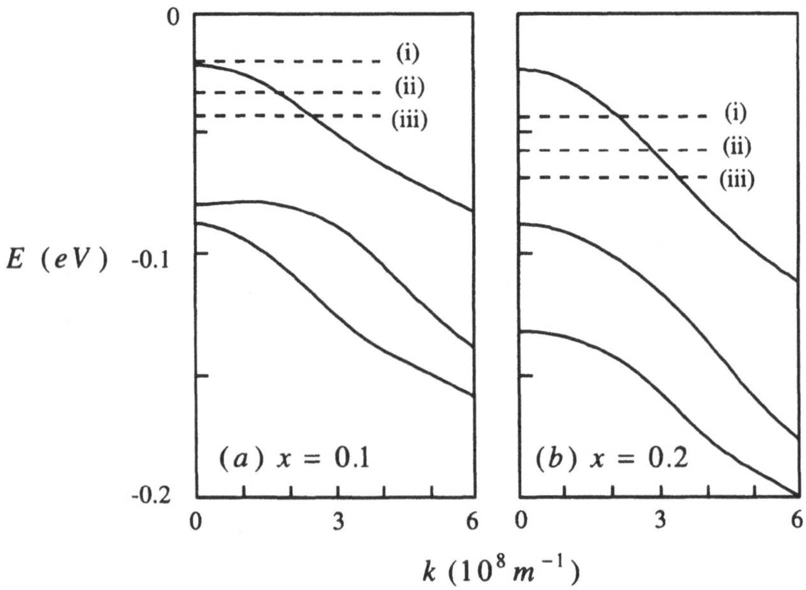


**Fig. 7.7a, b.** Hole chemical potentials (*dashed lines*) for 5 nm $In_xGa_{1-x}As$–$Al_{0.2}Ga_{0.8}As$, where the left figure (**a**) is for $x = 0.1$, and the right figure (**b**) is for $x = 0.2$. The densities are $N_{2d}$ = (i) 2×, (ii) 3×, and (iii) $4 \times 10^{18}\,cm^{-3}$. For the unstrained GaAs–AlGaAs, the hole chemical potentials for the same densities all lie inside the bandgap

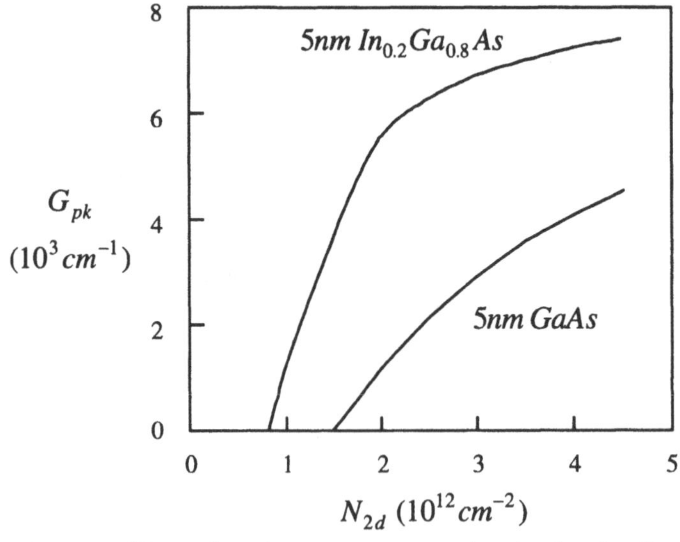


**Fig. 7.8.** Plots of peak gain versus carrier density for 5 nm $In_{0.2}Ga_{0.8}As$–$Al_{0.2}Ga_{0.8}As$ and 5 nm GaAs–$Al_{0.2}Ga_{0.8}As$

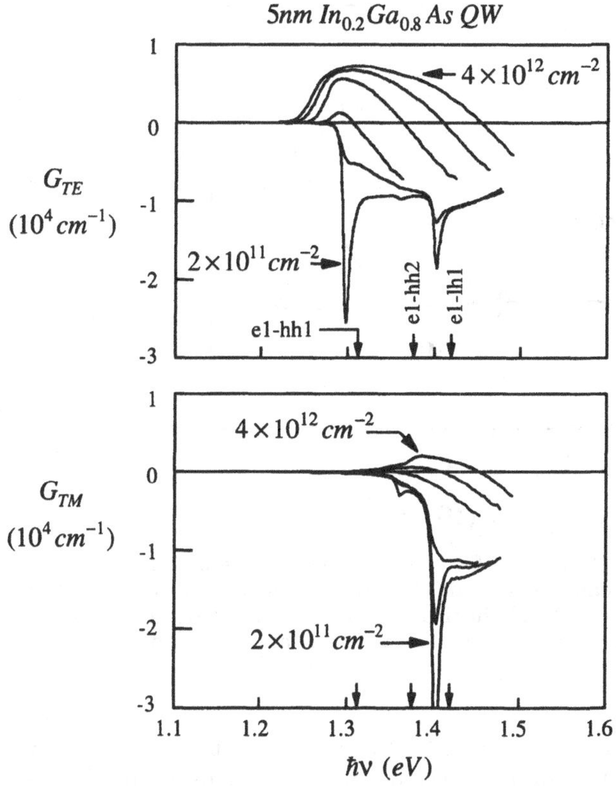


**Fig. 7.9.** TE (**top**)and TM (**bottom**) gain spectra for 5 nm $In_{0.2}Ga_{0.8}As$–$Al_{0.2}Ga_{0.8}As$ showing much lower TM gain for the same carrier density. The carrier densities are $N_{2d} = 2 \times 10^{11}$, $6 \times 10^{11}$, $10^{12}$, $2 \times 10^{12}$, $3 \times 10^{12}$ and $4 \times 10^{12}$ cm$^{-2}$ (from bottom to top). The arrows indicate the transition energies from the different bandedges

band it is harder to populate the lh1 band, which provides most of the TM gain. Furthermore, because of the larger hh1 and lh1 band separation, the TE and TM gain spectra are further apart in frequency than for an unstrained structure (compare Figs. 7.3, 9).

Figure 7.10 compares the linewidth enhancement factor at the gain peak for the 5 nm $In_{0.2}Ga_{0.8}As$–$Al_{0.2}Ga_{0.8}As$ and 5 nm GaAs–$Al_{0.2}Ga_{0.8}As$ structures. The InGaAs curve for TM polarization is not shown because there is negligible gain for experimentally realizable carrier densities. For low gain values, $\alpha(\nu_{pk})$ in the strained quantum well is below that of the unstrained structure. However, at high excitation levels the increase in peak gain results in a much larger value of $\alpha(\nu_{pk})$ for the strained quantum well. This is a result of gain rollover in the compressively strained system which reduces the denominator $dG/dN$ in $\alpha$ [see (2.110)].

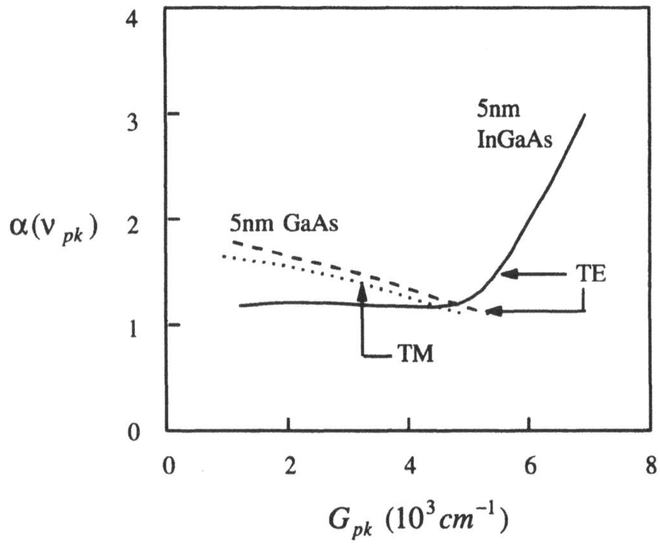


**Fig. 7.10.** Linewidth enhancement factor at gain peak as a function of peak gain for the TE mode in a compressive strained 5 nm $In_{0.2}Ga_{0.8}As$–$Al_{0.2}Ga_{0.8}As$ (*solid curve*). The TM gain is negligible in this case. *Dashed* and *dotted curve*, respectively, show the results of the TE and TM modes in an unstrained 5 nm GaAs–$Al_{0.2}Ga_{0.8}As$ quantum well

Replacing the barrier layers with GaAs leads to an interesting effect in InGaAs strained quantum wells. If the barriers are GaAs instead of AlGaAs, the confinement potential is too weak to yield a bound light hole state regardless of well width. Consequently, only the heavy hole bands are confined (Fig. 7.11a). However, this band is strongly influenced by the unconfined states. The dashed curves show the least squares fit using a parabolic approximation. The fit was performed for $0 \leq k \leq 1.28 \times 10^8\,\mathrm{m}^{-1}$, and gives for the effective masses $0.105m_0$ and $0.132m_0$ for hh1 and hh2, respectively.

Figure 7.11b shows that one may actually realize an almost ideal two-band system with a 5 nm $In_{0.1}Ga_{0.9}As$–GaAs structure. A similar least squares fit gives a hole effective mass of $0.147m_0$ in this case. One should note that because of mixing with the light hole unconfined states, the band curvature does not approach the value corresponding to transverse mass of the heavy hole.

One way to check the theoretical results is a detailed comparison with experiments. Figure 7.12 shows theoretical and experimental gain spectra for an 8 nm $Ga_{0.95}In_{0.05}As$–$Al_{0.2}Ga_{0.8}As$ single quantum-well gain region. The experimental gain spectra are extracted from transmission measurements using a spectrally broad low intensity laser probe-light, and injection currents of 3, 5, 8, 10 and 20 mA [*Ellmers* et al. (1998)]. The theoretical spectra are computed for the carrier densities, $N_{2\mathrm{d}} = 1.6$, 2.0, 2.4, 2.8 and $3.5 \times 10^{12}\,\mathrm{cm}^{-2}$.

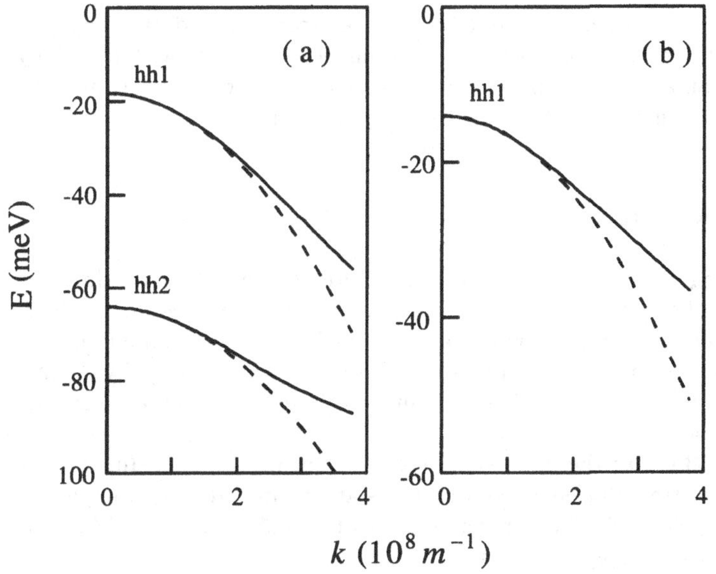


**Fig. 7.11a, b.** (**a**) With GaAs barriers the only confined state in the valence band is the heavy hole. The band structure is computed for a 5 nm $In_{0.2}Ga_{0.8}As$–GaAs structure. The effective hole mass is strongly influenced by the unconfined states. (**b**) With a 5 nm $In_{0.1}Ga_{0.9}As$–GaAs structure, one almost realizes the idealized two-band system. The *dashed curves* show the least squares fit with parabolic bands

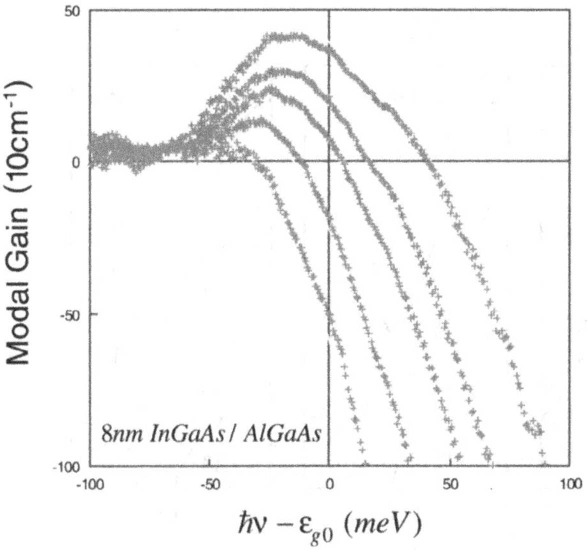


**Fig. 7.12.** Comparison of the experimental (*crosses*) and calculated (*solid curves*) TE spectra of an 8 nm $In_{0.05}Ga_{0.95}As$–$Al_{0.2}Ga_{0.8}As$ quantum well. This figure was originally published by [*Ellmers* et al. (1998)] where calculational details and material parameters are given

These carrier densities are chosen to best fit the experimental data. The figure shows that the theory reproduces the experimental data very well, especially the gradual bulk-like gain increase with increasing photon energy, and the blue shift in the gain peak for higher excitation levels.

## 7.3 InGaAs–InP

An interesting strained quantum-well structure consists of an $In_xGa_{1-x}As$ quantum well between InP barriers. The structure is unstrained for $x \simeq 0.53$, and under tensile (compressive) strain for smaller (larger) indium concentration. Lasers fabricated with InGaAs–InP quantum wells typically operate between 1.45 to 1.62 µm, which is a useful wavelength range for optical-fiber communications.

Figure 7.13 shows the band structures of 5 nm $In_xGa_{1-x}As$–InP for the three cases of strain. These band structures are computed using the parameters given in Chap. 6. The bandgap energy of InGaAs is given in (7.10–11) and the bandgap energy of InP is

$$\varepsilon_{g0} = 1.42667 - 0.000326436T \ , \tag{7.12}$$

which is determined using a linear fit to the data in Landolt-Börnstein (1982). For $x = 0.33$, there are three valence bands originating from one light and

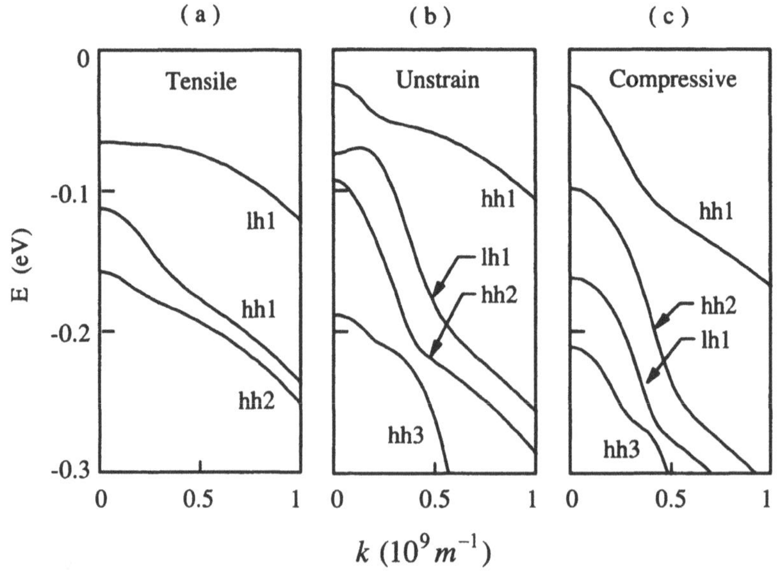


**Fig. 7.13a–c.** Band structures for 5 nm $In_xGa_{1-x}As$–InP with (**a**) $x = 0.33$, (**b**) $x = 0.53$ and (**c**) $x = 0.73$, showing the increase in the average hh1 band curvature with increasing indium concentration

two heavy hole states at $k = 0$, with the light hole state being associated with the lowest energy hole band. For $x = 0.53$, the number of conduction and valence bands increases to two and four, respectively, because of the deeper well. Note that now the energetically lowest hole band is predominately heavy hole in character and therefore has a higher average band curvature than for the case with $x = 0.33$. For $x = 0.73$, the light hole bandgap is shifted up in energy until two heavy hole states lie below it. This shift reduces the mixing of heavy and light hole states in the energetically lowest hole band, resulting in an even higher average hh1 band curvature. Not to be forgotten is the overall bandgap change with strain (from 0.73 eV with $x = 0.73$ to 1.01 eV with $x = 0.33$), which broadens the range of wavelengths accessible for lasing.

Figure 7.14 plots the gain spectra obtained using the band structures shown in Fig. 7.13. The carrier density is varied to give a peak gain of approximately 3000 $cm^{-1}$ for all the structures. The explanation for the variation of the gain spectra in relation to the corresponding band structures is similar to that for the InGaAs-AlGaAs system, especially for the $x = 0.73$ structure, which is under compressive strain. The gain spectrum is dominated by the e1–hh1 transition, since these bands have the highest electron and hole populations at the chosen density of $N = 8.0 \times 10^{11}$ $cm^{-2}$. There is no TM gain because the lh1 band is too far removed from the bandedge to be populated at this density.

For the unstrained structure with $x = 0.53$, the lowest energy hole band is still hh1 and the e1–hh1 transition dominates the gain spectra. However, the energy separation between hh1 and lh1 is reduced. This has two effects. One

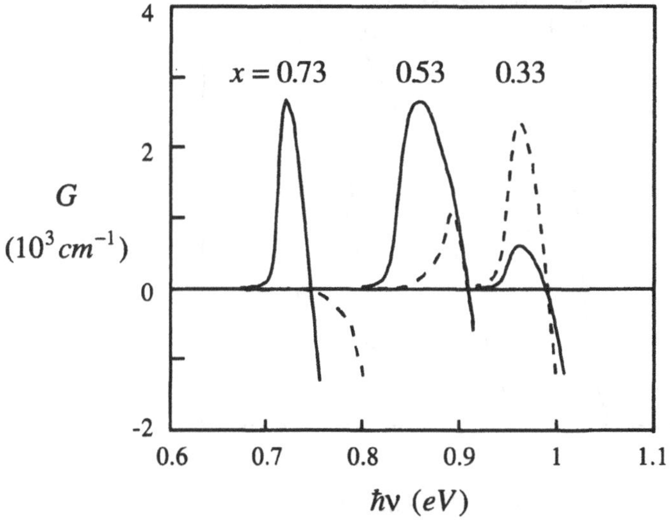


**Fig. 7.14.** $In_xGa_{1-x}As$–InP gain spectra for TE mode (*solid curves*) and TM mode (*dashed curves*) for the indium concentrations $x = 0.33$, 0.53 and 0.73. The densities, $N_{2d} = 2 \times 10^{12}$ $cm^{-2}$ for $x = 0.33$ and $x = 0.53$. For $x = 0.73$, we use $N_{2d} = 8 \times 10^{11}$ $cm^{-2}$ so that all three cases have similar peak gains of $\approx 3000$ $cm^{-1}$

is a higher hole population in lh1 and the other is a stronger heavy hole-light hole coupling, which increases the light hole content in hh1, and conversely the heavy hole content in lh1. This is responsible for the reduction in the difference of the peak gain values between the TE and TM modes seen in Fig. 7.14.

For indium concentrations $x < 0.53$ the quantum well is under tensile strain. At $x = 0.33$, the lowest hole band is lh1, which is mostly light hole in character around the zone center and therefore contributes more to the TM than to the TE mode. As we can see in Fig. 7.14, the TM mode dominates for $x = 0.33$. For indium concentrations $x \geq 0.53$ the TE gain spectra are red shifted with respect to the TM gain spectra because the lowest hole band has predominately heavy hole character around the zone center. On the other hand, for indium concentrations, $x < 0.53$, the lowest hole band has mostly light hole character around the zone center and therefore the TE gain spectra are blue shifted with respect to the TM gain spectra. There is, of course, an overall shift of the gain spectra to higher frequencies with decreasing indium concentration in InGaAs, due to the corresponding increase in the alloy bandgap.

More details on the effects of stain are summarized in Fig. 7.15. Here we plot the peak gain versus carrier density, for the three indium concentrations in the previous figure. As expected, the carrier density for a given TE peak gain is reduced by compressive strain. The curves also show that strain can be used for polarization selection. Tension favors the TM mode while compression favors the TE mode.

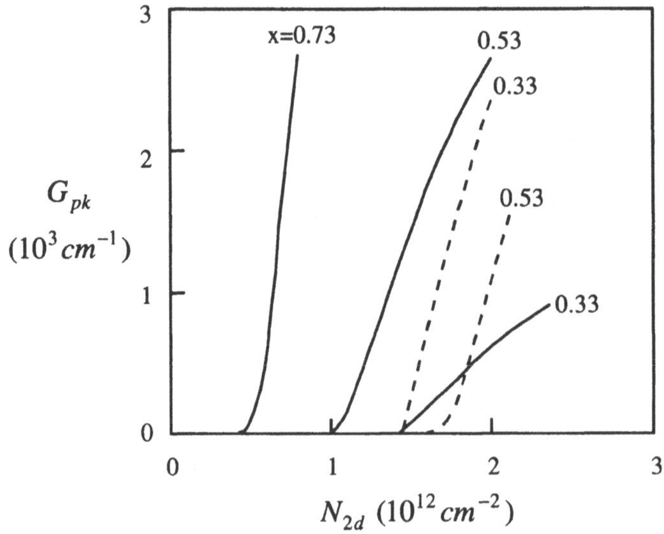


**Fig. 7.15.** Peak gain versus carrier density for 5 nm $In_xGa_{1-x}As$–InP with $x =$ 0.33, 0.53 and 0.73. The *solid (dashed) curves* are for the TE (TM) polarization. The TM gain for $x = 0.73$ is negligible

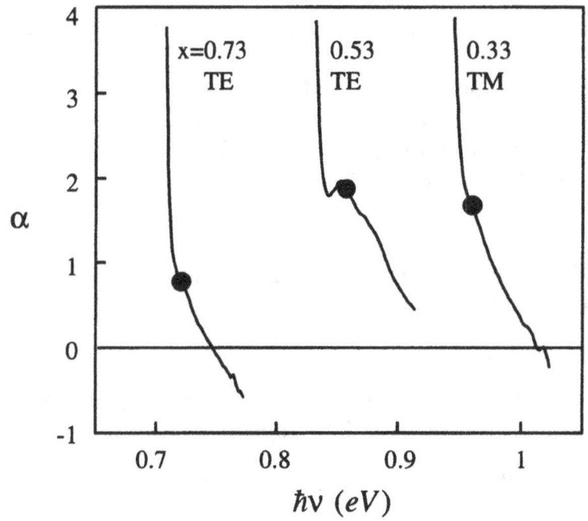


**Fig. 7.16.** Linewidth enhancement factor spectra for 5 nm $In_xGa_{1-x}As$–InP. For $x = 0.73$ and 0.53, the spectra are for the TE polarization and carrier densities $N_{2d} = 7.5 \times 10^{11}$ and $1.9 \times 10^{12}\,cm^{-2}$, respectively. The spectrum for $x = 0.33$ is for the TM mode and $N_{2d} = 1.9 \times 10^{12}\,cm^{-2}$

The slope of the curves in Fig. 7.15 gives the differential gain, $dG/dN$ at the gain peak. Unlike the free-carrier model, this derivative involves more than the Fermi-Dirac distributions. There are contributions from the Hartree-Fock and collision effects, in addition to band filling. From Fig. 7.15 we see that for the TE mode, the differential gain increases with compressive strain. However, a high differential gain can also be obtained under tensile strain in the TM mode.

Figure 7.16 shows the spectra of the linewidth enhancement factor. The dots indicate the values at the gain peak. In more extensive studies, $\alpha_{pk}$ is found to exhibit a complicated dependence on carrier density and strain. On the other hand, there are also common features among strained quantum-well structures. One is the increase in $\alpha_{pk}$ with decreasing compressive strain caused by band-structure effects.

## 7.4 InGaP–InAlGaP Red-Wavelength Lasers

Heterostructures made of InGaP quantum wells and InAlGaP barriers are important because they provide gain in the visible wavelength region below 700 nm. Light emission as short as 570 nm (yellow) is possible for material compositions where the quantum well is under tensile strain. Electrically injected vertical cavity surface emitting lasers (VCSELs) using these heterostructures demonstrated operation at wavelengths between 639 and

661 nm. Potential applications include optical displays and light sources for plastic fibers.

The bulk-material parameters for InP, GaP and AlP are given in Chap. 6. Parameter values for InGaP and InAlGaP are taken to be the properly weighed averages of those of InP, GaP and AlP. The bandgap of $In_{1-x}Ga_xP$ at room temperature as given by *Adachi* (1982) is

$$\varepsilon_{g0} = 1.35 + 0.643x + 0.786x^2 \quad . \tag{7.13}$$

However, the results of VCSEL experiments are found to fit better to the formula [*Stringfellow* et al. (1972)],

$$\varepsilon_{g0} = 1.421 + 0.73x + 0.7x^2 \quad . \tag{7.14}$$

The composition dependence of the bandgap for $In_{1-x}(Al_yGa_{1-y})_xP$ is even more uncertain. A reasonable fit to available data appears to be

$$\varepsilon_{g0} = \varepsilon_{g0}(In_{1-x}Ga_xP) + 0.6y \quad , \tag{7.15}$$

for $y \leq 0.6$ and

$$\varepsilon_{g0} = \varepsilon_{g0}(In_{1-x}Ga_xP) + 0.36 \quad , \tag{7.16}$$

otherwise. Reported values for the band-offset ratio ranges from $\simeq 0.39$ to $\simeq 0.67$. For the calculations in this section, we use the more recently reported band-offset ratio of 0.67 [*Dawson* and *Duggan* (1993)].

Figure 7.17a shows the hole band structure for a gain material composition often used in vertical-cavity surface-emitting lasers (VCSELs). It consists of 6 nm $In_{0.56}Ga_{0.44}P$ quantum wells and $(Al_{0.4}Ga_{0.6})_{0.5}In_{0.5}P$ barrier layers.

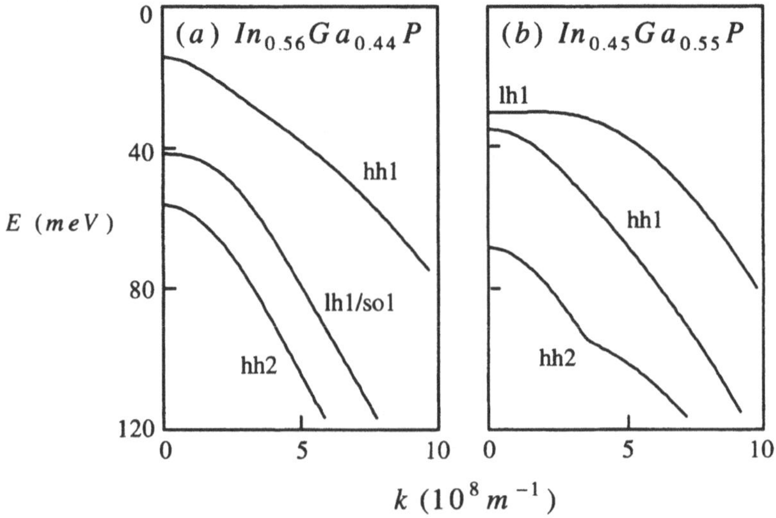


**Fig. 7.17a, b.** Band structures for 6 nm (**a**) compressive strained $In_{0.56}Ga_{0.54}P$–$(Al_{0.4}Ga_{0.6})_{0.5}In_{0.5}P$ and (**b**) tensile strained $In_{0.45}Ga_{0.55}P$–$(Al_{0.4}Ga_{0.6})_{0.5}In_{0.5}P$

The quantum wells are under compressive strain, with the lattice mismatch parameter, (6.34), $e_0 = -0.0044$. The bandgap energy is 1.882 eV which gives a transition wavelength of 660 nm. The calculation, which uses a $6 \times 6$ Luttinger Hamiltonian, shows that the lowest energy hole band has a heavy hole state at $k = 0$. Because of the coupling between $j = 3/2$ and $1/2$ states, the zone center state in the second band is a mixture of light hole and split-off state, with the approximate ratio of 0.84/0.16.

Shorter transition wavelengths are possible by reducing the In concentration in the quantum well. Experiments involving edge-emitting laser diodes have demonstrated 610 nm operation with tensile strained quantum wells. In this case, the TM polarized mode has the lower threshold. Figure 7.17b shows the band structure for a 6 nm $In_{0.45}Ga_{0.55}P$–$(Al_{0.4}Ga_{0.6})_{0.5}In_{0.5}P$ quantum well structure, which is under tensile strain. The bandgap energy is 2.261 eV which converts to 550 nm in transition wavelength. Here, the lowest energy hole band has a light hole state at $k = 0$. Unlike the earlier example for the compressively strained quantum well, there is little mixing with the split-off state because of the large energy separation between light hole and split-off hole states. The lower band curvature suggests that the threshold carrier density will be higher than for the compressively strained structure. In addition, we expect the threshold carrier density to be higher because of carrier leakage because of the smaller confinement potentials. For example, the state lh1 lies only about 60 meV below the top of the quantum-well confinement potential, whereas the hh1 state in the compressively strained structure of Fig. 7.17a is 115 meV below the top of the quantum-well confinement potential, which substantially reduces the leakage problem.

Carrier leakage in the InGaP–AlGaInP system places a practical limit on short wavelength operation. The cause of this limit is the $\Gamma - X$ conduction band crossing at which the conduction band minimum at the $X$ point (boundary of the Brillouin zone) becomes lower than that at the $\Gamma$ point (zone center). If the $X$ point minimum is below the $\Gamma$ point minimum the material is a semiconductor with an indirect bandgap, such as Si or Ge. This $\Gamma - X$ crossing in AlGaInP occurs at an Al mole fraction of approximately 0.56 to 0.7 in the quaternary alloy $(Al_xGa_{1-x})_{0.48}In_{0.52}P$ used in the barrier layers. Figure 7.18 shows the bandgap energy as a function of Al concentration for $(Al_xGa_{1-x})_{0.5}In_{0.5}P$ at the $X$ (dashed curve) and $\Gamma$ (solid curve) points of the Brillouin zone. The laser transition originates from the $\Gamma$ point, which has a direct energy gap. As long as $\varepsilon_{g0}(\Gamma) < \varepsilon_{g0}(X)$, most of the carrier population remains at the $\Gamma$ point. However, when $\varepsilon_{g0}(\Gamma) > \varepsilon_{g0}(X)$, electrons created at the $\Gamma$ point will escape to the $X$ point via the emission or absorption of phonons or by impurity and/or alloy scattering. According to Fig. 7.18, the bandgap crossing occurs at an Al concentration of 0.58. This crossing limits the size of the potential used for carrier confinement, with the limitation being more serious in structures with higher quantum-well bandgap energy. This in turn, places a practical limit on the shortest achievable lasing wavelength.

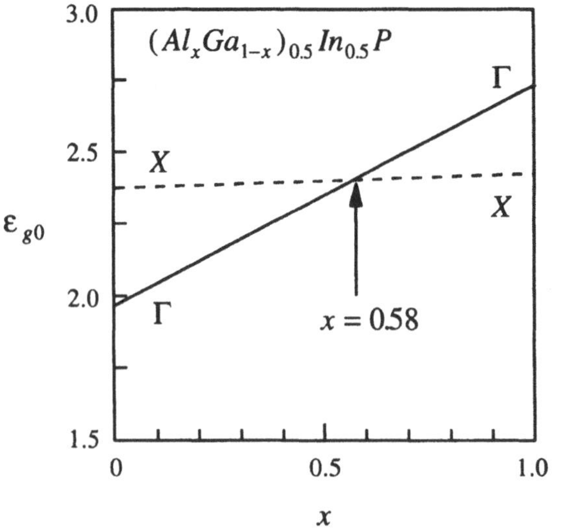


**Fig. 7.18.** $\Gamma$ and $X$ point bandgap energy versus Al concentration in $(Al_xGa_{1-x})_{0.5}In_{0.5}P$ showing a crossing at $x = 0.58$

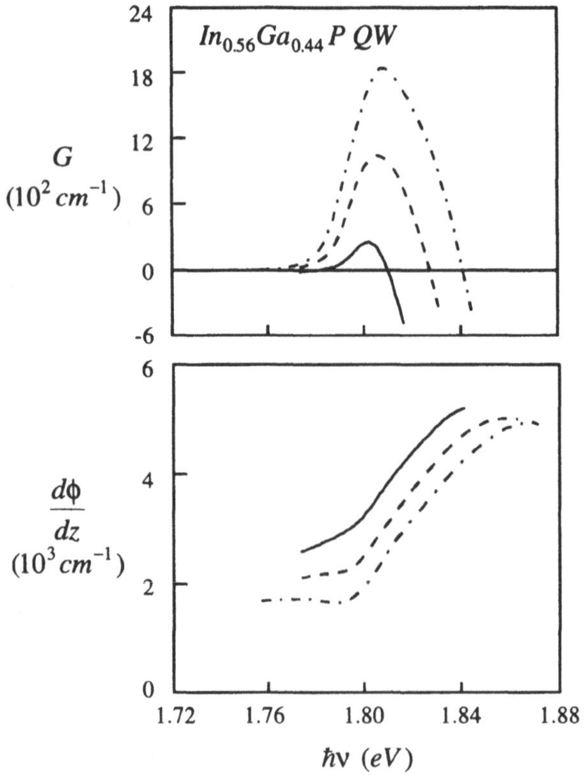


**Fig. 7.19.** TE gain (**top**) and phase shift (**bottom**) spectra for a 6.8 nm compressively strained $Ga_{0.41}In_{0.59}P$–$(Al_{0.5}Ga_{0.5})_{0.51}In_{0.49}P$ quantum-well structure. The carrier densities are $N_{2d} = 2.4 \times 10^{12}$ (*solid curves*), $2.8 \times 10^{12}$ (*dashed curves*) and $3.2 \times 10^{12}\,cm^{-2}$ (*dot-dashed curves*)

Figure 7.19 shows the computed TE gain and phase shift spectra for a 6.8 nm compressively strained $Ga_{0.41}In_{0.59}P$–$(Al_{0.5}Ga_{0.5})_{0.51}In_{0.49}P$ quantum-well structure. There is no gain for the TM polarization for the carrier densities considered. This quantum-well structure provides yet another opportunity to make comparison between theory and experiment. In many applications, it is important for the theory to correctly predict the peak gain and gain peak energy as functions of carrier density. Figure 7.20 shows the computed and measured peak gain and gain peak energy for the TE polarization versus chemical potential separation. The electron-hole chemical potential separation is a quantity that can be determined relatively unambiguously both theoretically and experimentally. The experimental points are extracted from spontaneous emission spectra measured through a 4 μm wide opening in the top contact of 50 μm wide oxide stripe lasers [*Chow* et al. (1997)]. The relationship between gain and emission (2.77) was used to determine the gain spectrum. The results show that agreement obtained with the full many-body theory extends over a wide range of excitation densities. The dashed curves are computed using the effective rate approximation, (3.100), with dephasing rate and carrier densities chosen to best fit the experimental

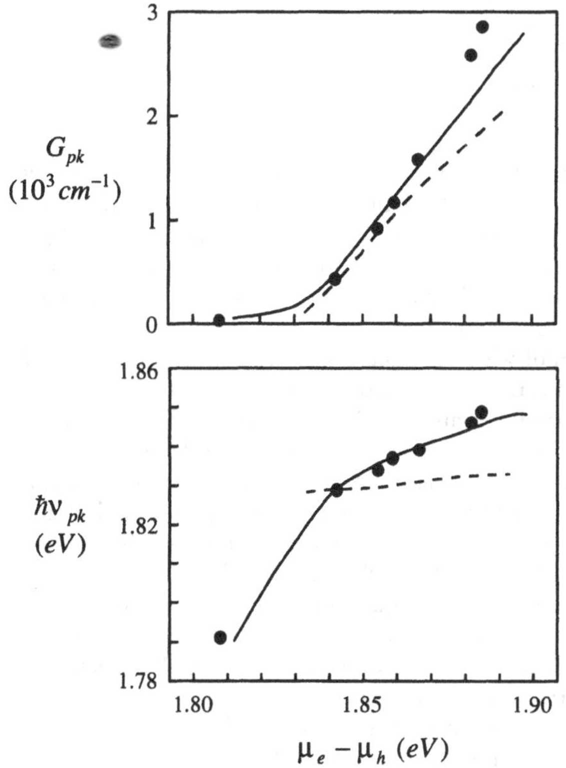


**Fig. 7.20a, b.** (**a**) Peak gain and (**b**) gain peak energy versus chemical potential separation for structure in Figs. 7.19. The points are from experiment [*Chow* et al. (1997)], the *solid curve* is from the full many-body theory, and the *dashed curve* is from effective relaxation rate approximation, with $\gamma = 10^{13}\,s^{-1}$, chosen to best fit the experimental data

data. The results impressively demonstrate that the more detailed treatment of Coulomb collision effects is crucial, especially at low excitations, or when the accurate location of the gain peak is desired.

## 7.5 II–VI Wide-Bandgap Systems

Semiconductor compounds with large bandgap energies are under intensive investigation because of the many potential electro-optical applications for visible and ultraviolet wavelength light sources. These applications include optical data-storge, optical displays and chemical sensing. The wide bandgap laser compounds are either II–VI compounds or group-III nitrides. The former includes ZeSe and CdSe, while the latter includes InGaN and GaN, to list just a few representative examples.

Table 7.1 below summarizes the approximate bulk-material exciton binding energies and Bohr radii for GaAs, GaN and ZnSe.

**Table 7.1.** Exciton binding energies and Bohr radii for GaAs, GaN, and ZnSe

| | GaAs | GaN | ZnSe |
|---|---|---|---|
| $\varepsilon_R^{3d}$ | 4 meV | 23 meV | 19 meV |
| $a_0^{3d}$ | 12 nm | 3 nm | 5 nm |

The values for the exciton Bohr radius $a_0^{3d}$ are calculated from $\varepsilon_R^{3d}$ using (1.15). The large binding energies in bulk-materials point to the importance of excitonic effects (and of Coulomb effects, in general) in the wide bandgap compounds. The case is even more convincing for quantum wells, where the exciton binding energies are typically increased in comparison to the respective bulk values by a factor of two or more.

We begin the discussion of wide bandgap laser structures with the II–VI compounds. An example of a II–VI laser heterostructure is a CdZnSe quantum well between ZnSe-barriers. Similar to GaAs, the spin-orbit splitting of the valence bands in II–VI compounds is significant (of the order of 420 meV). Therefore, we can neglect the influence of the split-off states and return to using the $4 \times 4$ Luttinger-Kohn theory for calculating the band structure. Figure 7.21 shows the band structure for a 7 nm compressively strained $Cd_{0.25}Zn_{0.75}Se$–ZeSe, which consists of two electron, four heavy-hole and one light hole bands that are confined in the quantum well. The four lowest energy hole bands have heavy hole character around $k = 0$. The calculations for Fig. 7.21 use a bulk $Cd_{0.25}Zn_{0.75}Se$ bandgap energy of 2.520 eV, which also takes into account the contribution from strain.

The TE gain spectra of the 7 nm $Cd_{0.25}Zn_{0.75}Se$–ZeSe quantum well are shown in Fig. 7.22. The carrier density ranges from $N_{2d} = 5 \times 10^{11}$ to

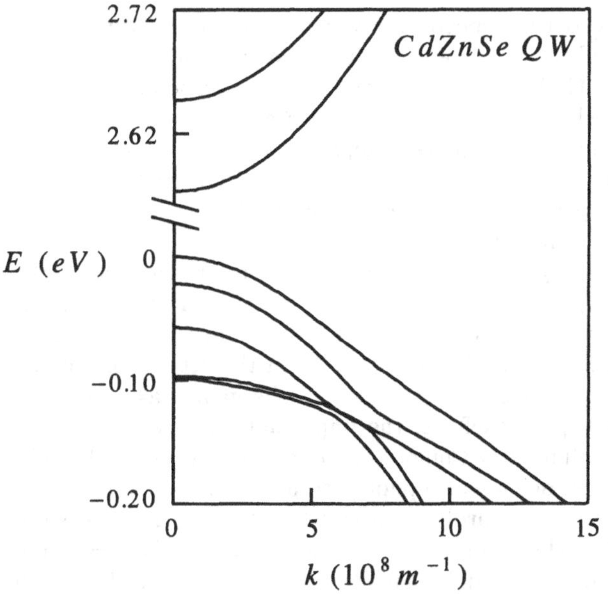


**Fig. 7.21.** Band structure for a 7 nm $Cd_{0.25}Zn_{0.75}Se$–ZeSe quantum well showing the two energetically lowest conduction bands and the five top valence bands

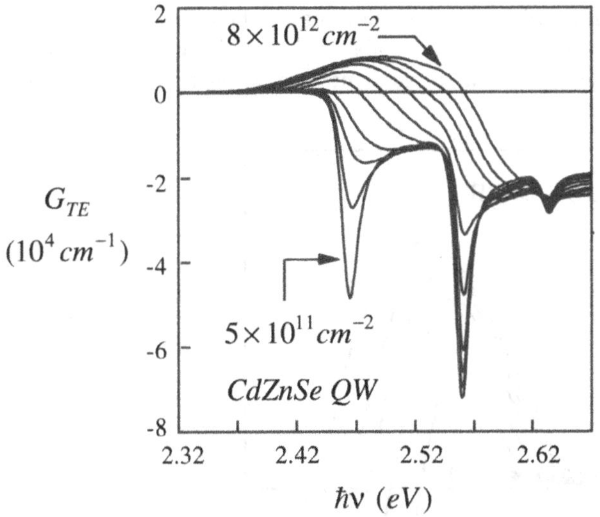


**Fig. 7.22.** TE spectra of 7 nm $Cd_{0.25}Zn_{0.75}Se$–ZeSe quantum well at $T = 300$ K and for carrier densities $N_{2d} = 5 \times 10^{11}$, $10^{12}$, $1.5 \times 10^{12}$, $2.0 \times 10^{12}$ to $8.0 \times 10^{12}$ cm$^{-2}$, from bottom to top, in increments of $10^{12}$ cm$^{-2}$

$8 \times 10^{12}\,\mathrm{cm}^{-2}$. The low density spectra show two excitonic absorption lines due to the hh1 → e1 and the hh2 → e2 transitions. While conventional III–V semiconductors such as GaAs display clear excitonic absorption lines only at low carrier densities and temperatures, these features dominate the spectra in II–VI materials even at room-temperature, and for carrier densities up to $N_{2d} \approx 10^{12}\,\mathrm{cm}^{-2}$. With increasing carrier density, the computed absorption resonances are gradually bleached, but remain basically at their respective spectral position, in agreement with experiment.

For carrier densities $N_{2d} \geq 1.5 \times 10^{12}\,\mathrm{cm}^{-2}$, gain is present due to the e1 → hh1 transition. Figure 7.23 shows the gain portion of the spectra in more detail. The energetic position of the gain maxima shifts slightly to higher energies when the carrier density is increased. As in the other semiconductor gain media the carrier density dependence of the gain peak is a result of a combination of competing effects, the important ones being bandgap renormalization, band filling and Coulomb correlations. Although the subbands other than the first one are also populated at higher carrier densities, the spectral shape of the gain hardly changes. Only for very high densities of $N_{2d} > 6 \times 10^{12}\,\mathrm{cm}^{-2}$, when an inversion develops between the $n = 2$ subbands, one notices qualitative modifications in the spectral shape of the gain spectra, e.g., at $N_{2d} = 8 \times 10^{12}\,\mathrm{cm}^{-2}$ a shoulder in the gain develops.

The spectra for TM-polarization are plotted in Fig. 7.24. For this polarization, the first conduction band is only weakly coupled to the three energetically lower hole bands, which have predominately heavy hole character. As a result, the spectra show only weak absorption and gain at low energies ($\hbar\nu < 2.52\,\mathrm{eV}$). The dipole matrix elements are larger between the first

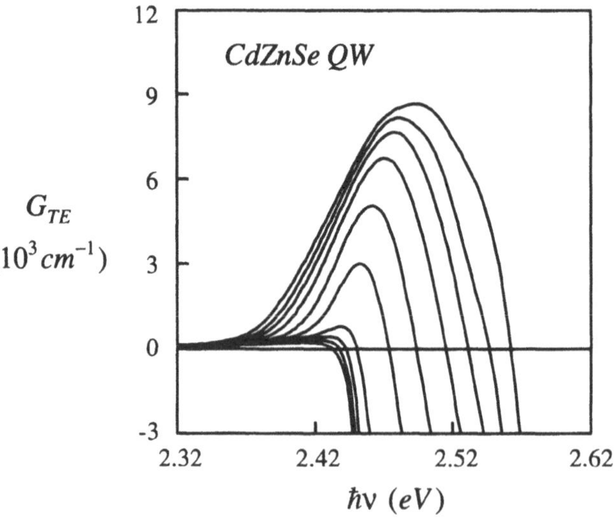


**Fig. 7.23.** Gain portion of the spectra in Fig. 7.22 on a larger scale

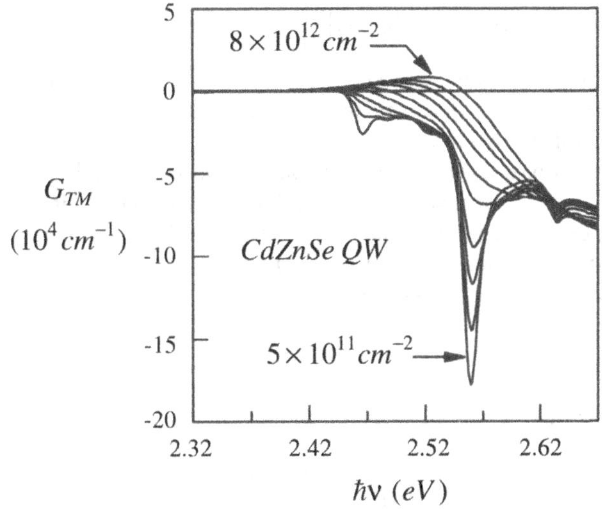


**Fig. 7.24.** TM spectra of 7 nm $Cd_{0.25}Zn_{0.75}Se$–ZeSe quantum well at $T = 300\,K$. The carrier densities are $N_{2d} = 5 \times 10^{11}$ (bottom), $10^{12}$, $1.5 \times 10^{12}$, $2 \times 10^{12}$ up to $8 \times 10^{12}\,cm^{-2}$ (top) in increments of $10^{12}\,cm^{-2}$

conduction band and the fourth and fifth hole bands, which are superpositions of heavy and light hole states. These interband transitions lead to the strong excitonic TM-absorption resonance apparent in the figure. The comparison with Fig. 7.22 shows that an absorption resonance at approximately the same energy exists also in the low density TE spectra. This coincidence is a consequence of the less than 3 meV separation between the energies for the hh2 → e2, hh4 → e1 and hh5 → e1 transitions. Similar to the TE-spectra, the TM exciton resonance is bleached with increasing carrier density. Because the TM optical transitions involve the higher energy hole bands, the onset of gain occurs at very high plasma densities. Even then, the value of TM gain achievable is significantly smaller than that of the TE polarization.

Thus far in this book we have considered only homogeneously broadened structures, where the quantum-well thicknesses and compositions are precisely known. This may be a reasonably good assumption for the red or near-infrared wavelength lasers, where growth techniques are well developed. However, for the relatively new II–VI and group-III nitride laser structures, the experimental spectra are likely to be inhomogeneously broadened by localized regions of varying quantum-well thicknesses and compositions.

In a simple phenomenological way, we approximate the inhomogeneously broadened spectrum $g_{\rm inh}(\omega)$ as a statistical average of the homogeneously broadened spectra

$$G_{\rm inh}(\omega, N, T) = \int {\rm d}x\, P(x)\, G(x, \omega, N, T) \quad , \tag{7.17}$$

where a typical weighting function is a normal distribution

$$P(x) = \left(\sqrt{2\pi}\sigma\right)^{-1} \exp\left\{- \left[(x - x_0)/\left(\sqrt{2}\sigma\right)\right]^2\right\} , \tag{7.18}$$

representing the variation in $x$, which can either be the quantum-well thickness or indium concentration. The distribution (7.18) is characterized by an average $x_0$, and a standard deviation $\sigma$.

Figure 7.25 shows examples of the effects of inhomogeneous broadening as predicted by (7.17). In this case, we assume that the net effect of either quantum-well thickness or composition variations is a distribution of bandgap energies, whose standard deviation is given by $\sigma$. Figure 7.25 (top) shows broadening of the exciton resonances as well as reduction in the peak ab-

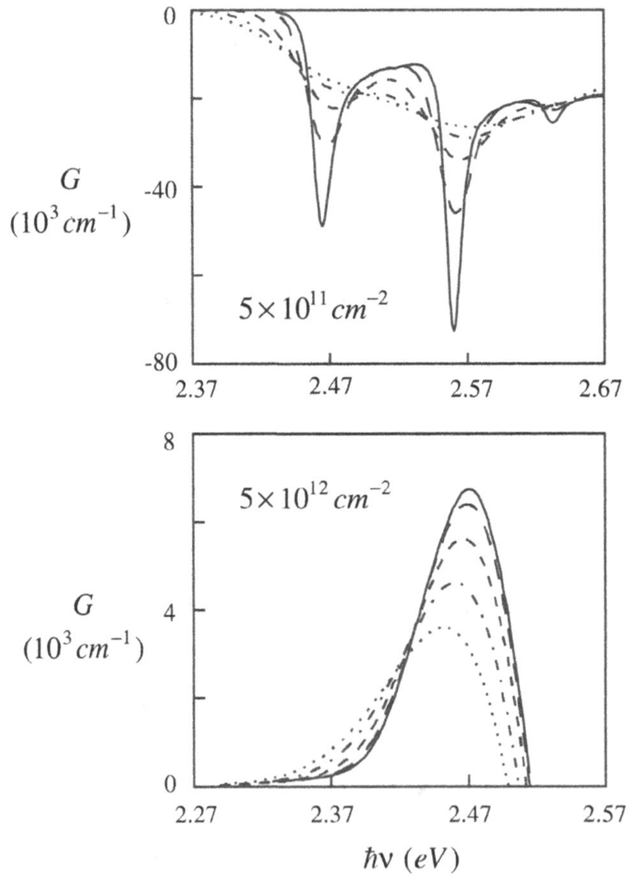


**Fig. 7.25.** Effects of inhomogeneous broadening on absorption (**top**) and gain (**bottom**) spectra. The *solid curve* is are for homogeneous broadening, the other curves are for quantum-well thickness or composition fluctuations leading to a variation in the bandgap energy of standard deviation $\sigma = 10$ (*long-dashed curve*), 20 (*short-dashed curve*), 30 (*dot-dashed curve*) and 40 meV (*dotted curve*)

sorption with increasing inhomogeneous broadending. Figure 7.25 (bottom) presents the results for high carrier densities where gain exists. We clearly see how the inhomogeneous broadening reduces the gain and shifts the spectrum to lower energies. This red shift may be understood by noting that the inhomogeneously broadened spectrum consists of a superposition of homogeneously broadened spectra, each shifted with respect to another because of their different bandgap energies. In this superposition, the gain portion of a higher lying spectrum overlaps the absorption portion of a lower lying spectrum, such that the lower lying spectra contribute more to the net gain.

Using our simple inhomogeneous broadening model, we are in the position to make comparisons with experiments. The points in Fig. 7.26 shows experimental gain spectra for a 6.9 nm $Cd_{0.27}Zn_{0.73}Se$ quantum well between $Zn_{0.94}S_{0.6}Se$ barrier and ZnMgSSe cladding layers. The entire structure is lattice matched to GaAs. The different spectra are obtained by varying the injection current from 45 mA to 60 mA. The solid curves show the calculated spectra for the same structure and carrier densities $N_{2d} = 2.8 \times 10^{12}$, $3.2 \times 10^{12}$, $3.5 \times 10^{12}$, $3.6 \times 10^{12}$, $3.8 \times 10^{12}$ and $4 \times 10^{12}\,cm^{-2}$. We also assumed an inhomogeneous broadening of $\sigma = 10\,meV$. The good agreement between experiment and theory is significant because it shows that the room temperature experimental gain spectra in II–VI lasers may be well described by our model of an interacting electron-hole plasma. Many of the differences in the optical properties between these lasers and conventional near-infrared

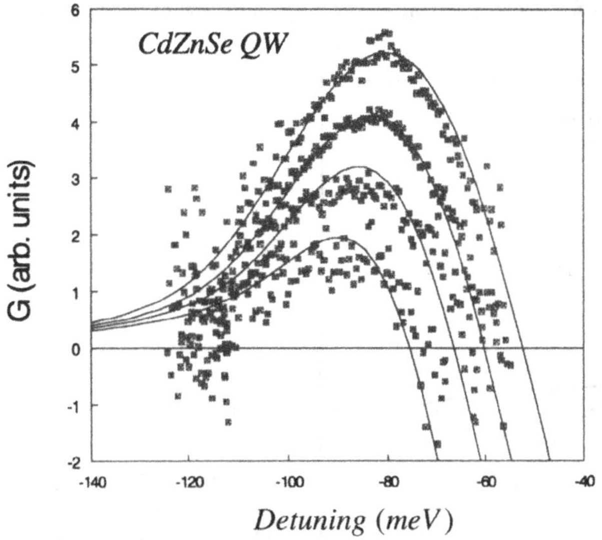


**Fig. 7.26.** TE spectra of 6.9 nm $Cd_{0.27}Zn_{0.73}Se$–ZeSe quantum well at $T = 300\,K$ and for different carrier densities. The points are from experiment [*Yoshida* et al. (1996)] and the curves are from the theory [from Girndt et al. (1998), where also the material parameters can be found]

lasers are a result of the significantly stronger Coulomb correlation effects in the II–VI materials.

## 7.6 Group-III Nitrides

This section discusses the wide bandgap group-III nitrides. Laser structures of these compounds show considerable promise for light emitting diodes and lasers operating in a wide spectral range from the visible up to ultraviolet wavelength regions. Like the II–VI lasers, the optical properties of group-III nitride structures are significantly influenced by strong Coulomb effects. In addition, there are differences in behavior from the red and near-infrared wavelength lasers because nitride structures are typically grown with the hexagonal or wurtzite crystal symmetry. The most relevant aspects of the wurtzite band-structure properties are discussed in Chap. 6.

To illustrate some of the unique gain and absorption features, we choose the example of a 2 nm compressively strained $In_{0.2}Ga_{0.8}N$–GaN quantum-well structure. The confined bands consist of one electron, two heavy hole and two light hole bands. Figure 7.27 shows the conduction band $e$ and the two energetically lowest hole bands, hh and lh. The other bands are more than 200 meV separated in energy at $k = 0$, and therefore have negligible hole populations. The band structure is computed with a sufficently low ($\leq 10^{10}\,cm^{-2}$) carrier density so that the piezoelectric effect is essentially unscreened. (See Fig. 6.13 for a comparison of the high and low carrier density

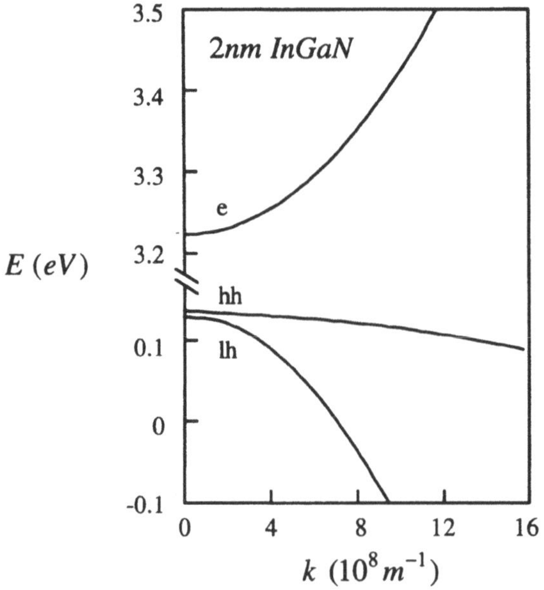


**Fig. 7.27.** Band structure for 2 nm $In_{0.2}Ga_{0.8}N$–GaN quantum-well structure. The carrier density is $N_{2d} = 10^{10}\,cm^{-2}$, so that the piezo electric field is essentially unscreened

cases). In the present case we obtain over 100 meV reduction in the bandgap energy due to the piezoelectric effect. Comparison with the computed band structures of earlier examples (e.g., Fig. 7.21) shows the much larger effective mass mismatch between the conduction and lowest energy hole band in the group-III nitrides. The large hh effective mass is one contribution to the difficulty in creating a population inversion in nitride gain structures.

Figure 7.28 shows the TE and TM dipole matrix elements versus carrier moment. As discussed in Chap. 6, because of the hexagonal crystal symmetry, both el–hh and el–lh transitions contribute to the TE polarization. On the other hand, their contributions to the TM polarization are negligible. The very small but still discernible TM dipole for the e–lh transition is due to mixing between the confined light hole and unconfined crystal-split states. As a consequence of the results shown in Fig. 7.28, the TM gain is negligible for the carrier densities realizable in experiments.

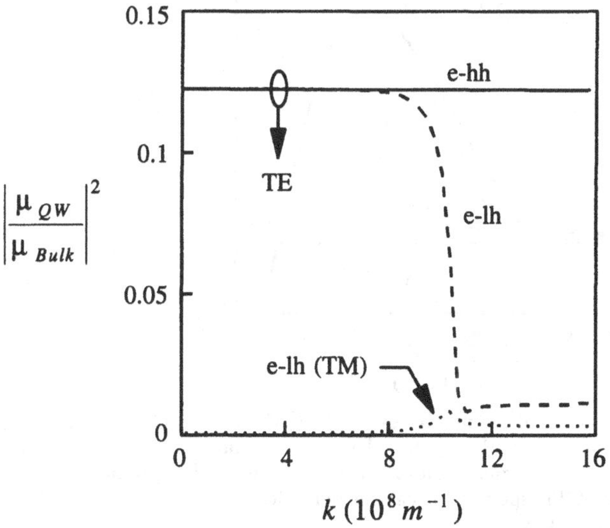


**Fig. 7.28.** TE and TM dipole matrix elements versus carrier momentum for the 2 nm $In_{0.2}Ga_{0.8}N$–GaN quantum-well structure. The transitions are between the one electronic state *e*, and the two hole states, hh and lh. The e–hh TM dipole matrix element is essentially zero and therefore not shown

Figure 7.29 (top) shows the TE gain/absorption spectra for the 2 nm $In_{0.2}Ga_{0.8}N$–GaN quantum well at carrier densities ranging from $N_{2d} = 2\times10^{12}$ to $6\times10^{12}\,cm^{-2}$. The spectra display the transition from the exciton absorption resonance at low electron and hole densities, to gain from an interacting electron-hole plasma at high carrier densities. From these spectra, we again recognize several indications of strong Coulomb effects. Concentrating first on the spectra for lower carrier densities, we extract from the energy dif-

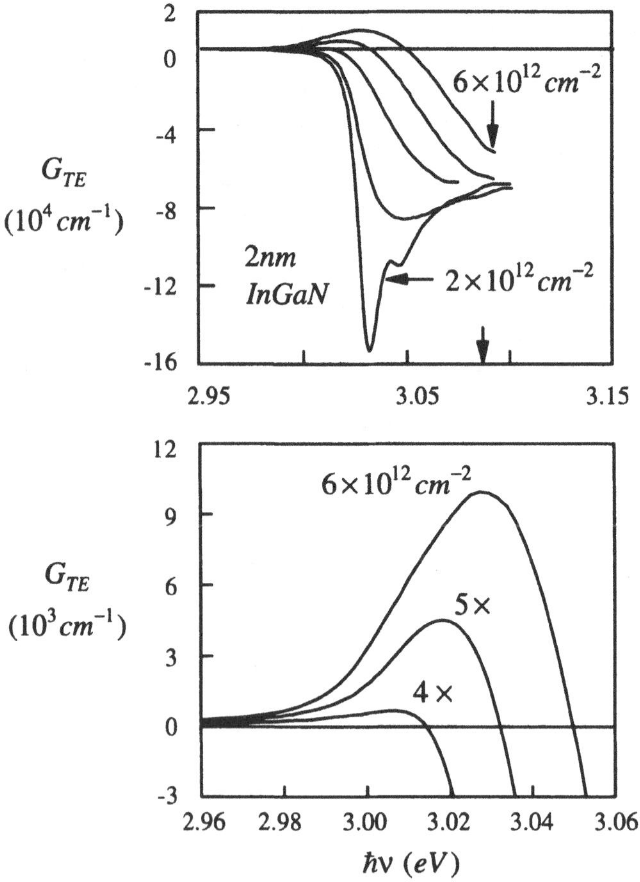


**Fig. 7.29.** (**Top**) TE spectra of a 2 nm $In_{0.2}Ga_{0.8}N$–GaN quantum-well structure for carrier densities $N_{2d} = 2 \times 10^{12}$ to $6 \times 10^{12}\,cm^{-2}$, from bottom to top, in increments of $10^{12}\,cm^{-2}$. The *arrow* indicates the quantum-well bandgap energy. (**Bottom**) Gain portion of TE spectra for carrier densities $N_{2d} = 4 \times 10^{12}$, $5 \times 10^{12}$ and $6 \times 10^{12}\,cm^{-2}$, from bottom to top

ference between the exciton resonance and the quantum-well bandgap energy (indicated by arrow), a quantum-well exciton binding energy of 55 meV. This high binding energy is a clear evidence of strong attractive Coulomb effects. It is also the reason why an exciton resonance still exists at the relatively high carrier density of $N_{2d} \simeq 2 \times 10^{12}\,cm^{-2}$. Figure 7.29 (bottom) shows in more detail the gain portion of the spectra for the carrier densities $N_{2d} = 4 \times 10^{12}$, $5 \times 10^{12}$, and $6 \times 10^{12}\,cm^{-2}$. The curves clearly exhibit an even more gradual rise in gain from the bandedge than in the earlier examples involving other III–V lasers (compare, e.g., Fig. 7.12). Another indication of strong Coulomb effects is the significant shift to lower energy of the gain peak relative to the quantum-well bandgap at 3.087 eV. This energy shift is largely due to bandgap renormalization.

Similar to the case of II–VI gain structures, experimental data suggest that the nitride quantum-well gain region is inhomogeneously broadened by spatial variations in quantum-well thickness or composition. Using (7.17), we obtain the curves in Fig. 7.30 which show the effects of inhomogeneous broadening due to an In variation in the quantum well. Figure 7.30 (top) shows that even a 5 % fluctuation in indium concentration (standard deviation $\sigma_{In} = 0.01$) leads to a significant reduction in the absorption resonance. With $\sigma_{In} \geq 0.04$, the exciton resonance becomes unobservable. Also present

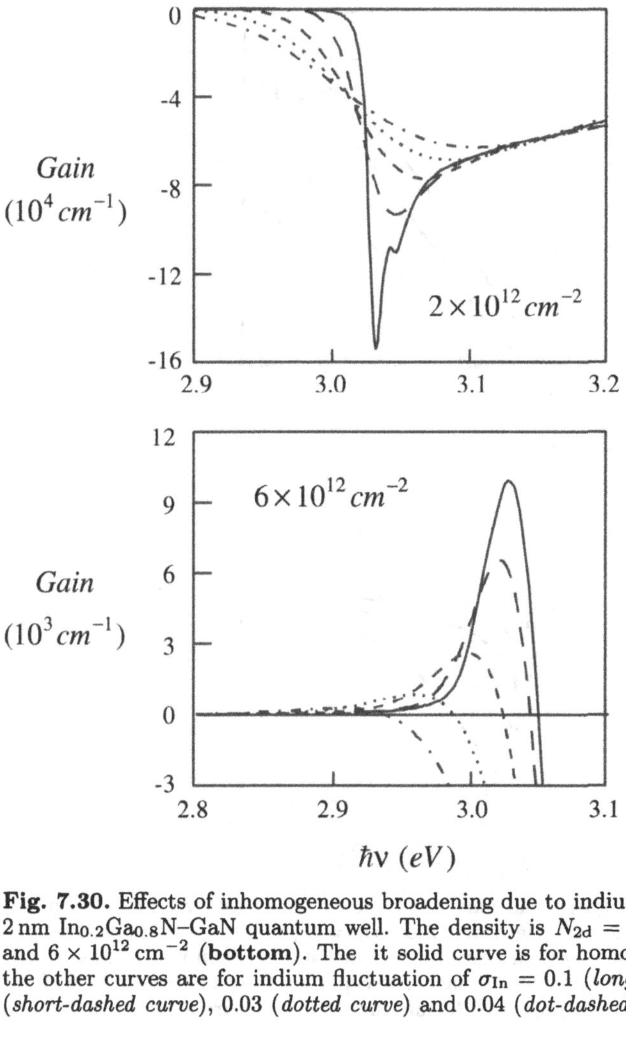


**Fig. 7.30.** Effects of inhomogeneous broadening due to indium fluctuations in the 2 nm $In_{0.2}Ga_{0.8}N$–GaN quantum well. The density is $N_{2d} = 2 \times 10^{12}$ cm$^{-2}$ (**top**) and $6 \times 10^{12}$ cm$^{-2}$ (**bottom**). The it solid curve is for homogeneous broadening, the other curves are for indium fluctuation of $\sigma_{In} = 0.1$ (*long-dashed curve*), 0.02 (*short-dashed curve*), 0.03 (*dotted curve*) and 0.04 (*dot-dashed curve*)

is a blue shift in the absorption resonance with increasing inhomogeneous broadening due to the asymmetry in the homogeneously broadened spectrum. For carrier densities sufficently high to produce gain, inhomogeneous broadening reduces the gain and shifts the gain peak to lower energy, as shown in Fig. 7.30 (bottom). With $\sigma_{\mathrm{In}} = 0.04$, the gain peak is red shifted by 122 meV, compare the discussion in Sect. 7.5. Together with the 60 meV shift in gain peak of the homogeneously broadened spectrum, we get a significant 181 meV red shift of the gain peak relative to the unrenormalized quantum-well bandgap. Significant red shifts of the gain peak relative to the bandedge have also been observed in experiments. Figure 7.31 summarizes the two pri-

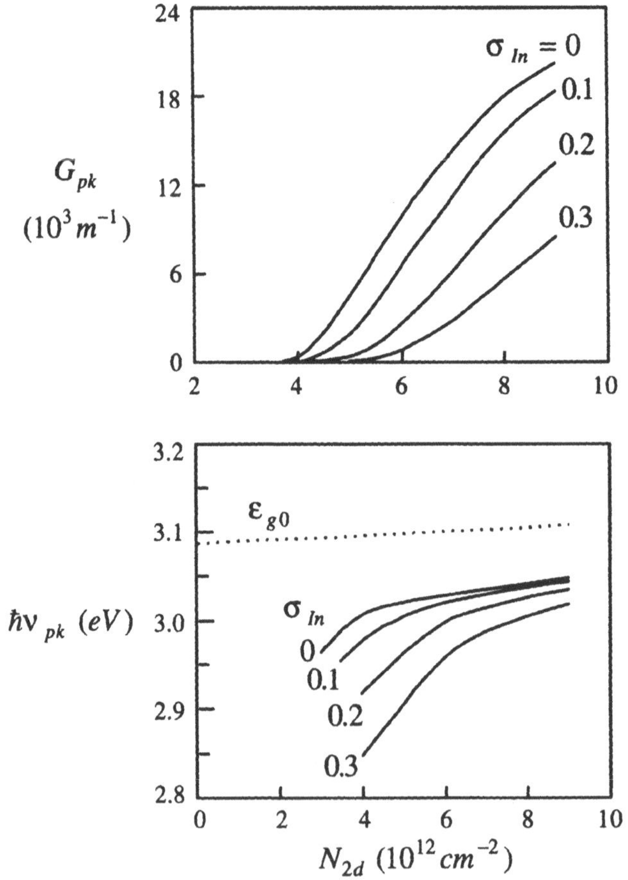


**Fig. 7.31.** Peak gain (**top**) and gain peak energy (**bottom**) vs. carrier density for indium concentration standard deviation $\sigma_{\mathrm{In}} = 0$ (*homogeneously broadening*), 0.01, 0.02 and 0.03. The *dotted curve* shows the quantum-well bandgap energy which increases with carrier density due to plasma screening of the piezoelectric field. The results are for a 2 nm $\mathrm{In_{0.2}Ga_{0.8}N}$-GaN quantum-well structure at $T = 300\,\mathrm{K}$

mary effects of inhomogeneous broadening. The top part of Fig. 7.31 shows the degradation of the peak gain, and the bottom part shows the red shift of the gain peak with increasing inhomogeneous broadening.

We conclude this section with a discussion on how one may estimate the laser threshold current density. This discussion is particularly appropriate for the group-III nitride lasers because systematic experimental investigations aimed at determining lasing threshold properties are particularly difficult to perform, currently mainly due to uncertainties in sample growth. The calculation involves first using (2.77) to compute the spontaneous emission spectra $SE(\omega)$ from the corresponding gain/absorption spectra. Figure 7.32 shows the spontaneous emission spectra for the 2 nm $In_{0.2}Ga_{0.8}N$–GaN quantum-well structure and carrier densities ranging from $N_{2d} = 4\times10^{12}$ to $8\times10^{12}\,cm^{-2}$. Each spectrum is changed by inhomogeneous broadening, as shown in Fig. 7.33. However, all the spectra in Fig. 7.33 have the same area. In other words the total spontaneous emission rate,

$$w_{sp} = \int_0^\infty d\omega\, SE(\omega) \ , \tag{7.19}$$

is independent of inhomogeneous broadening.

Two experimentally useful parameters can be derived from the spontaneous emission rate. One is an effective radiative lifetime,

$$\tau_{sp} = \frac{N}{w w_{sp}} \ , \tag{7.20}$$

where $N$ is the two-dimensional carrier density and $w$ is the quantum-well width. Figure 7.34 shows the spontaneous emission rate and the corresponding radiative lifetime as a function of carrier density. The second parameter is the spontaneous emission contribution to the injection current density

$$J_{sp} = e w w_{sp} \ , \tag{7.21}$$

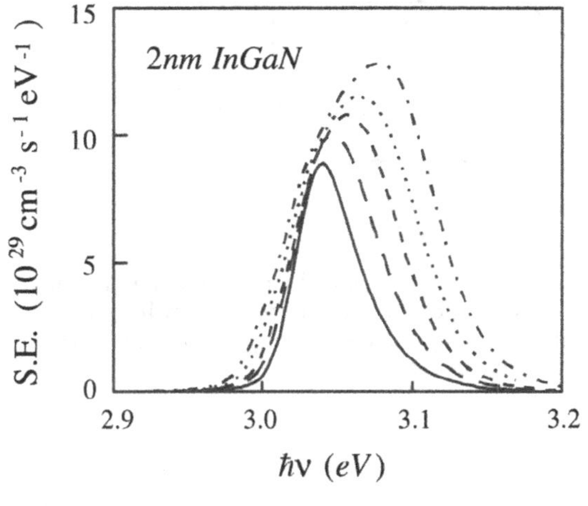


**Fig. 7.32.** Spontaneous emission spectra for a 2 nm $In_{0.2}Ga_{0.8}N$–GaN quantum-well structure and carrier densities $N_{2d} = 4\times10^{12}$ (*solid line*) to $8\times10^{12}$ $cm^{-2}$ (*dash-dot line*) in $10^{12}\,cm^{-2}$ increments. The results are calculated using gain spectra such as those in Fig. 7.29

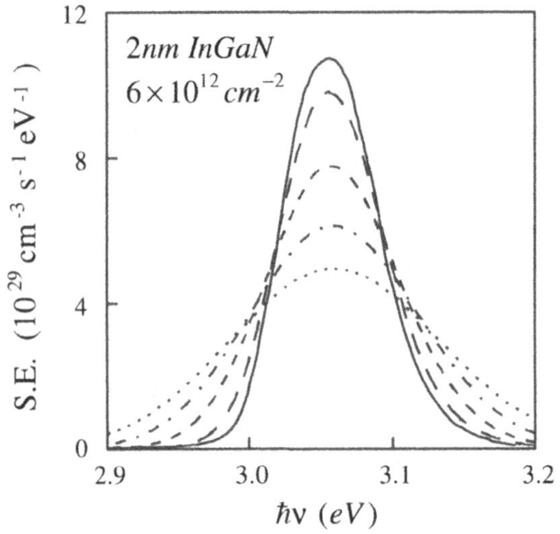


**Fig. 7.33.** Effects of inhomogeneous broadening on spontaneous emission spectrum. The *solid curve* shows the spontaneous emission spectrum for the 2 nm $In_{0.2}Ga_{0.8}N$–GaN quantum well and carrier density $N_{2d} = 6 \times 10^{12}\,cm^{-2}$. The other curves are for Indium fluctuation of $\sigma_{In} = 0.1$ (*long-dashed curve*), 0.02 (*short-dashed curve*), 0.03 (*dotted curve*) and 0.04 (*dot-dashed curve*). Each spectrum has an area of $w_{sp} = 9.1 \times 10^{28}\,cm^{-3}\,s^{-1}$

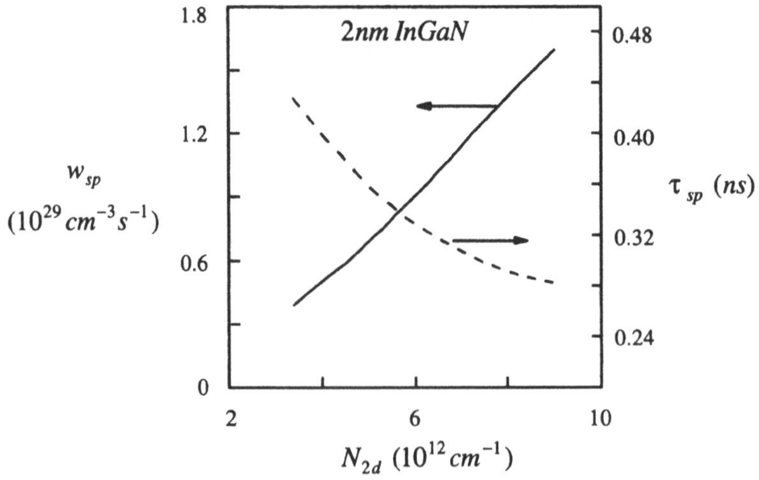


**Fig. 7.34.** Radiative decay rate $w_{sp}$ (*solid curve*) and the corresponding radiative lifetime $\tau_{sp}$ (*dashed curve*) as a function of carrier density for a 2 nm $In_{0.2}Ga_{0.8}N$–GaN quantum-well structure and $T = 300\,K$. As discussed in the text, the results are independent of inhomogeneous broadening

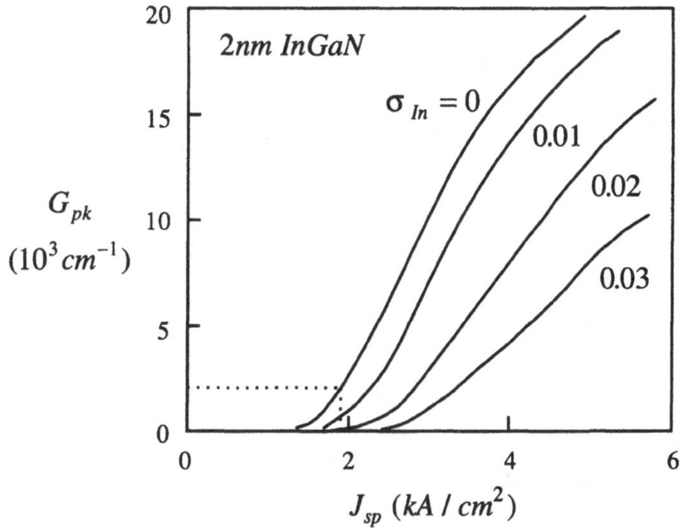


**Fig. 7.35.** Peak gain versus spontaneous emission current for a 2 nm $In_{0.2}Ga_{0.8}N$–GaN quantum-well structure and $T = 300\,K$. The different curves corresponds to different In concentration fluctuations in the quantum well

where $e$ is the electron charge. Figure 7.35 shows the peak gain as a function of the spontaneous emission current density. The different curves are for different inhomogeneous broadening due to composition fluctuations in the quantum well.

The threshold current density may be estimated using one of the curves in Fig. 7.35. For example, choosing a nominal material threshold gain of $G_{th} = 2 \times 10^3\,cm^{-1}$ and a homogeneously broadened sample, the figure shows a threshold current density of $J_{th} \approx 2\,kA/cm^2$. Of course, this is a prediction for the fundamental limit to the current density because we have neglected extrinsic contributions, such as those due to nonradiative recombination and current leakage.

A similar exercise can also be carried out to see if lasing in a particular GaN–AlGaN quantum well is feasible. Figure 7.36 shows gain/absorption spectra of a 2 nm GaN–$Al_{0.2}Ga_{0.8}N$ quantum-well structure. From the lower figure, we see that transparency occurs at $N_{2d} \simeq 4 \times 10^{12}\,cm^{-2}$. At higher densities, the spectra show gain in the ultraviolet wavelength region. To learn more about the emission wavelength, we plot in Fig. 7.37 the gain and spontaneous emission peak energies as functions of carrier density. The spontaneous emission peak ranges from 348 to 350 nm. The corresponding stimulated emission peak is about 3 nm longer in wavelength. Both sets of peaks show a blue shift with increasing carrier density.

The solid curve in Fig. 7.38 shows the peak gain as a function of spontaneous emission current for the 2 nm GaN–$Al_{0.2}Ga_{0.8}N$ quantum-well struc-

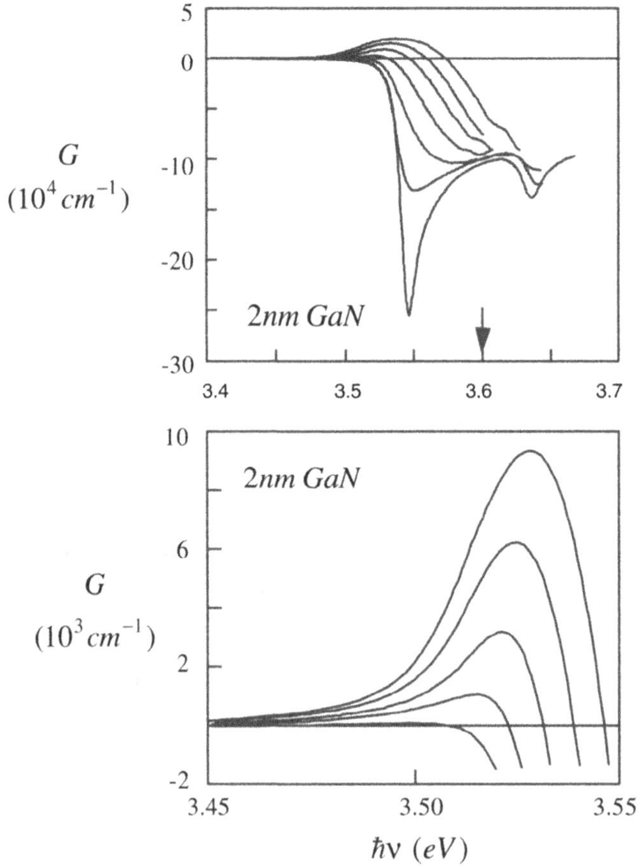


**Fig. 7.36.** TE gain for 2 nm GaN–$Al_{0.2}Ga_{0.8}N$. The **top** figure is for carrier densities $N_{2d} = 2 \times 10^{12}$ to $8 \times 10^{12}\,cm^{-2}$ (from bottom to top) in $10^{12}\,cm^{-2}$ increments. The **bottom** figure is for carrier densities $N_{2d} = 4 \times 10^{12}$ to $6 \times 10^{12}\,cm^{-2}$ (from bottom to top) in $5 \times 10^{11}\,cm^{-2}$ increments

ture. For comparison, we also plotted the curve for the 2 nm $In_{0.2}Ga_{0.8}N$–GaN quantum well. The theory predicts GaN-$Al_{0.2}Ga_{0.8}N$ quantum-well threshold current densities that are twice as high as those for the InGaN quantum well. This higher threshold current density is due primarily to band-structure differences.

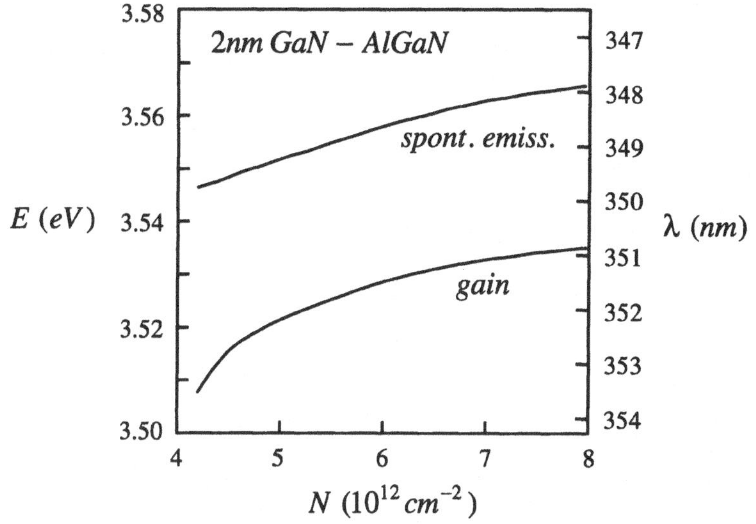


**Fig. 7.37.** Gain and spontaneous emission peak energies versus carrier density for a 2 nm GaN–$Al_{0.2}Ga_{0.8}N$ quantum well. The corresponding wavelengths are shown on the right axis

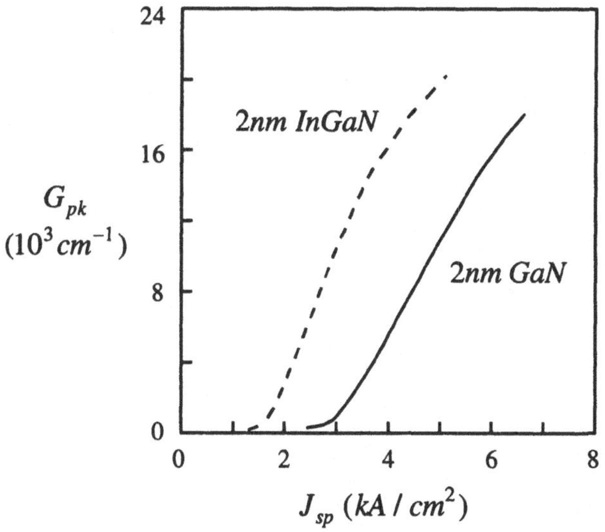


**Fig. 7.38.** Peak gain versus spontaneous emission current for a 2 nm GaN–$Al_{0.2}Ga_{0.8}N$ quantum well. For comparison the dashed curve shows the plot for the 2 nm $In_{0.2}Ga_{0.8}N$–GaN quantum-well structure

# References

## Chapter 1


Basov, N. G., O. N. Kroklin, and Y. M. Popov (1961), Pis'ma Zh. Eskp. Theor. Fiz. **40**, 1879 (see also Sov. Phys. JETP **13**, 1320).

Haug, H. and S. W. Koch (1994), *Quantum Theory of the Optical and Electronic Properties of Semiconductors*, 3rd Edition, World Scientific Publ., Singapore.

Chow, W. W., S. W. Koch, and M. Sargent III (1994), *Semiconductor-Laser Physics*, Springer-Verlag, Berlin.

Meystre, P. and M. Sargent III (1991), *Elements of Quantum Optics*, 2nd Edition, Springer-Verlag, Heidelberg.

Agrawal, G. A. and N. K. Dutta (1986), *Long-Wavelength Semiconductor Lasers*, Van Nostrand Reinhold Co., New York.

Streifer, W., R. D. Burnham, T. L. Paoli, and D. R. Scifres (1984), Laser Focus/Electro Optics, June, 1984.

Thompson, G. H. B. (1980), *Physics of Semiconductor Lasers*, John Wiley, New York.

Yariv, A. (1975), *Quantum Electronics*, 2nd Edition, John Wiley, New York.


For general information on semiconductor lasers, see


Agrawal, G. A. and N. K. Dutta (1993), *Semiconductor Lasers*, 2nd Edition, Van Nostrand Reinhold Co., New York.

Chuang, S. L. (1995), *Physics of Optoelectronic Devices,* Wiley, New York.

Coldren, L. A. and S. W. Corzine (1995), *Diode Lasers and Photonic Integrated Circuits*, Wiley, New York.

Ebeling, K. L. (1993), Integrated Opto-Electronics, Springer-Verlag, Berlin.

Kressel, H. and J. K. Bulter (1977), Semiconductor Lasers and Heterojunction LEDs, Academic Press, San Diego.

Streifer, W., R. D. Burnham, T. L. Paoli, and D. R. Scifres (1984), Laser Focus/Electro Optics, June, 1984.

Thompson, G. H. B. (1980), *Physics of Semiconductor Lasers*, John Wiley, New York.


For laser theory and quantum optics, see


Louisell, W. H. (1973), *Quantum Statistical Properties of Radiation*, Wiley, New York.

Meystre, P. and M. Sargent III (1991), *Elements of Quantum Optics*, 2nd Edition, Springer-Verlag, Heidelberg.

Sargent, M. III, M. O. Scully, and W. E. Lamb, Jr. (1974), *Laser Physics*, Addison-Wesley, Reading.

Siegman, A. (1986), *Lasers*, University Science Books, Mill Valley.

Yariv, A. (1975), *Quantum Electronics*, 2nd Edition, John Wiley, New York.

For more information about properties of low-dimensional structures, see

Arakawa, Y., K. Vahala, and A. Yariv (1986), Surf. Sci. **174**, 155.
Asada, M., Y. Miyamoto, and Y. Suematsu (1986), IEEE J. Quantum Electron. **22**, 1915.
Banyai, L. and S. W. Koch (1993), *Semiconductor Quantum Dots*, World Scientific Series in Atomic, Molecular and Optical Physics – Vol. 2, World Scientific Publ., Singapore.
Jacak, L., P. Hawrylak, and A. Wójs, (1997), *Quantum Dots*, Springer, Berlin.
Kapon, E., J. P. Harbison, R. Bhat, and D. M. Hwang (1989), p. 49 in Optical Switching in Low-Dimensional Systems, H. Haug and L. Banyai, eds., NATO ASi Series B, Vol. 194, Plenum, New York.
Vahala, K. (1988), IEEE J. Quantum Electron. **24**, 523.
Woggon, U. (1997), *Optical Properties of Semiconductor Quantum Dots*, Springer Tracts in Modern Physics **136**, Springer-Verlag, Berlin.
Zory, P. S. (1993), *Quantum Well Lasers*, Academic Press, San Diego.

The above texts and reviews give also references to the original papers.

For more information about properties of quantum wires and quantum dots, see

Banyai, L. and S. W. Koch (1993), *Semiconductor Quantum Dots*, World Scientific Series in Atomic, Molecular and Optical Physics – Vol. 2, World Scientific Publ., Singapore.
Arakawa, Y., K. Vahala, and A. Yariv (1986), Surf. Sci. **174**, 155.
Asada, M., Y. Miyamoto, and Y. Suematsu (1986), IEEE J. Quantum Electron. **22**, 1915.
Vahala, K. (1988), IEEE J. Quantum Electron. **24**, 523.
Kapon, E., J. P. Harbison, R. Bhat, and D. M. Hwang (1989), p. 49 in Optical Switching in Low-Dimensional Systems, H. Haug and L. Banyai, eds., NATO ASi Series B, Vol. 194, Plenum, New York.
Woggon, U. (1997), *Optical Properties of Semiconductor Quantum Dots*, Springer Tracts in Modern Physics **136**, Springer-Verlag, Berlin.

These reviews give also references to the original papers.

## Chapter 2

For free-carrier treatments, see

Chow, W. W., S. W. Koch, and M. Sargent III (1994), Chap. 3 in *Semiconductor-Laser Physics*, Springer-Verlag, Berlin.
Chuang, S. L. (1995), *Physics of Optoelectronic Devices*, Wiley, New York.
Coldren, L. A. and S. W. Corzine (1995), *Diode Lasers and Photonic Integrated Circuits*, Wiley, New York.
Thompson, G. H. B. (1980), *Physics of Semiconductor Lasers*, John Wiley, New York.
Zory, P. S. (1993), *Quantum Well Lasers*, Academic Press, San Diego.

Early references of the linewidth enhancement and antiguiding factors include:

H. Haug and H. Haken (1967), Z. Phys. **204**, 262.
Lax, M. (1968), in Brandeis University Summer Institute Lectures (1966), Vol. II, ed. by M. Chretien, E. P. Gross, and S. Deser, Gordon and Breach, New York.
Thompson, G. (1972), Opto-Electron. **4**, 257.
Henry, C. (1982), IEEE J. Quantum Electron. **QE-18**, 259.

## Chapter 3

For more details on the semiconductor Bloch equations and for further references see:

Binder, R. and S. W. Koch, Progress in Quantum Electronics **19**, 307 (1995).
Koch, S. W., N. Peyghambarian, and M. Lindberg (1988), J. Phys. **C21**, 5229.
Haug, H. and S. W. Koch (1989), Phys. Rev. **A39**, 1887.
Haug, H. (1988), Ed., *Optical Nonlinearities and Instabilities in Semiconductors*, Academic, New York (1988).
Stahl, A. and I. Balslev (1987), *Electrodynamics of the Semiconductor Band Edge*, Springer Tracts in Modern Physics **110**, Springer-Verlag, Berlin.
Haug, H. and S. W. Koch (1994), *Quantum Theory of the Optical and Electronic Properties of Semiconductors*, 3rd ed., World Scientific, Singapore.
Lindberg, M. and S.W. Koch (1988), Phys. Rev. **B38**, 3342.

Discussions of the two-level Bloch equations can be found in:

Allen, L. and J. H. Eberly (1975), *Optical Resonances and Two-Level Atoms*, John Wiley, New York; reprinted (1987) with corrections by Dover, New York.
Meystre, P. and M. Sargent III (1991), *Elements of Quantum Optics*, 2nd Ed., Springer-Verlag, Heidelberg.
Sargent III, M., M. O. Scully, and W. E. Lamb (1977), *Laser Physics*, Addison Wesley, Reading, MA.

The classical theory of plasma screening is discussed in:

Ashcroft, N. W. and N. D. Mermin (1976), *Solid State Theory*, Saunders College, Philadelphia.
Harrison, W. A. (1980), *Solid State Theory*, Dover Publ. New York.
Haug, H. and S. W. Koch (1994), Op. Cit.

General many-body theory and sum rules are discussed in:

Lundquist, B. I. (1967), Phys. Konden. Mat. **6**, 193 and 206.
Mahan, G. D. (1981), *Many Particle Physics*, Plenum Press, New York.

For the modifications of the plasmon-pole approximation in an electron-hole plasma see:

Haug, H. and S. Schmitt-Rink (1984), Op. Cit.
Zimmermann, R. (1988), *Many-Particle Theory of Highly Excited Semiconductors*, Teubner, Berlin.

For the Padé approximation, see:

Gaves-Morris, P. R. (1973), Ed., *Padé Approximants and Their Application*, Academic Press, N.Y.
Haug, H. and S. W. Koch (1989), Phys. Rev. **A39**, 1887.

We have used the integral tables in:

Gradshteyn, I. S. and I. M. Rhyzhik (1980), *Tables of Integals, Series and Products*, Academic Press, New York.

## Chapter 4

Many references listed in Chap. 3 are also relevant to this chapter. Additional references include:

Binder, R. and S. W. Koch (1995), Progress in Quantum Electronics **19**, 307.
Jahnke, F., M. Kira and S. W. Koch (1997), Z. Physik **B104**, 559.
Binder, R., D. Scott, A. E. Paul, M. Lindberg, K. Henneberger, and S. W. Koch (1992), Phys. Rev. **B45**, 1107.
Press, W. H., B. P. Flannery, S. A. Teukolsky, and W. T. Vetterling (1988), *Numerical Recipes*, Cambridge University Press, Cambridge.

The solutions of the Boltzmann equation in Figs. 4.2, 4.3, and 4.4 are presented in

Jahnke, F. and S. W. Koch (1995), Appl. Phys. Lett. **67**, 2278.
Jahnke, F. and S. W. Koch (1995), Phys. Rev. **A52**, 1712

## Chapter 5

Much of the material discussed in this chapter can be found in many solid state physics textbooks and review articles, e.g.,

Altarelli, M. (1985), p. 12 in Heterojunctions and Semiconductor Superlattices, Eds. G. Allan, G. Bastard, N. Boccara, M. Lannoo, and M. Voos, Springer-Verlag, Berlin.
Ashcroft, N. W. and N. D. Mermin (1976), Solid State Physics, Saunders College (HRW), Philadelphia.
Bastard, G. (1988), Wave Mechanics Applied to Semiconductor Heterostructures, Les Editions de Physique, Paris.
Callaway, J. (1974), Quantum Theory of the Solid State, Part A, Academic Press, New York.
Kane, E. O. (1966), Semiconductors and Semimetals, edited by R. K. Willardson and A. C. Beer, Academic, New York, p. 75.
Kittel, C. (1971), Introduction to Solid State Physics, Wiley & Sons, New York; Kittel, C. (1967) Quantum Theory of Solids, Wiley & Sons, New York.
The block diagonalization of the Luttinger Hamiltonian has been done by: Broido, D. A. and L. J. Sham (1985), Phys. Rev. **B31**, 888.

The spin-orbit coupling scheme is discussed, e.g., in

Schiff, L. (1968), Quantum Mechanics, McGraw-Hill, New York. Chap. 12.

A large number of material parameters for many semiconductors can be found in:

Landolt-Börnstein (1982), Numerical Data and Functional Relationships in Science and Technology, ed. K. H. Hellwege, Vol. 17 Semiconductors, edited by O. Madelung, M. Schulz, and H. Weiss, Springer-Verlag, Berlin.

## Chapter 6

The Hamiltonian for strained semiconductors has been derived by

Bir, G. L. and G. E. Pikus (1974), *Symmetry and Strain-Induced Effects in Semiconductors*, Wiley & Sons, New York.
Pikus, G. E. and G. L. Bir (1960), Sov. Phys. – Solid State 1, 1502 [Fiz. Tverd. Tela (Leningrad) **1**, 1642 (1959)].
Kittel, C. (1971), *Introduction to Solid State Physics*, Wiley & Sons, New York; Kittel, C. (1967), *Quantum Theory of Solids*, Wiley & Sons, New York.
Landolt-Börnstein (1982), *Numerical Data and Functional Relationships in Science and Technology*, ed. K. H. Hellwege, Vol. 17 Semiconductors, edited by O. Madelung, M. Schulz, and H. Weiss, Springer-Verlag, Berlin.

For papers and reviews dealing with bandstructure calculations and optical properties of strained superlattices see, e.g.,

Ahn, D. and S. L. Chuang (1988), IEEE J. Quantum Electron. **24**, 2400.
Chuang, S. L. (1991), Phys. Rev. **B43**, 9649.
Dawson, M. D. and G. Duggan (1993), Phys. Rev. **B47**.
Duggan, G. (1990), SPIE **1283**, 206.
Marzin, J. Y. (1986), *Heterojunctions and Semiconductor Superlattices*, eds. G. Allan, G. Bastard, and M. Voos, Springer, Berlin, p. 161.
Chuang, S. L. (1995), *Physics of Optoelectronic Devices*, Wiley & Sons, New York.

For references on the composition dependence of the $In_{1-x}Ga_xP$ bandgap see, e.g.,

Adachi, S. (1982), J. Appl. Phys. **53**, 8775.
Stringfellow, G. B., P. F. Lindquist, and R. A. Burmeister (1972), J. Electron. Mater. **1**, 437.

For the wurtzite group-III nitrides see, e.g.,

Pearton, S. J., Ed. (1997), *GaN and Related Materials*, Vol. 2, (Gordon and Breach, Netherlands).
Nakamura, S., et al. (1996), Appl. Phys. Letts. **69**, 4056 (1996).
Miles K. and I. Akasaki, Eds. (1998), *GaN-Based Lasers: Materials, Processing, and Device Issues*, IEEE Journal of Selected Topics in Quantum Electronics **4**.
Chuang, S. L. and C. S. Chang (1996), Phys. Rev. **B54**, 2491.

## Chapter 7

For AlGaInP bandstructure and optical properties see:

Dawson, M. D. and G. Duggan (1993), Phys. Rev. **B47**.
Duggan, G. (1990), SPIE **1283**, 206.
Adachi, S. (1982), J. Appl. Phys. **53**, 8775.
Stringfellow, G. B., P. F. Lindquist, and R. A. Burmeister (1972), J. Electron. Mater. **1**, 437.

For nitride lasers see e.g. the reviews:

Nakamura, S. and G. Fasol (1997), The Blue Laser Diode (Springer, Berlin).
Pearton, S. J., editor, (1997), GaN and related materials (Gordon and Breach, Netherlands).

Calculations of quantum well bandstructures can be found in

Sirenko, Y. M., J.-B. Jeon, K. W. Kim, M. A. Littlejohn, and M. A. Stroscio (1996), Phys. Rev. **B53**, 1997.

Chuang, S. L. and C. S. Chang (1996), Phys. Rev. **B54**, 2491.

Additional references for the bulk material parameters include

Suzuki, M., T. Uenoyama, and A. Yanase (1995), Phys. Rev. **B 52**, 8132.

Nykhovshi, A., B. Gelmonst, and M. Shur (1993), J. Appl. Phys. **74**, 6734.

Examples of the gain calculations have been published in

Chow, W. W., A. F. Wright, A. Girndt, F. Jahnke, and S. W. Koch (1997), Appl. Phys. Lett. **71**, 2608.

Experimental results for II-VI lasersare presented e.g. in

Ding, J., M. Hagerott, T. Ishihara, H. Jeon, and A. V. Nurmikko (1993), Phys. Rev. **B47**, 10528.

Theory/experiment comparisons have been published

Chow, W. W., P. M. Smowton, P. Blood, A. Girndt, F. Jaknke, and S. W. Koch (1997), Appl. Phys. Lett. **71**, 157.

Ellmers, S., M. Hofmann, W. Ruehle, A. Girndt, F. Jahnke, W. Chow, A. Knorr, S.W. Koch, H. Gibbs, G. Khitrova, and M. Oestreich (1998), phys. stat. sol. **b206**, 407; Ellmers et al., Appl. Phys. Lett. **72**, 7647 (1998).

Girndt, A., S. W. Koch, and W. W. Chow (1998), Appl. Phys. **A66**.

The experimental data for the II-VI gain spectra have been reported by

Yoshida, H., Y. Gonno, K. Nakano, S. Taniguchi, T. Hino, A. Ishibashi, M. Ikeda, S. L. Chuang, and J. Hegarty (1996), Appl. Phys. Lett **69**, 3893.

# Subject Index